Production
Operations

**Well Completions, Workover,
and Stimulation
Volume 1**

Production Operations

Volume 1

Well Completions, Workover, and Stimulation

Thomas O. Allen
and
Alan P. Roberts

Oil & Gas Consultants International, Inc.
Tulsa

Library of Congress Catalog Card Number: 77-13239
International Standard Book Number: 0-930972-00-7
Printed in U.S.A.

Preface

The subject of well completion, workover, and stimulation seemingly occupies only a small part in an overview of the oil industry. The same is true even if we limit ourselves to the exploration and production phases of the industry.

From our vantage, however, the operation of completing a well so as to obtain, and maintain, effective communication with the desired reservoir fluids is the focal point of exploration and production activities. The technology required for effective well completion involves many disciplines and many different types of talents. A well completion is not merely a mechanical process of drilling a hole, setting casing, and perforating a hydrocarbon section.

The importance of total reservoir description; the role of effective communication between the reservoir and the wellbore; the hazards of flow restrictions around the wellbore; the importance of knowing where fluids are and where they are moving to; the rigors of excluding undesirable fluids all become more and more evident as we move deeper into the areas of enhanced methods of maximizing recovery of increasingly valuable hydrocarbon fluids.

In preparing Production Operations Volume 1 and Volume 2, we have tried to logically separate well completion and well operation technology into packages to permit detailing the more important facets. The interrelation of most subjects makes this somewhat difficult and results in some duplication for clarity. Effective well completion and recompletion operations require consideration of specific problems using all available technology.

Volumes 1 and 2 are the product of some twelve years of conducting training programs throughout the world for industry groups, including engineers, managers, geologists, technicians, foremen, and others.

The question is often asked, "What's new in well completion technology?" Our answer must be that new technology per se is not the real issue in considering improvement in producing operations. "The key to optimizing oil and gas recovery and profits is the *effective application* of *proved* technology." This has been the theme of our Production Operations courses since our first effort in 1966, and is the theme of these two new books on Production Operations. A primary objective of our technical training has been to assist operating groups reduce the length of time required for "proved techniques" to become routine field practice.

The business of well completion is continually changing. The learning process continues, technology improves; and just as important, the rules of the game change with the times and with the area. In many areas, effective and economic recovery of hydrocarbons from more and more marginal

v

reservoirs is the name of the game. In other areas where costs are tremendous due to the complications of offshore activities or geographic location, high production rates needed to provide sufficient return on the uncomprehensible investment required, provide the winning combination.

Thomas O. Allen
Alan P. Roberts

Tulsa, Oklahoma
September, 1977

Acknowledgments

Authors of a book always receive many different kinds of assistance from many different people. Ours is no exception. We wish to acknowledge a number of these contributions by name, and also give our thanks to many others, as well.

We owe a special debt of gratitude to the various oil-well service companies. Without their help, there would be no oil industry. Nor would there be any book on PRODUCTION OPERATIONS. We particularly want to acknowledge extensive help and counsel from Baker Oil Tools, Dowell Division of Dow Chemical Co., Dresser Industries, Halliburton Company, and Schlumberger.

Special recognition is due Dr. Scott P. Ewing (deceased), who prepared the original writeup of the chapter on Corrosion (Vol. 2), and to Wallace J. Frank, who contributed a great deal to the current revision.

C. Robert Fast, OGCI; and C. C. McCune, Chevron Research, reviewed the chapter on Acidizing (Vol. 2) and made valuable suggestions. Dr. D. A. Busch, Dr. P. A. Dickey, Dr. G. M. Friedman, and Dr. Glenn Visher, OGCI, rendered valuable assistance in the preparation of the chapter on Geology (Vol. 1).

C. P. Coppel, Chevron Research, reviewed the chapter on Surfactants (Vol. 2). Norman Clark (deceased) and Dr. Charles Smith made valuable technical contributions to the Reservoir Engineering chapter (Vol. 1). Ray Leibach, OGCI, assisted in the preparation of the chapter on Sand Control (Vol. 2). G. W. Tracy, Amoco Production Research, reviewed the section on production testing, Well Testing chapter, Vol. 1. John E. Eckel, OGCI, contributed to the chapter on Downhole Production Equipment.

Our thanks and appreciation go, too, to the various operating companies which have participated in our PRODUCTION OPERATIONS course sessions. They have helped hone and refine many generations of lecture notes into this two-volume series of textbooks.

We appreciate the assistance of Gerald L. Farrar, long-time Engineering Editor of The Oil and Gas Journal, and now OGCI's Publications Editor, who served as Editor for this series.

Finally, we acknowledge the valuable contribution of Jewell O. Hough, who in the past 12 years has prepared dozens of revisions, leading up to the publication of this book.

—The Authors

Contents Vol. 1

Contents Vol. 2

Chapter 1

Geologic Considerations In Producing Operations

The role of geology
Sandstone reservoir properties
Application of concepts in specific sandstone reservoirs
Reservoir description, Elk City
Reservoir description, Niger Delta
Carbonate reservoir properties
Application of concepts in specific carbonate reservoirs
San Andres stimulation
Monahans Clearfork and Judy Creek waterfloods
Golden Spike miscible flood
Glossary of geological terms

INTRODUCTION

Geologic studies have provided significant data for the finding, development, and operation of oil and gas reservoirs during the more than 100 years of oil and gas field operation.

Engineers, as well as geologists, have long employed isopachs, structure maps, isobaric maps, core and log information, production tests, and other data as a guide in decisions relative to the development and operation of oil and gas fields. These tools, however, often proved to be inadequate for correct extrapolation or prediction of the vertical and horizontal continuity of porosity and permeability beyond the wellbore. Transient pressure tests provide additional clues, but only represent average values of both vertical and lateral variations in reservoir properties of porous zones open to the wellbore.

Multiple well tests, particularly "pulse" tests, are useful in determining continuity of porosity between wells, and drawdown tests may help determine reservoir limits. However, the results of all types of transient pressure tests are questionable in layered zones, particularly due to the possibility that some of the zones may be plugged or partially plugged by formation damage effects.

Research during the past 20 years by various groups has focused attention on the importance to reservoir development and operation of more comprehensive analysis of the reservoir geology, particularly the interrelation of fluid flow in the reservoir with depositional environment and post-depositional history. These studies can provide a much improved basis for extrapolating and predicting reservoir rock properties and reservoir fluid flow away from the wellbore.

The aim of this section is to highlight geologic technology useful in solving production problems, and to present several reservoir case studies to illustrate the value of improved reservoir description for both sandstone and carbonate reservoirs. The total geologic description of the reservoir must then be incorporated with reservoir fluid flow data, reservoir production history and reservoir performance predictions to provide information required to plan effective depletion of the reservoir.

Harris[18] in his 1974 treatise on "The Role of Geology in Reservoir Simulation" outlined geologic controls on reservoir properties and pointed out the significance of these controls on reservoir performance. Figure 1-1 from Harris shows combined data input required for good reservoir description from both geologic and reservoir engineering studies.

Some of the functions of the geologist in this type of study may be:

1

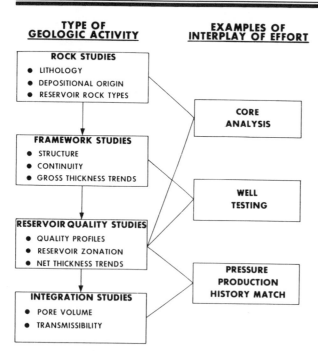

FIG. *1-1—Interrelated geologic and engineering activities in reservoir description.* [18] *Permission to publish by The Society of Petroleum Engineers.*

—Select core samples for both geologic and reservoir studies.

—Identify the depositional environment and source rocks.

—Develop a depositional model modified by post-depositional changes.

—Construct a structure map.

—Develop cross sections or other representations to show changes in rock properties throughout the reservoir.

—Develop porosity and permeability trends including both horizontal and vertical barriers to fluid flow—particularly those trends predictable from depositional and post depositional history of the reservoir.

Figure 1-2 illustrates typical data obtained from combined studies of geologic and reservoir engineering aspects.

The Habitat of Oil and Gas

Most oil and gas reservoirs are found in sandstones or carbonates. There are very limited occurrences in shale, volcanic rock, and fractured basement rock (basalt). Comparing the significance of sandstone and carbonate reservoirs, sandstones are more abundant, yet limestones are more important as reservoirs for hydrocarbons, as shown in Figure

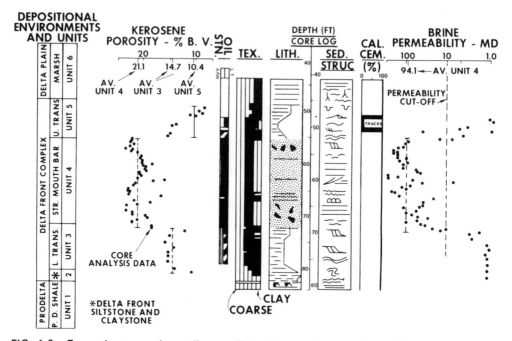

FIG. *1-2—Example reservoir quality profile.* [18] *Permission to publish by The Society of Petroleum Engineers.*

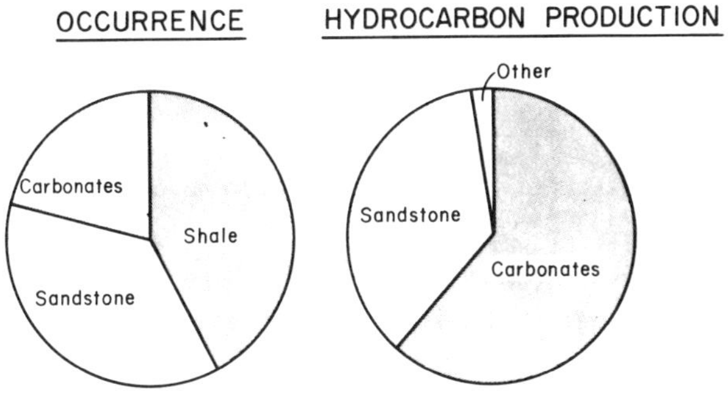

OCCURRENCE   HYDROCARBON PRODUCTION

FIG. *1-3—Significance of sandstone and carbonates.*[15]

1-3. The greater percentage of hydrocarbons in carbonates is greatly influenced by the numerous reservoirs in the Middle East area.

Traps for Oil and Gas Accumulation

The three prerequisites to a commercial accumulation of oil and/or gas are:

1. Source rock.
2. Porous and permeable container rock.
3. Impermeable caprock.

A trap may be structural, stratigraphic, or a combination of the two. Some of the more common types of traps are illustrated in Figures 1-4, 1-5, 1-6, 1-7, 1-8, and 1-9.

Fractures and Joints in Reservoir Rocks

Mechanical characteristics of reservoir rocks can vary considerably. Some may be plastic or semiplastic and bend and deform without fracturing. Others are hard and brittle and fracture or break rather than bend. Reservoir rocks may be fractured by either tensional or compressive forces. Shales found at depths to about 15,000 ft in the Gulf Coast of Texas and Louisiana are examples of plastic rocks.

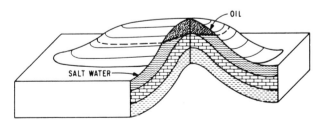

FIG. *1-4—Oil accumulation in an anticlinal structure.* Elements of Petroleum Reservoirs. *Permission to publish by The Society of Petroleum Engineers.*

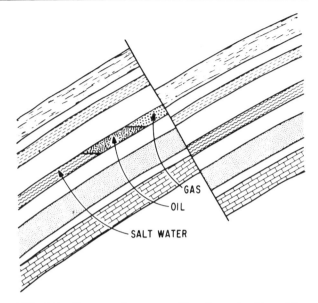

FIG. *1-5—Structural trap resulting from faulting.* Elements of Petroleum Reservoirs. *Permission to publish by The Society of Petroleum Engineers.*

Most dolomites have a high strength with little plasticity and frequently fracture rather than bend when subjected to high stress. Fracture systems, such as those found in the huge oil fields of the Middle East, may result from regional stress. Fractures provide major flow channels in these reservoirs while matrix permeability of the carbonates varies considerably.

Faults associated with salt domes in the Texas and Louisiana Gulf Coast are representative of

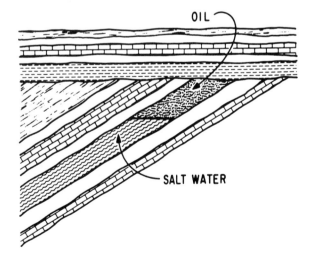

FIG. *1-6—Oil accumulation under an unconformity.* Elements of Petroleum Reservoirs. *Permission to publish by The Society of Petroleum Engineers.*

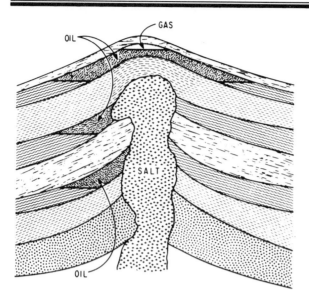

FIG. *1-7—Oil accumulation in the vicinity of a piercement-type salt dome. Elements of Petroleum Reservoirs. Permission to publish by The Society of Petroleum Engineers.*

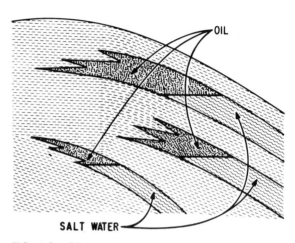

FIG. *1-8—Oil accumulation in sand lenses of the sand bar type. Elements of Petroleum Reservoirs. Permission to publish by The Society of Petroleum Engineers.*

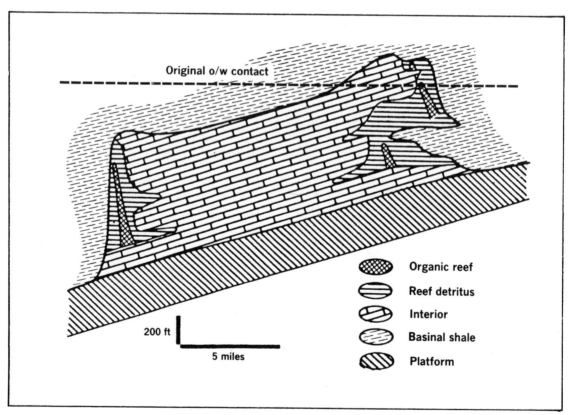

FIG. *1-9—Reef with oil accumulation in top.[20] Permission to publish by The Society of Petroleum Engineers.*

fracturing caused by local rather than regional stress patterns, although regional stress may always contribute to total fracturing and faults around salt domes. These faults are usually closed and seldom provide flow paths along the fracture face.

Joints are closed, vertical or near vertical, smooth straight fractures. They occur in parallel sets with spacings ranging from a few centimeters to several meters apart. The direction of fracture planes is constant over large areas. Joints are usually closed at depth. However, joints are frequently propped open during hydraulic fracturing operations or etched during fracture acidizing to form flow channels. During drilling, cementing, or well killing operations if joints are forced open by hydraulic pressure, the result may be partial loss of circulating fluid or complete loss of returns under more extreme cases.

It is very common for joints to be held open during waterflood operations, resulting in loss of sweep efficiency. Joints are found to some degree in most oil and gas producing reservoirs throughout the world; however, they occur most frequently in high compressive strength rock, and especially in carbonates. Parallel joints may be continuous for many miles, and joint systems may cover hundreds of square miles.

Linears or joints can sometimes be seen on the surface, and their direction is the same as joints in the subsurface. Joints are often seen in cores as vertical fractures; however, widely spaced fractures may not always be penetrated by a core bit. The Spraberry trend in the Permian Basin near Midland, Texas and the Marmaton trend in North Texas and Western Oklahoma are examples of extensively jointed oil-bearing reservoirs.

SANDSTONE RESERVOIRS

Most oil and gas production from sandstone is derived from deposits originating from river-borne sediments. Environments and depositional models for river-borne sediments forming continental, transitional, and marine deposits are shown in Figures 1-10, 1-11, and 1-12 (from Geometry of Sandstone Reservoir Bodies by Rufus J. LeBlanc, AAPG Memoir 18[22]). Examples of deposits having river-borne sediments as their source material are the hundreds of reservoirs laid down by ancestral rivers during Tertiary time, including those related to the Mississippi River, and the Niger River.

The Latrobe River deposited sediments in Bass

Straits Fields off Australia, and the Orinoco River accounts for the prolific sandstone deposits off Trinidad. The McKenzie delta deposits in Canada, the Prudhoe Bay field in Alaska, Burgan field in Kuwait, and the many reservoirs in Indonesia are primarily river-borne deposits.

A model of types of deposits found in a delta complex is shown in Figure 1-13, which depicts the alluvial, distributary channels, deltaic, and marine barrier bar deposits. This model of the types of deposits found in the Elk City field, Oklahoma is illustrative of river-borne deposits found all over the world.

Figure 1-14 shows how the current in a river erodes the outside of a bend, and deposits sediment on the inside to form point bars, which are scattered all over the flood plain of a meandering river. Many significant oil and gas fields produce from point bars, including a large number of the reservoirs in the Niger delta area.

Significant wind deposited dune sand reservoirs are relatively rare. A notable exception is the Rotliegendes formation of Permian age extending from Germany through the Netherlands to Southeastern England where gas production is obtained primarily from ancient sand dune deposits. The most significant reservoir producing from the Rotliegendes is the Groningen gas field in Holland, which contains the largest gas reserve in Europe.

Rotliegendes deposits under the North Sea generally exhibit a lower permeability than in the Groningen field. The Leman field and others off Southeastern England are representative of this lower permeability group with permeabilities averaging 1 md or less where pores have been filled primarily with secondary silica cement.

GEOLOGIC FACTORS AFFECTING RESERVOIR PROPERTIES IN SANDSTONE RESERVOIRS
Porosity and Permeability

Porosity is the pore volume divided by the bulk volume of the rock, and is usually expressed in percent. It provides the container for the accumulation of oil and gas, and gives the rock characteristic ability to absorb and hold fluids. The ease with which fluids move through inter-connected pore spaces of the reservoir rock is defined as permeability.

Compared with carbonates, sandstone porosity appears to be relatively consistent and easy to

ENVIRONMENTS   DEPOSITIONAL MODELS

CONTINENTAL				
ALLUVIAL (FLUVIAL)	ALLUVIAL FANS (APEX, MIDDLE & BASE OF FAN)	STREAM FLOWS	CHANNELS / SHEETFLOODS / "SIEVE DEPOSITS"	ALLUVIAL FAN
		VISCOUS FLOWS	DEBRIS FLOWS / MUDFLOWS	
	BRAIDED STREAMS		CHANNELS (VARYING SIZES)	BRAIDED STREAM
			BARS: LONGITUDINAL / TRANSVERSE	
	MEANDERING STREAMS (ALLUVIAL VALLEY)	MEANDER BELTS	CHANNELS / NATURAL LEVEES / POINT BARS	MEANDERING STREAM
		FLOODBASINS	STREAMS, LAKES & SWAMPS	
EOLIAN	DUNES	COASTAL DUNES	TYPES: TRANSVERSE / SEIF (LONGITUDINAL) / BARCHAN / PARABOLIC / DOME-SHAPED	COASTAL DUNES
		DESERT DUNES		DESERT DUNES
		OTHER DUNES		

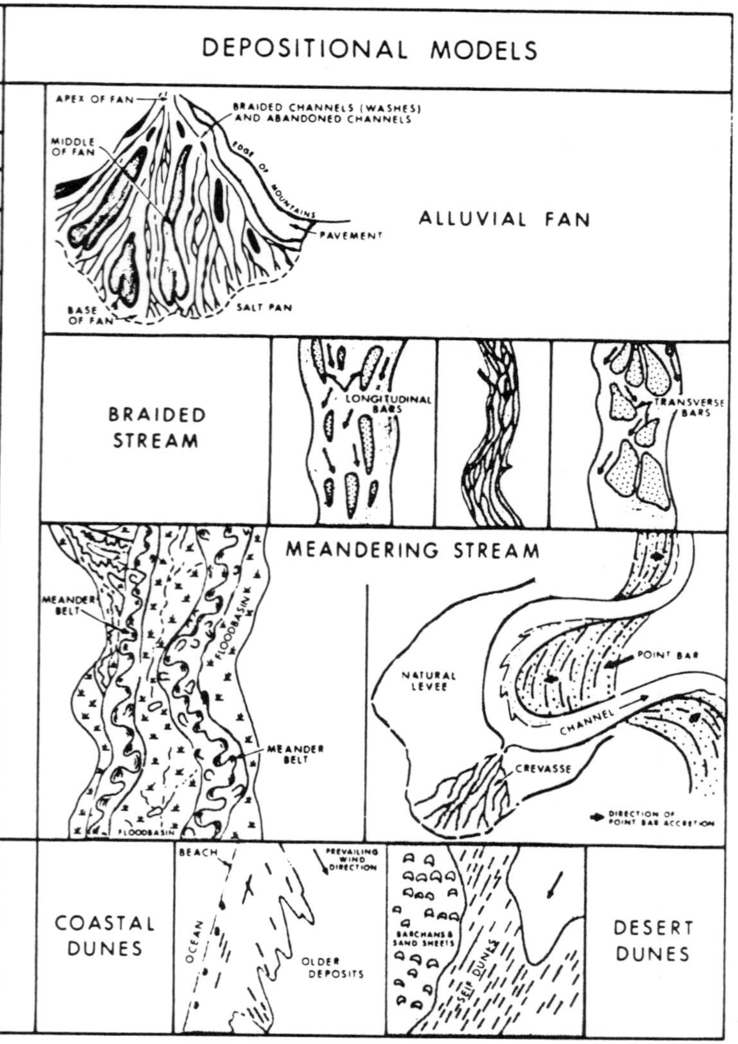

FIG. 1-10—Alluvial (fluvial) and eolian environments and models of clastic sedimentation.[22] Permission to publish by AAPG.

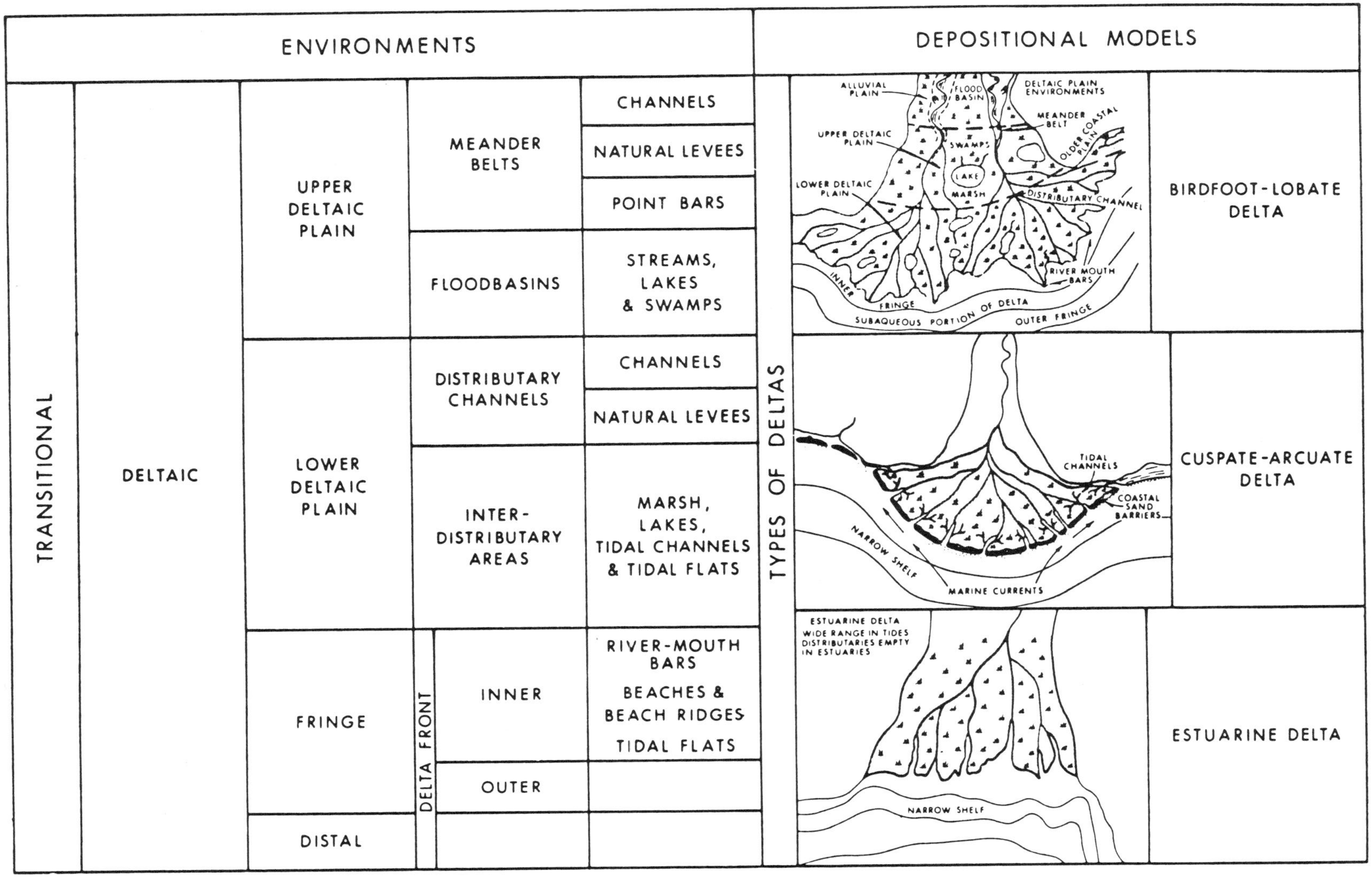

ENVIRONMENTS

TRANSITIONAL				
DELTAIC	UPPER DELTAIC PLAIN	MEANDER BELTS	CHANNELS	
			NATURAL LEVEES	
			POINT BARS	
		FLOODBASINS	STREAMS, LAKES & SWAMPS	
	LOWER DELTAIC PLAIN	DISTRIBUTARY CHANNELS	CHANNELS	
			NATURAL LEVEES	
		INTER-DISTRIBUTARY AREAS	MARSH, LAKES, TIDAL CHANNELS & TIDAL FLATS	
	DELTA FRONT	FRINGE	INNER	RIVER-MOUTH BARS
				BEACHES & BEACH RIDGES
				TIDAL FLATS
			OUTER	
		DISTAL		

DEPOSITIONAL MODELS

TYPES OF DELTAS

- BIRDFOOT-LOBATE DELTA
- CUSPATE-ARCUATE DELTA
- ESTUARINE DELTA

(Diagram labels — Birdfoot-Lobate Delta: ALLUVIAL PLAIN, UPPER DELTAIC PLAIN, LOWER DELTAIC PLAIN, DELTAIC PLAIN ENVIRONMENTS, MEANDER BELT, OLDER ABANDONED DISTRIBUTARY CHANNEL, RIVER-MOUTH BARS, INNER FRINGE, OUTER FRINGE, SUBAQUEOUS PORTION OF DELTA, SWAMP, MARSH, LAKE, LEVEE/FLOOD BASIN)

(Cuspate-Arcuate Delta: TIDAL CHANNELS, COASTAL SAND BARRIERS, MARINE CURRENTS, NARROW SHELF)

(Estuarine Delta: ESTUARINE DELTA WIDE RANGE IN TIDES DISTRIBUTARIES EMPTY IN ESTUARIES, NARROW SHELF)

FIG. 1-11—Deltaic environments and models of clastic sedimentation.[22] *Permission to publish by AAPG.*

7

	ENVIRONMENTS				DEPOSITIONAL MODELS
TRANSITIONAL	COASTAL INTER-DELTAIC	COASTAL PLAIN (SUBAERIAL)	BARRIER ISLANDS	BACK BAR, BARRIER, BEACH, BARRIER FACE, SPITS & FLATS, WASHOVER FANS	BARRIER IS COMPLEX
			CHENIER PLAINS	BEACH & RIDGES	CHENIER PLAIN
				TIDAL FLATS	
			TIDAL	TIDAL FLATS	
				TIDAL DELTAS	
		SUBAQUEOUS	LAGOONS	SHOALS & REEFS	
			TIDAL CHANNELS		
			SMALL ESTUARIES		
MARINE	SHALLOW MARINE	SHELF (NERITIC)	INNER	SHOALS & BANKS	SHALLOW MARINE
			MIDDLE		
			OUTER		
	DEEP MARINE	CANYONS			DEEP MARINE
		FANS (DELTAS)			
		SLOPE & ABYSSAL			
		TRENCHES & TROUGHS			

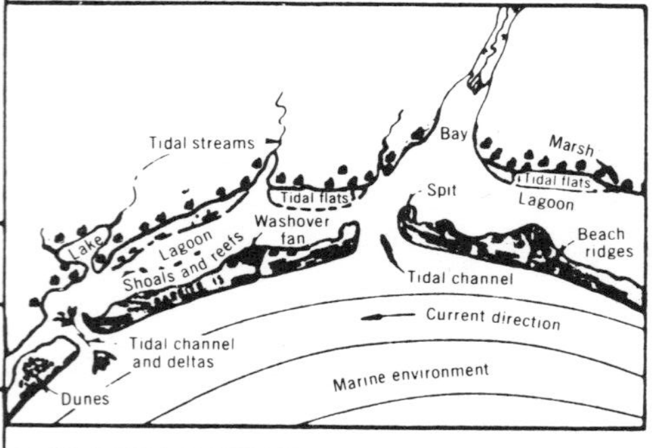

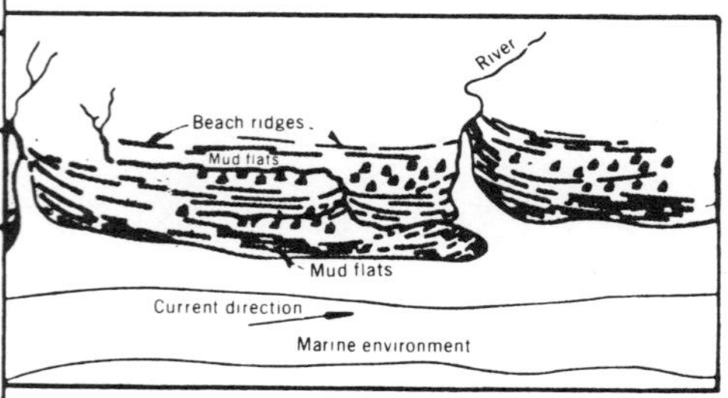

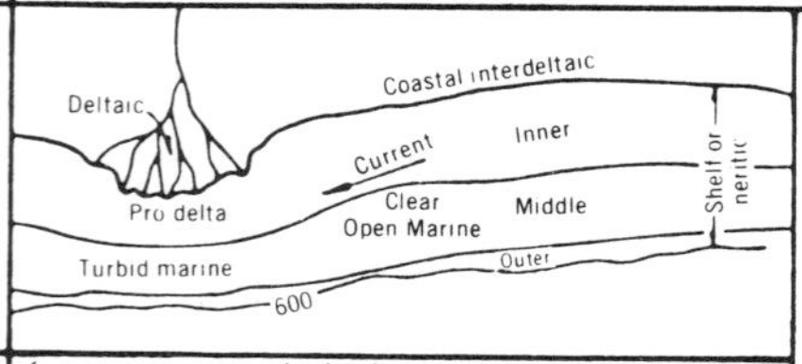

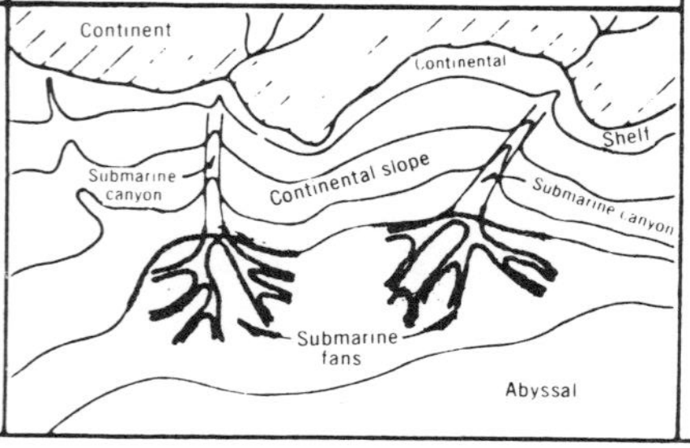

FIG. 1-12—*Coastal-interdeltaic and marine environments and models of clastic sedimentation.*[22] *Permission to publish by AAPG.*

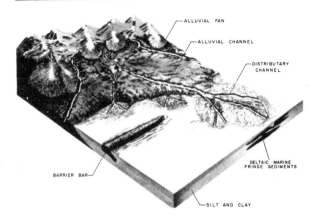

FIG. *1-13—Types of sandstone deposits, Elk City field, Oklahoma.*[28] *Permission to publish by The Society of Petroleum Engineers.*

predict. Primary factors controlling porosity are texture and sorting of the grains, type and manner of cementation of grains, amount and location of clay or other minerals associated with the sand, and the degree of layering or intrusion of clay, carbonates, or other materials in the interstices of the sandstone deposit. Cementing material is usually silica, but may be calcite, clay or other minerals.

Table 1-1 provides a brief analysis of sandstone porosity. Porosities in sandstone reservoirs are usually in the 15–30% range; however, sandstones having much lower porosity are currently being economically produced, primarily through the use of hydraulic fracturing. Because of sedimentary layering in sandstone deposits, continuous homogeneity in a vertical direction is often limited to a few inches or a few feet.

It is evident that 1-in. core plugs, frequently used in evaluating sandstones, are inadequate for the measurement of a vertical porosity and permeability. Continuity of vertical permeability is especially significant in studies to determine whether vertical movement of unwanted fluids within a producing oil or gas reservoir is due to coning, fingering, a poor cement job, or other mechanical problems.

Shale Break Prediction From a Study of Outcrops

Planning of well completions, including selection of the zones to be perforated, requires knowledge of the continuity of reservoir permeability and the

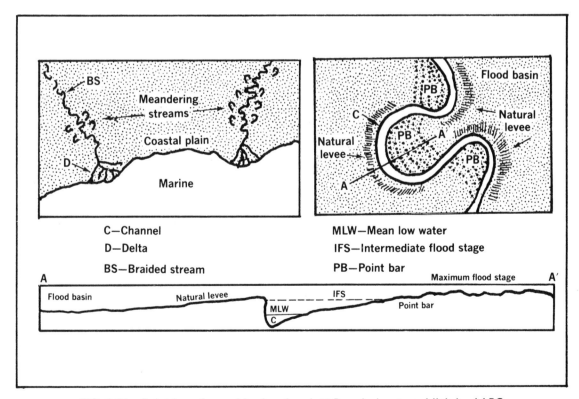

FIG. *1-14—Point bars formed in river bends.*[22] *Permission to publish by AAPG.*

TABLE 1-1
Porosity in Sandstone Reservoirs[7]

Aspect	Description
Primary porosity in sediments	Commonly 25–40%.
Ultimate porosity in rocks	Commonly 10–30%.
Types of primary porosity	Almost exclusively interparticle.
Types of ultimate porosity	Primary interparticle, greatly modified by precipitation of arthigenic clay minerals and silica.
Sizes of pores	Diameter and throat sizes closely related to sedimentary particle sizing and sorting.
Shape of pores	Initially dependent on particle shape, but greatly modified by secondary clay and silica.
Uniformity of size, shape, and distribution	Commonly fairly uniform within homogeneous body.
Influences of diagenesis (changes in rock since deposition)	Often large reduction of primary porosity by compaction and cementation.
Influence of fracturing	Generally not of major importance in reservoir properties.
Adequacy of core analysis for reservoir evaluation	Core plugs of 1-in. diameter commonly used for "matrix" porosity.
Permeability-porosity interrelations	Relatively consistent; commonly dependent on particle size and sorting.

TABLE 1-2
Lateral Continuity of Shale Breaks in Marine Sandstone[33]

% of shale breaks	Lateral extent, ft	Confidence
96	>250	99%
89	>500	86%
83	>1,000	52%
80	>2,000	48%

—Shale breaks in channel and deltaic sandstone tend to be converging, randomly distributed and unpredictable.

—Shale breaks in deltaic sands tend to be more continuous than in channel sands and less continuous than in marine deposits.

Implications of Outcrop Studies—In channel and deltaic sands, the tendency for a shale stringer to converge upon an adjacent shale stringer has serious connotations in regard to trapping oil or gas. Converging shales can reduce recovery efficiency in waterflood and other improved recovery operations. Selection of perforation interval can be critical in channel and deltaic sands. Crossbedding tends to place a directional restriction on sweep efficiency during waterflooding, and other improved recovery operations.

The lateral continuity of shale breaks and sandstone members in marine deposits provides favorable depositional conditions for enhanced recovery operations.

Effect of Silt and Clay Content on Sandstone Permeability[19]

Figure 1-15 relates permeability to percent of silt and clay for several sandstone formations in the United States and Canada. In all nine reservoirs an increase in silt and clay decreased permeability. However, direct correlation of clay content with permeability is usually restricted to a single genetic sand unit as illustrated by comparing the Stevens and Bow Island sandstones.

In the Stevens sand, 9% silt and clay reduced permeability to 1 md. In the Bow Island sand, permeability was about 1000 md with 9% clay and silt. This wide difference in permeability of the two sandstones, each with 9% silt and clay, is primarily due to the much coarser texture of the Bow Island sandstone.

location of reservoir barriers. Shale breaks have been correlated for miles through log and core studies of producing reservoirs[16]. Zeito's 1965 study[33] sandstone of marine, deltaic and channel origin provides valuable guidelines in this regard.

Table 1-2 shows that shale breaks in marine sandstone may be reliably predicted to have lateral continuity for considerable distances.

Zeito found that shale deposited in channel or fluviatile sands usually converges on adjacent shales; thus, sands tend to be pinched out by shales. Because of converging shales, only 17% of the sand deposits in Zeito's study were continuous for over 500 ft., and only 34% were continuous for over 250 ft. Further Zeito found that:

—Shale breaks in marine sandstone are generally parallel, and concentrated near the top and bottom of a sandstone unit.

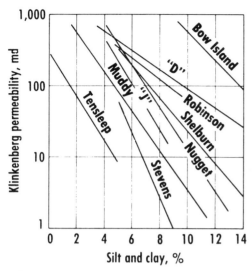

FIG. *1-15—Effect of silt and clay on sandstone
permeability.*[19] *Permission to publish by The
Petroleum Publishing Co.*

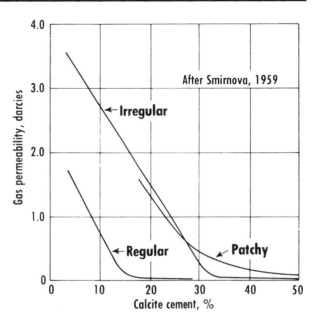

FIG. *1-16—How calcite cement alters permeability.*[19]
*Permission to publish by The Petroleum Publishing
Co.*

Effect of Mineral Cement on Sandstone Permeability[19]

Figure 1-16 shows a relation developed by Smirnova for the effect of calcite cement on fine-grained quartz sandstones from the Ugersko suite of the Miocene in Russia. With uniform distribution of calcite within the reservoir rock, approximately 20% calcite reduced gas permeability to less than 50 md. If 20% calcite was irregularly distributed within the rock, this Miocene sand had a gas permeability of 1500 md.

Permeability Variations with Texture Changes

Figure 1-17[19] shows data from a fluviatile sandstone reservoir having a fine-grained, ripple-laminated upper zone and medium-grained, cross-stratified lower zone.

The finer grained upper zone had lower water-oil relative permeability ratios at all water saturations.

Texture contrast causes some variation in permeability and relative permeability. A 1961 study by

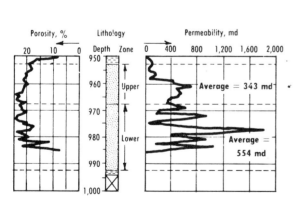

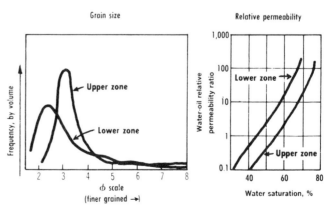

FIG. *1-17—Data from a layered Pennsylvanian sandstone reservoir.*[19] *Permission to publish by The Petroleum
Publishing Co.*

Hutchinson, et. al., of a shallow marine sandstone showed that sand texture caused a maximum variation in permeability of 5:1. By comparison, however, the degree and type of cementation caused a 100:1 variation in permeability.

Relation of Permeability to Irreducible Water Saturation

Figure 1-18[19] provides data developed by White, et. al., in 1960, showing the relationship between permeability and irreducible water saturation for a number of formations. In general, permeability decreases as irreducible water saturation increases. For a single genetic formation, the irreducible water relationship is sufficiently precise for calculation and prediction purposes.

Identification of Channel or Fluviatile Sandstones

Hewitt[19] reported detailed studies of a Pennsylvania sandstone reservoir, described in following

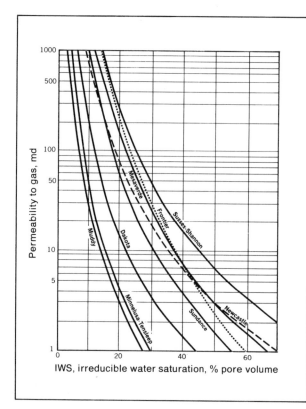

FIG. 1-18—How permeability and irreducible water saturation are related.[19] Permission to publish by The Petroleum Publishing Co.

paragraphs, which can be used to help identify channel or fluviatile reservoirs. Core and log data illustrated in Figure 1-19, along with production tests, structure maps, and isopach maps, are frequently all that is available to guide engineers in the development and operation of oil and gas fields.

Usual interpretation of log and core data would provide this information.

—Lithology log—sand from reservoir top at 885 ft to 942 ft. Neither the SP log or resistivity logs provide reliable data on continuity of vertical permeability and porosity in this sand-shale sequence.

—Core porosity—15% to 25%, average 20%.

—Core horizontal permeability—lower two-thirds erratic, varying from 50 to 750 md. In upper one third, permeability gradually decreased upward to zero at 885 ft.

These core and log data per se are insufficient in this case to allow extrapolation or prediction of reservoir values beyond the wellbore in which they were measured. Transient pressure analysis could indicate the extent of the reservoir, but provides only an average of vertical and lateral variations in reservoir properties.

Better methods are required to detail the continuity of shale and sandstone members. Determination of the depositional environment is a big step in this direction.

Examination by Hewitt, et. al., of slabbed, continuous cores of the Pennsylvanian sandstone of Figure 1-19, along with previously available data, revealed these characteristics of the deposit.

—Gradational upper contact of sand body.

—Sharp erosional basal contact.

—High-angle cross stratification in the middle part of the reservoir.

—Common slumping and distortion of the cross strata.

—Trough-shaped ripples in the upper part of the reservoir.

—Fragments of carbonized wood.

—Few burrows by organisms in the overlying shales.

These observations indicate this reservoir to be non-marine fluviatile sandstone.

These characteristics should generally be representative of any fluviatile or channel sandstone

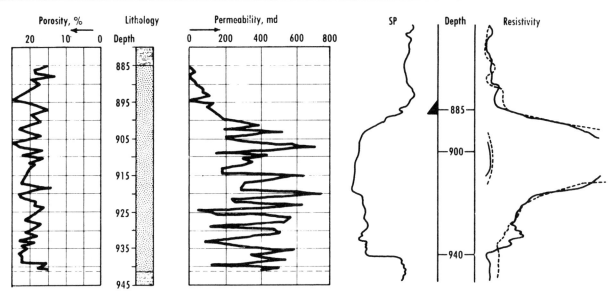

FIG. *1-19—Properties of a Pennsylvanian sandstone reservoir.*[19] *Permission to publish by the Petroleum Publishing Co.*

reservoir and should be useful to reservoir or production engineers involved with channel sand deposits:

—Trend of deposit: Long dimension parallel with paleoslope. Large angle to paleoshore.

—Shape: Commonly lenticular and elongate, with length many times the width.

—Boundaries: Lower boundary—sharp and erosional. Upper boundary—gradational. Lateral boundaries—sharp lateral boundary if erosional depression completely filled with sand; gradual lateral boundary if sandstone body is small relative to size of a large alluvial plane.

—Texture: Grain size may decrease upward in reservoir.

—Structures: Trough-shaped ripples; high angle cross-stratification; slump common; burrows rare.

—Permeability: Decreases upward if grain size decreases. Maximum horizontal permeability is parallel with direction of sediment transport, which usually parallels long dimension of sand body. This directional permeability results from (1) trough-shaped ripples, (2) high angle cross-stratification, and (3) the long axis of sand grains being oriented in direction of stream flow during transport and deposition.

The most significant single criterion for identifying a fluviatile sandstone is downward thickening,

at the expense of the subjacent (eroded) strata. Although no single criterion is diagnostic, various combinations of the above criteria usually are adequate for precise determination.

Geologic Control Summary for Sandstone Reservoirs[19]

Figure 1-20 outlines the relation of depositional environment, source material, and post depositional history to various sandstone reservoir parameters,

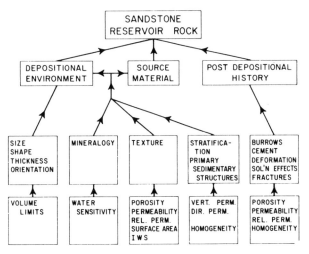

FIG. *1-20—Potential value of geologic studies of sandstones in reservoir evaluation.*[19] *Permission to publish by The Petroleum Publishing Co.*

such as volume limits of the sand deposit, water sensitivity, porosity, horizontal and vertical permeability, relative permeability, surface area of deposited material, irreducible water saturation, and homogeneity. This information, plus log studies, analysis of core permeability and porosity, provide most of the basic information required for reservoir studies, and the planning, operation and maintenance of wells.

APPLICATION OF GEOLOGIC CONCEPTS IN SPECIFIC SANDSTONE RESERVOIRS

Reservoir Description in the Niger Delta

Extensive work on reservoir description of the Niger Delta was reported by K. J. Weber in 1971[31] and again in 1975[32] by Weber and Daukoru, of Shell BP Nigeria Ltd. This work covers a 20-year period and is being continued and expanded. Techniques used in these studies have increased recoverable reserves, decreased the number of dry holes, and increased the profitability of operations.

Commercial oil was discovered by Shell-BP in the Niger River Delta in 1955. By 1975, Niger delta production made Nigeria the seventh largest oil producer in the world.

The Niger delta sequence starts with a thick deposit of under-compacted marine clays overlain by near shore deposits, in turn overlain by progradational continental sands. All of this is a composite of sediments deposited during repetitious cycles of prograding sedimentation. Basement faulting affects delta development and sediment thickness distribution. In the near-shore sequence, roll-over structures associated with growth faults trapped the hydrocarbons. Various depositional environments of these reservoir sands strongly influence well completions, well productivity, and oil and gas recovery.

Figure 1-21 is a block diagram illustrating the depositional environments and growth faulting in the Niger delta. Producible sand deposits are found in barrier bars, river mouth bars, tidal channels, distributary channels, and point bars.

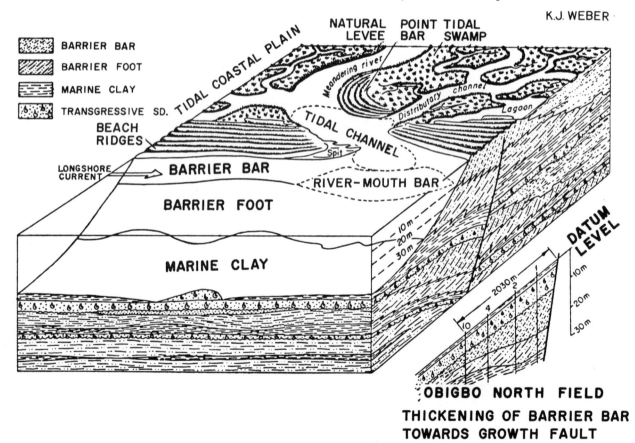

FIG. 1-21—Block diagram of typical Niger delta sediments.[31]

These different types of sand bodies can usually be distinguished by electric logs, gamma-ray logs, and by grain-size distribution obtained from cores. The permeability, porosity, and reservoir characteristics differ markedly from one type to another.

Barrier bar sands are deposited in a marginal marine environment on top of the finer grained barrier-foot deposits, as illustrated in Figure 1-21. General characteristics of barrier bars are:

—Sand grain size and permeability increases from the bottom to the top of the bar.

—Clay "breaks" in barrier bars can be correlated over long distances. These clay "breaks" may limit or stop fluid flow across them in a vertical direction.

—A series of barrier-bar sands may be found on top of each other with only a thin marine clay and/or thin interval of barrier foot between clean bar sands.

—Barrier bars may be eroded in places by cross-trending river channels which, in turn, may be filled with reservoir sandstone.

Tidal channel fills often consist of a series of thin cross-bedded sequences fining upwards with a clay pebble or gravelly lag deposit at the base, and separated by thin clay beds. Grain size distribution is similar to fluviatile or point bar deposits. Clay breaks generally are difficult to correlate in tidal channel fills. River-mouth bars may be associated with tidal channels but are usually similar

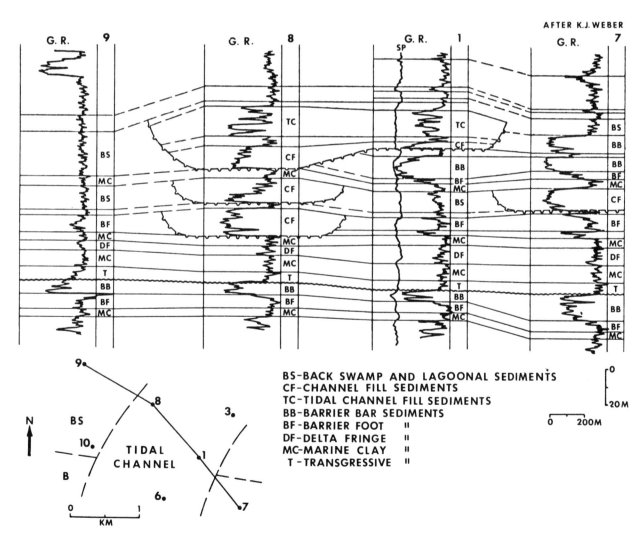

AFTER K.J. WEBER

BS—BACK SWAMP AND LAGOONAL SEDIMENTS
CF—CHANNEL FILL SEDIMENTS
TC—TIDAL CHANNEL FILL SEDIMENTS
BB—BARRIER BAR SEDIMENTS
BF—BARRIER FOOT   "
DF—DELTA FRINGE   "
MC—MARINE CLAY   "
T—TRANSGRESSIVE   "

FIG. 1-22—Stratigraphic section, Egwa field, Niger Delta.[31]

to barrier bars in depositional character.

Tidal channel locations are shown in Figure 1-21. Log correlation of various zones for part of the Egwa field is illustrated in Figure 1-22 with the various types of deposits shown for each well. A tidal channel is the topmost oil productive zone in wells 1 and 8. Well 9 penetrated back swamp and lagoon deposits and lacked permeable sands.

Fluviatile or Channel Deposits—These are formed by meandering rivers, and thus form point bars in a brickwork fashion covering much of the ancient river flood plain. River channel fills are shown below the tidal channel deposit in Wells 1 and 8, Figure 1-22. Fluviatile sediments are characterized by upward fining of grain-size distribution in the upper part of the fills, which are composed of laminated wavy-bedded clay and silty sand. However, the clayey part on top of the point bars frequently is eroded off, leaving coarse-grained sand on top.

Fluviatile sand bodies often have a sharp erosive base, which is easy to identify on SP and Gamma ray logs. Oval point bars are often bounded on three sides by clay plugs of the oxbow type lakes. Distributary channel fills are usually formed nearer the coast than point bars and frequently erode their course through barrier sediments.

Growth Faults or Step Faults—Almost all of the hydrocarbon accumulations in the Niger delta are contained in roll-over structures associated with growth faults. These growth faults were caused by rapid sedimentation on top of undercompacted marine clays along the delta edge. Figure 1-23 illustrates roll-over structures, trapping hydrocarbons in the closures on the downthrown side of a boundary (growth) fault. Combinations of crestal and antithetic faults in Figures 1-23c and 1-23d produced collapsed graben-like structures with smaller and more numerous traps.

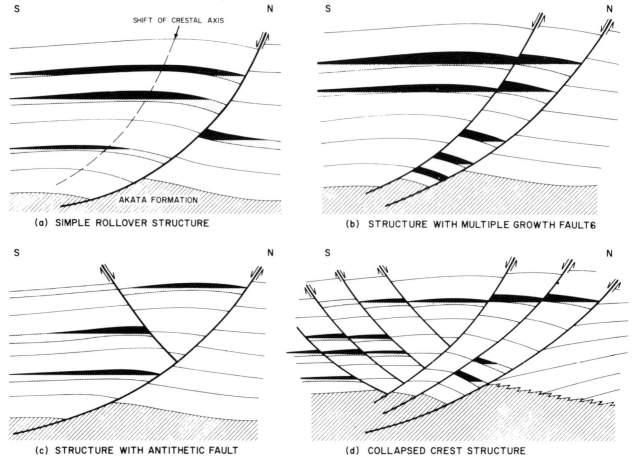

(a) SIMPLE ROLLOVER STRUCTURE

(b) STRUCTURE WITH MULTIPLE GROWTH FAULTS

(c) STRUCTURE WITH ANTITHETIC FAULT

(d) COLLAPSED CREST STRUCTURE

NOTE: ONLY A FEW RESERVOIR SANDS ARE SHOWN IN THE SCHEMATICAL SECTIONS AND THE SAND THICKNESS HAS BEEN ENLARGED

FIG. *1-23—Types of traps in the Niger delta.*[31]

Practical Use of Sedimentological Information in Oil and Gas Field Development

Figure 1-24 was prepared to illustrate the practical use of sedimentological information. It is readily apparent that well locations and well completions must be made to take advantage of sedimentary conditions if reservoirs are to be efficiently produced. In differentiating between formation damage in these fields and naturally low permeability, the type of deposit may be critical.

For example, typically the completion in a barrier-bar sand may produce only a tenth as much oil per day as a channel-fill completion with the same length of interval. The higher productivity and permeability of channel fills usually is due to larger grained sand deposited in the channels.

Following are some of the significant highlights of the Niger delta studies.

1. Continuity of clay breaks or other impermeable barriers can be predicted from depositional environment. This information is very significant in predicting gas and water coning. The success of squeeze cementing in shutting off unwanted water or gas depends on knowledge of barriers to vertical movement of fluids. This can affect ultimate recovery appreciably.

2. Channel fills, point bars, and barrier bars have characteristic distributions of permeability both vertically and laterally. This can be critical in selecting the location of perforations to drain each sand member.

3. Sand grains in the Niger delta are generally large with many sands being quite unconsolidated. Identification of type of deposit is made from side-wall cores and conventional cores plus petrophysical data obtained from logs.

4. Channel-fills, eroding across barrier bars, may

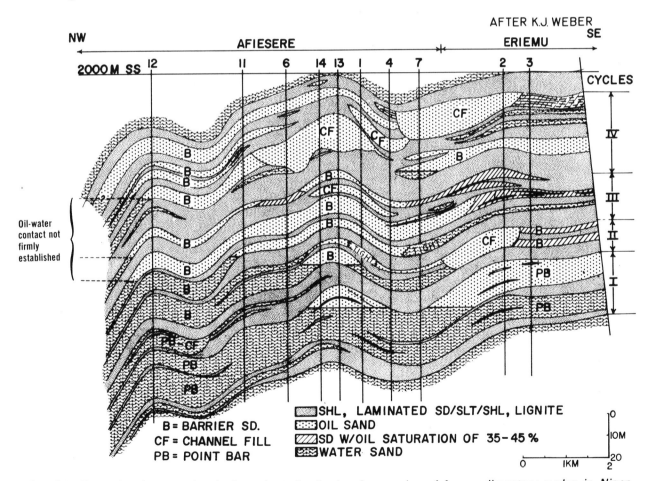

FIG. 1-24—*Example of a complex hydrocarbon distribution in a series of four sedimentary cycles in Niger Delta.*[31]

have higher productivities than barrier bars because of larger sand grains; however, sand grains may be poorly sorted in channel sands.

5. Grain size analysis from cores and side wall cores has been a most significant tool to identify the various types of deposits and also to select the high productivity intervals for completion.

Reservoir Description—Elk City Field, Oklahoma [28]

Detailed knowledge of the distribution of porosity and barriers influencing fluid flow must be integrated with reservoir engineering data to select, plan, and implement any enhanced recovery project. Studies must adequately account for inhomogeneities or variations in reservoir and nonreservoir rock properties. Better understanding of sand body genesis—how it is deposited and the environment of deposition—make possible a more accurate description of the reservoir, which is required for predicting recovery performance.

A reservoir description study of the Elk City field was carried out by the Shell group as part of the planning phase for waterflooding. A total of 310 wells were drilled on 40-acre spacing in this 100-million bbl field, producing from a ±500-ft. thick zone of Pennsylvanian sandstones and conglomerates. The structure is anticline with eight reservoir zones between 8,800 and 11,000 ft.

The study is an excellent illustration of the step by step process required for effective reservoir description. It includes many useful techniques and methods of presenting data.

Approach and Methods of Study—In this investigation, data were integrated from genetic sand studies, petrologic-petrophysical analysis, and rock-log calibrations to characterize and map pore-space distribution, and to help predict fluid-flow response of the waterflood. Thickness maps, based on sand genesis, of net permeable sand were prepared to represent floodable sand volume. Predictions were made, from genetic facies maps, of the spatial distribution of permeability and permeability barriers, as well as anticipated flood performance.

Interrelated studies covered are:

1. Lithology and petrophysical properties, with logs being calibrated with lithology in cored intervals.

2. Interpretation of genesis of reservoir and reservoir rocks.

3. The establishment of formation correlations.

4. Mapping by subzones of net sand and sequences of rock types.

Most of the 310 wells were logged with the SP log, the 8- and 16-in. Normal Resistivity logs, the 24-ft. Lateral log, and the Microlog. Gamma ray-neutron logs were run on 16 wells in place of the Microlog. In addition, over 1700 feet of cores from 26 wells were available representing 3 to 9 in. of section from each foot cored.

One of the starting points for this study was a structure map, Figure 1-25, on a limestone bed easily correlated across the field, and located between the primary zones in the field.

For the lithologic study, mineralogy, grain size, sorting, and sedimentary structures were examined visually and microscopically. Pore sizes and geometry, and the type and amount of pore filling material were obtained from thin sections of cores. Grain size and sorting were determined either by sieve analysis or with a grain size-sorting comparator. Porosity and permeability were measured on a large number of cores and capillary pressure measurements were made on 35 cores.

Lithology and Petrophysical Properties and Relations—Reservoir rocks in this field are conglomerates and sandstones. Nonreservoir rocks are siltstones, shales, and limestone. Sand grain sizes range from very fine (0.062 mm diameter) to very coarse (2.0 mm). Most conglomerates range from granule (2.0 mm to 4.0 mm) to pebble size (4.0 to 64.00 mm). The finer grained rocks are the best sorted. With increase in median grain size, the sorting becomes progressively poorer.

Porosity, pore size, and permeability of any sandstone or conglomerate depend primarily on (1) grain size and sorting, and (2) the amount of cementation and compaction. Reservoir rocks in the Elk City field are all compacted. Cement or pore-fill material is less than 7% by volume. In these rocks, pore space correlates with grain size and sorting. Figure 1-26 shows a plot of grain size and sorting versus porosity. The fine grained rocks have highest porosities, and with increase in grain size, sorting is poorer and porosity decreases.

Pore size was estimated from pore-size measurements in thin sections, and from capillary pressure

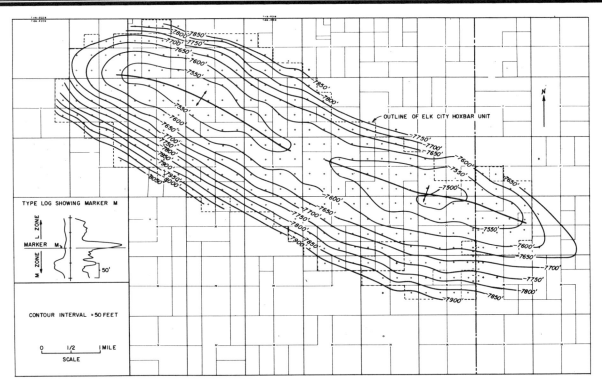

FIG. *1-25—Structure contour map on marker M, Elk City field.*[28] *Permission to publish by The Society of Petroleum Engineers.*

curves. Capillary pressure curves of representative reservoir rock types are shown in Figure 1-27. The fine-grained rocks have predominantly fine and very fine pores. With an increase in grain size, the average pore size and the number of medium and large pores increase.

From Figure 1-27, it is interesting to note the rather high permeability of 1638 md in relatively low porosity of 12.5% in pebble conglomerate with well-sorted coarse sand matrix. On the other end of the grain-size scale, the very fine, very well sorted sandstone had a permeability of 1.6 md and a porosity of 20.0%.

Log Calibration—Logs were correlated with cores so that logs could be used to supplement core data. The SP and 8-in. Normal Resistivity curves and the Microlog were used to determine rock type and sequences of rock type as illustrated in Figure 1-28.

Impermeable siltstones, calcareous shales and limestones are distinguished from sandstones and conglomerates on the basis of SP development. For the sandstones and conglomerates, which exhibit moderate to well-developed SP, grain size is indi-

cated by the 8-in. Normal Resistivity curve. The fine-grained rocks exhibit low resistivity, and with increase in grain size, the values of resistivity increase. Grain size determinations of the very silty or shaly or cemented rocks present a problem. The measured resistivity of these rocks is higher than clean or noncemented rocks of equivalent grain size.

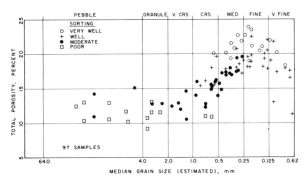

FIG. *1-26—Relationship among sorting, grain size, and porosity in Elk City sandstones and conglomerates.*[28] *Permission to publish by The Society of Petroleum Engineers.*

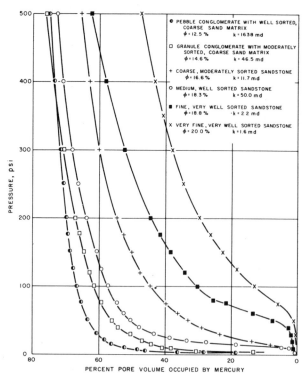

FIG. *1-27—Capillary-pressure curves of typical reservoir rock types.*[28] *Permission to publish by The Society of Petroleum Engineers.*

The Microlog opposite these rocks shows no separation or a characteristically "hashy" appearance.

Correlation between grain size and log resistivity is apparent because of the relationship between rock types and pore space. The better sorted fine-grained rocks have the highest porosity and largest number of fine pores. This results in low values of formation factor and resistivity. Formation factor and resistivity increase as grain size increases, sorting becomes poorer, and porosity and number of fine pores decrease. Hydrocarbon saturation has only a minor influence on resistivity values from the 8-in. Normal because this curve seldom probes beyond the invaded zone adjacent to the well bore.

The distribution of porosity, permeability, and pore size follows in general the distribution of rock types. Correlation of rock types, permeability, and porosity is considered later in this discussion for various types of sandstone deposits.

Genesis of Sand Bodies—From core studies, including thin sections, the reservoirs consist principally of one or more of the following genetic types: barrier bar, alluvial channel, deltaic distribu-

tary channel, and deltaic marine fringe. Figure 1-13 illustrates schematically the trend and distribution of the various deposits.

Barrier Bar Deposit—Barrier bar deposits are composed of rocks ranging from siltstone to pebble conglomerate with progressive increase in grain size upward. A typical barrier bar sequence is illustrated by Figure 1-29. It may be noted that the resistivity gradually increases as grain size increases in an upward direction. Figure 1-30 shows that permeability follows the porosity trend and generally increases in an upward direction within the sand body.

The barrier bar deposits trend parallel with the depositional strike of the marine strata. Figure 1-31 shows the barrier bar deposits in relation to adjacent deltaic deposits and a lower marine limestone marker.

The coarsest-grained material is concentrated on the landward side and at the top of the deposits. The overall grain-size and sand thickness decreases in a seaward direction across the trend of the deposit. The permeability profile is essentially the same, parallel with the sand-body trend and decreases uniformly toward sea side of the bar and is highest on the shore or back side of the deposit.

No shale breaks exist in the bar deposits except near the base. The lower and lateral boundaries are gradational with adjacent impermeable siltstones and shale.

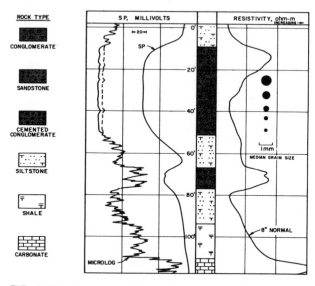

FIG. *1-28—General relations between rock type and log response.*[28] *Permission to publish by The Society of Petroleum Engineers.*

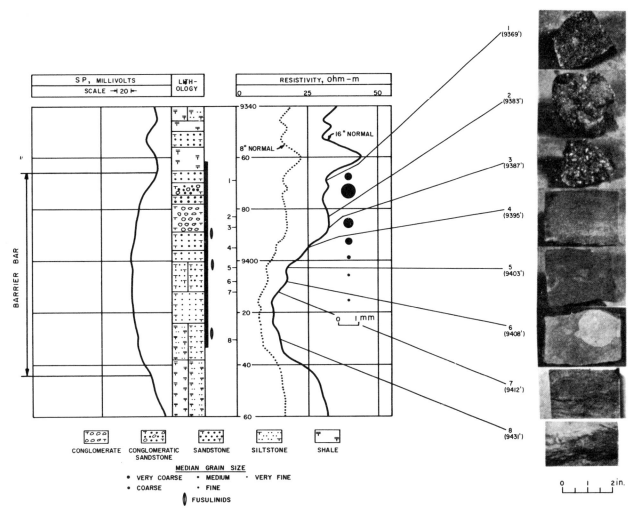

FIG. *1-29—Vertical sequence of rock types and log response of a barrier bar deposit, Shell, G. Slatten No. 1.*[28] *Permission to publish by The Society of Petroleum Engineers.*

Conglomerates, coarse sandstones, and some fine and medium-grained sandstones are massively-bedded. The fine- and medium-grained rocks have faintly developed, slightly inclined or horizontal bedding. The fine sandstones and siltstones are laminated and ripple bedded and show some reworking by marine organisms.

Alluvial Channel Deposits—This type of channel deposit is predominantly coarse grained rock with a cyclical vertical variation in grain size due to being deposited by periodic floods from high gradient braided streams. Conglomerates, conglomeratic coarse-grained sandstones are the dominant rocks. Fine to very fine grained sandstones, siltstones, and silty shales are interbedded with coarser rocks but are not so abundant.

Typical log response of the SP, Microlog and 8-in Normal Resistivity curves for the channel deposits is shown in Figure 1-32.

The SP curve is well developed opposite the sandstones and conglomerates, with the SP being depressed slightly toward the shale baseline opposite silty sands. Opposite the permeable sandstones and conglomerates, the Microlog has positive separation. In the low permeability silty members, the Microlog is "hashy" and shows no positive separation, thus indicating small grain size.

The alluvial channel deposits are linear in trend and tend to be oriented perpendicular to the depositional strikes of the marine strata. Figure 1-33 shows a channel sandstone and conglomerate deposited in a valley cut into older marine siltstones and shales.

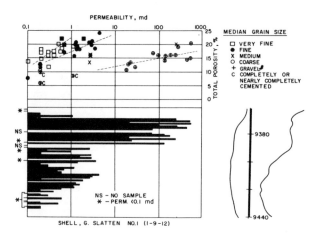

FIG. *1-30—Plot of porosity vs. permeability and vertical permeability profiles, Shell, G. Slatten No. 1.*[28] *Permission to publish by The Society of Petroleum Engineers.*

The distribution of deposits is controlled by size and shape of the stream-cut valley. Sand and gravel filled much of the valley, with silt and clay filling the remainder.

Rocks composing the channel deposits have relatively low porosity (10 to 15%), but high permeabilities (75 to 1500 md). Because of the relative frequency of depositional cycles, vertical and lateral permeabilities may be quite variable.

Distributary Channel Deposits—These deposits are similar to the alluvial channel deposits except that the rocks are not as coarse.

Deltaic Marine Fringe Deposits—Deltaic deposits in the Elk City field consist of either one or two stacked genetic units. Rock types range from shale to pebble conglomerate.

The vertical sequence of rock types in these deposits is characterized by (1) interbedding of sandstone with shale and siltstone, (2) a general upward decrease in the number of shale and siltstone beds and in the amount of interstitial clay- and clay-size particles in the sandstones, and (3) a general upward increase in grain size for each depositional sequence.

The deltaic marine fringe deposits are irregular in geometry and distribution. Toward the lateral and basinward edges of the deposits, sandstones become progressively finer and grade into siltstone and shale.

For each sequence in the delta buildup, permeability increases in an upward direction. The lower part of each sequence may be shaly or silty with low permeability. Each sequence is separated by impermeable or low permeability rocks.

Toward the lateral or basinward edges of the fringe deposits, the permeability gradually decreases as sand grades into impermeable silt or clay.

Permeability distribution in delta marine fringe deposits is similar to that found in barrier bars. However, the thin siltstone/shale beds that cap each cycle, are widespread and are effective barriers to vertical flow.

Genetic Sand-Body Identification and Delinea-

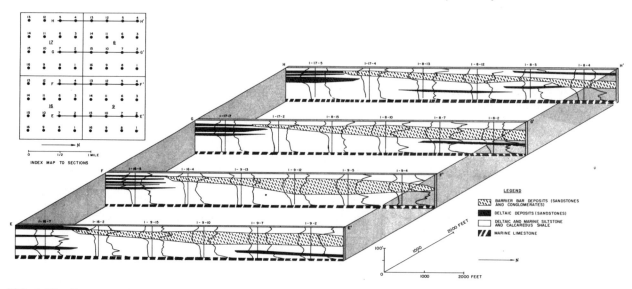

FIG. *1-31—Cross sections through a barrier bar deposit.*[28] *Permission to publish by The Society of Petroleum Engineers.*

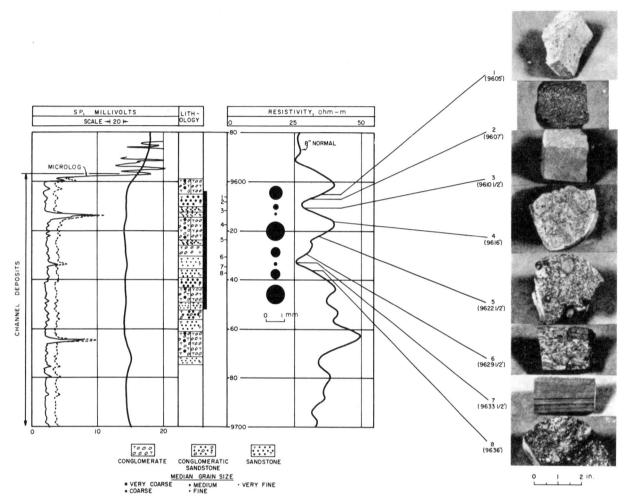

FIG. *1-32—Vertical sequence of rock types and log response, Shell, B. Pinkerton No. 1.*[28] *Permission to publish by The Society of Petroleum Engineers.*

tion—Most sandstone deposits may be divided into individual genetic units deposited during a single cycle of sedimentation. For example, a genetic unit or cycle for a fluviatile sand may be the sand, siltstone, or shale layers deposited during a single flood cycle. Such a cycle can be correlated across a field unless interrupted or cut into by other deposits. Figure 1-34 shows the lateral distribution of genetic sand units for one zone within the reservoir. It may be noted that all types of sand found in the field are represented in this single zone.

Figure 1-35 is a sand thickness map of the L_3 Zone. This map used with other data aids in understanding the heterogeneity of the reservoir and well problems associated with both primary recovery, pressure maintenance, and enhanced recovery. The type of map presented here is particularly helpful in providing better communication between various personnel involved in technical studies and in field operations.

Permeability distribution in delta marine fringe deposits is similar to that found in barrier bars. However, the thin siltstone/shale beds that cap each cycle, are widespread and are effective barriers to vertical flow.

Lithology and pore space in many types of sand are distributed in a systematic and predictable manner. The distribution and continuity of porosity, permeability and lithology can be predicted with reasonable certainty after identifying the genesis of sand members in a field.

From the identification and delineation of genetic sand bodies in the L_3 subzone, Figures 1-34 and

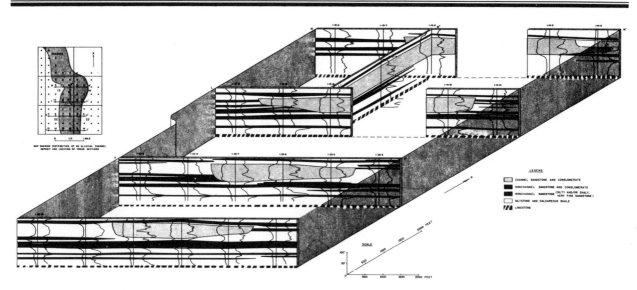

FIG. *1-33—Areal distribution of a portion of an alluvial channel deposit and cross sections through the deposit.*[28] *Permission to publish by The Society of Petroleum Engineers.*

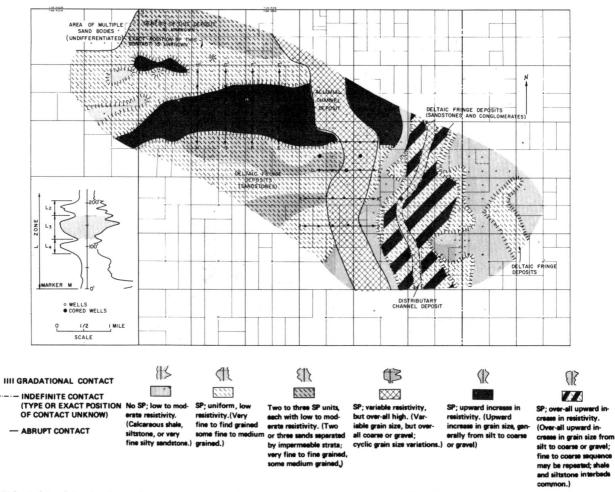

FIG. *1-34—Distribution of sequence of rock types, L_3 subzone, showing the distribution of the genetic types of sand bodies.*[28] *Permission to publish by The Society of Petroleum Engineers.*

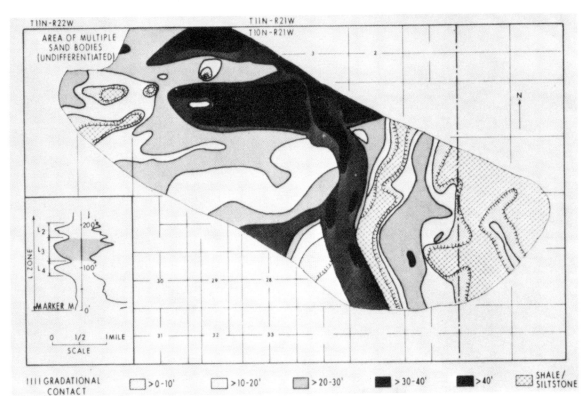

FIG. *1-35—Thickness map of net sand, L₃ subzone.*[28] *Permission to publish by The Society of Petroleum Engineers.*

1-35, it can readily be seen how important this type of study can be from the standpoint of selecting and completing wells for the waterflood project. There is no doubt that this type of study would have been quite useful in planning operations under primary recovery.

CARBONATE RESERVOIRS

Carbonate deposits are usually laid down as calcium carbonate ($CaCO_3$), primarily originating from a mixture of ground-up shells and excrement of marine organisms. Because of the heterogeneous nature of carbonates, a detailed description of any carbonate trap is required.

Limestones are composed of granular particles, a fine matrix, and cementing material. The grains may vary in size from a few microns to recognizable fragments of shells or corals. There are several different kinds of grains, of which four are the most important. These are (1) shell fragments, called "bio"; (2) fragments of previously deposited limestone called "intraclasts"; (3) small round pellets,

the excreta of worms; and (4) ooliths spheres formed by rolling lime particles along the bottom.

The matrix is usually calcareous mud containing clay-size calcite particles called micrite. The clear secondary calcite cement is called sparite. A rock consisting primarily of clear secondary calcite with intraclast grains is called "intrasparite".

A rock consisting mainly of micrite (lime mud) with grains of broken shell fragments would be called "biomicrite". Biomicrite and pelmicrite are the most common limestone types.

A modification of Folk's classification of limestone is shown in Figure 1-36.

In addition to these classifications, some limestones consist only of micrite; others consist of the remains of upstanding reef-building organisms.

Pores in limestone may be either primary or secondary. The primary pores are those that were formed during deposition in environments where there were strong waves or currents to remove the fine muds. In such cases the limestones may consist mostly of grains with a minimum of micrite or sparite.

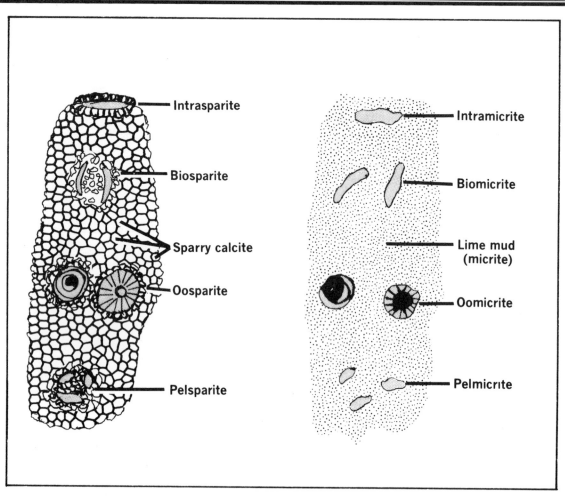

FIG. *1-36—Modified Folk's classification of limestone.*[15]

Secondary pores are formed by solution and reprecipitation of limestone after deposition. The most common types are "molds" and "channels". In the case of molds, certain grains are dissolved preferentially, leaving cavities. In the case of channels, the solutions moving through the rock dissolved out small tunnels. These are usually small in dimension (0.1 to 10 mm) and are lined with calcite crystals.

Dolomite, $CaMg(CO_3)_2$, is formed by recrystallizing micritic limestone. In the dolomitization process, water containing magnesium dissolves $CaCO_3$ and reprecipitates $CaMg(CO_3)_2$. Dolomitization may completely alter the texture of original limestone, and it may destroy such structures as bedding and fossils. The final texture is medium to coarsely crystalline, with a general massive overall structure.

Carbonates, being heterogeneous, may change from dense to porous within a few feet or even inches. It is very important to determine pore structure of carbonates. This is primarily controlled by the type of reservoir rock.

Carbonate Porosity

A few limestone reservoirs contain their original porosity. Chalks, oolitic limestone, and fragmented limestone have high porosity, and many have some continuous porosity, see Table 1-3. A major problem in nearly all limestone is the lack of continuous porosity over a great distance. Fractures and joints often provide the best source of continuous porosity and permeability, unless fractures have been recemented with a secondary deposition.

Joints developed under a given stress pattern will be usually near vertical and parallel. If the reservoir rock is later subjected to a different stress pattern,

**TABLE 1-3
Porosity in Carbonate Rocks[7]**

Aspect	Significance
Primary porosity in sediments	Commonly 40-70%.
Ultimate porosity in rocks	Commonly none or only small fraction of initial porosity; 5-15% common in reservoir facies.
Types of primary porosity	Interparticle commonly predominates, but intraparticle and other types are important.
Types of ultimate porosity	Widely varies because of post-depositional modifications.
Sizes of pores	Diameter and throat sizes commonly show little relation to sedimentary particle size or sorting.
Shape of pores	Greatly varied, ranges from strongly dependent "positive" or "negative" of particles to form completely independent of shapes of depositional or diagenetic components.
Uniformity of size, shape, and distribution	Variable, ranging from fairly uniform to extremely heterogeneous, even within body made up of single rock type.
Influence of diagenesis (change in rock since deposition)	Major; can create, obliterate, or completely modify porosity; cementation and solution important.
Influence of fracturing	Of major importance in reservoir properties if present.
Adequacy of core analysis for reservoir evaluation	Core plugs commonly inadequate; whole cores may be inadequate for large pores.
Permeability—porosity interrelations	Greatly varied; commonly independent of particle size and sorting.

a different set of joints may develop and may intersect previously formed joints.

The great variations in limestones are due primarily to the fact that limestones may be modified or changed continuously as long as there is some porosity filled with water. When the water is at rest, solution and precipitation of carbonates may occur to develop larger crystals. If the water is circulating, recrystallization, solution, or precipitation of introduced materials will occur.

The original, primary porosity of most limestone reservoirs has been greatly altered by solution and reprecipitation performed by water moving through the rock. Table 1-3 provides general information on carbonate porosity.

Porosity Through Dolomitization—Porosity may be developed during the process of converting limestone ($CaCO_3$) to dolomite $CaMg(CO_3)_2$. Continuous porosity sufficient to produce oil or gas may depend on the degree of dolomitization in many carbonate reservoirs. Larger continuous flow channels or cavities are probably due to leaching out of limestone or calcitic fossils.

Porosity and pore size increase as complete dolomitization is approached. However, about 70% of the limestone must be dolomitized before appreciable porosity is formed. Saccharoidal or "grainey" dolomite usually is a good reservoir rock with well sorted intercrystalline porosity. Vugs or fractures usually are required for dense carbonates to produce oil and gas. In such instances oil or gas in "pinpoint" porosity may "bleed" into the fractures.

Types of Porosity—To properly develop and operate carbonate reservoirs, the basic type of porosity must be known. Because of the great variations in porosity in both the vertical and horizontal directions in a carbonate reservoir, detailed geologic, and chemical description of reservoir rocks should be available from a representative number of wells penetrating the reservoir.

As an aid in understanding geologic description of carbonates, the seven most abundant types of carbonate porosity are:

1. *Interparticle Porosity*—This is porosity between grains or particles of depositional origin. This is the dominant type of porosity in most carbonate sediments. Secondary interparticle porosity may be developed by selective dissolution of fine particles between larger particles.

2. *Moldic Porosity*—A mold is a pore formed by selective removal, normally by solution, of a rock constituent such as a shell or oolith.

3. *Fenestral Porosity*—Mud-supported and grain-supported fabrics, where the openings are larger than interparticle openings give this porosity.

4. *Intraparticle Porosity*—Internal chambers or other openings within individual or colonial skeletal

organisms are the most common of intraparticle
pores.

5. *Intercrystal Porosity*—Porosity between crystals of equal size, as is found in porous dolomite,
is of this type.

6. *Fracture Porosity*—This is formed by fracturing or breaking the reservoir rocks, normally by
regional or localized tectonic stress.

7. *Vugular Porosity*—An opening in the rock
large enough to be visible with the unaided eye.
The opening does not conform in position, shape,
or boundaries to particular fabric elements of the
reservoir rock. Most vugs were created by dissolution.

Types of Carbonate Reservoir Traps—The single
most important facies that serves as a trap for
hydrocarbons is the reef. A reef is a stratigraphic
trap, with some being enclosed in shale. A reef
has drape and flank beds which commonly dip away
from the reef. Reefs may occur in large complexes,
or may be scattered over an area as patch reefs.
If oil or gas, localized in the upper part of a
spire-shaped reef, alone or cresting as a summit,
it is often called a pinnacle reef.

Figure 1-37 shows the relative location of three
very common types of carbonate traps, namely,
reef, backreef, and lagoon. The reef is normally
deposited where waves are strong and original
porosity may be high. Lagoon type carbonates are
formed from reef debris which has become calcareous mud, with deposition in a low energy
environment. Backreef deposits are formed in a
slightly higher energy environment than lagoonal
deposits.

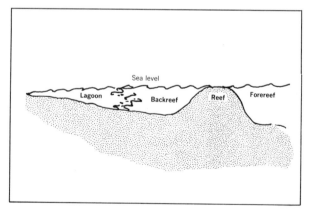

FIG. *1-37—Cross section of carbonate depositional
environment.*[2]

Fractured carbonates with either high or low
matrix permeability can contain huge quantities of
oil and gas if fractures are open. Shells and other
carbonate particles may be carried by long shore
current and deposited in the same manner as sandstone barrier bars. Some of the better quality
carbonate reservoirs are reefs, shoals, and carbonate banks.

APPLICATION OF GEOLOGIC CONCEPTS IN CARBONATE RESERVOIRS

Geologic concepts have obvious application in
any study of a carbonate reservoir. The real question
is how much and what type of geologic effort should
be exerted to optimize economic recovery from
the reservoir. Case histories discussed here show
that detailed reservoir description was, or would
have been, very beneficial.

In one case, improved well stimulation resulted.
In two reservoirs, improved reservoir description
showed the reasons for poor waterflood efficiency
and indicated required remedial action. The fourth
example involves a pinnacle reef, which was be-
lieved to be relatively homogeneous; however, a
horizontal barrier prevented vertical sweep of the
reef with a miscible fluid, appreciably reducing
ultimate recovery.

Variations in the San Andres Reservoirs Significant in Well Completions and Well Stimulation [2,3]

The San Andres in the Permian Basin of West
Texas is an important carbonate deposit penetrated
by many thousands of wells. One of the most
significant parts of this study was the identification
of the basic types of carbonate present in the porous
interval of each well through core analyses and
thin section studies.

The San Andres consists of three basic rock types:
lagoon, backreef and reef, resulting from variations
in the depositional environment during the Permian
Period of the Paleozoic Era.

An important characteristic of the San Andres
is its gradation from a full dolomitic section in the
southern part of the Central Basin Platform to an
almost entirely evaporitic section of anhydrite,
gypsum, and rock salt with a thin dolomite bed
in North Central Texas, the Texas Panhandle, and
Western Oklahoma.

When the seas covered the Permian Basin area,

larger particles of coral and fossil fragments settled and formed reef-banks in the high energy zone away from the shoreline. These reef-banks later formed high porosity, high permeability formations classified as reef type rock.

Shelfward from these banks, the quiet water environment allowed finer particles to deposit which later resulted in the lower porosity backreef type rock. In the shallow water near the shoreline, even finer grained sedimentation eventually caused the formation of lagoon type rock.

Rock Type Description in San Andres—The *reef rock* of the San Andres formation has a massive, non-bedded appearance and may contain large nodules of anhydrite. It has a granular, sucrosic appearance with intergranular porosity. Permeabilities range from 4 to 32 md and porosities from 10 to 18%. Grain size in this rock is generally larger than 100 microns. Hydrochloric acid solubilities averaged about 85% for reef cores from the large Wasson field, but acid solubilities may be higher in some areas.

Permeabilities in the San Andres *backreef rock* usually range from less than 0.1 md to 3.0 md with porosities in the 4 to 10% range. The backreef rock type has intergranular porosity in addition to porosity from shell fragments. Grain particle size ranges from 10 to 250 microns. Backreef cores average about 75% solubility in hydrochloric acid.

The *lagoon rock type* is characterized by very fine, micritic grains, low intergranular porosity, poor permeability and well developed bedding planes. Lagoon type rock usually has a particle size less than 10 microns with permeabilities less than 0.1 md. Due to its low porosity and permeability it is generally not considered commercially productive of hydrocarbons unless natural fractures or vugs are present. Vugs in lagoon type rock are formed by burrowing of marine life in the calcareous mud near marine channels.

Productivity Improvement in San Andres Wells— After identifying the rock types in specific depositional environments, laboratory and field studies can be made to plan well completion, workover, and well stimulation. Many failures of well stimulation probably have resulted from the lack of knowledge of the reservoir rock penetrated by the well being stimulated.

Figure 1-38 is an example of type of laboratory data obtained from backreef cores during a study

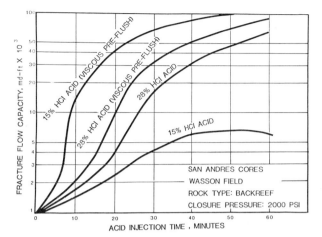

FIG. *1-38—Etched-fracture flow capacity vs. acid injection time on Backreef cores.*[2] *Permission to publish by Southwestern Petroleum Short Course.*

involving stimulation of the San Andres reported in 1975[3].

Fracture flow efficiency tests were run, as illustrated in Figure 1-38, to determine the optimum type of acid treatment required for specific types of formations in each well. These etching tests indicated that a viscous preflush followed by 15% HCl acid was the most efficient acid fracturing treatment for San Andres backreef cores tested.

Table 1-4 shows stimulation results of all wells completed in essentially 100% backreef type rock[2] in this test program. As may be noted, hydraulic fracturing has been quite successful in the backreef type San Andres dolomite using water base frac fluids. Laboratory tests indicated the release of excessive quantities of fines when the backreef cores were reacted with acid.

Results of acid fracturing using a viscous preflush followed by HCl acid tends to verify the acid etching test results shown in Figure 1-38. Also the value of using a fines suspending agent in acidizing San Andres backreef is indicated in Table 1-4. Although more data is needed, results shown on this table suggest that acid fracturing may be competitive with hydraulic fracturing and propping, provided sufficient data is available on the type of rock, including the anticipated quantity of fines resulting from acid etching.

Various types of reservoir rock were stimulated; however, the results of stimulation of backreef provide an example of the value of both geologic and acid etching studies in planning well stimulation.

TABLE 1-4[2]
Stimulation Results in San Andres Backreef

Well No.	Fluid type	Rate bpm	Volume gal × 10³	Sand, Sks	Production, bpd		Remarks
					Before	After	
Hydraulic Fracturing Jobs							
1	Viscous water gel	15	27.5	180 (20–40) 550 (10–20)	20 (oil) 0 (water)	45 (oil) 10 (water)	2.25 Fold increase
2	Gelled water with 25 lb F.L.A. per 1000 gallons	5	20.0	300 (20–40)	30 (oil) 9 (water)	52 (oil) 18 (water)	1.7 Fold increase six months after frac.
3	Gelled 1% KCl water	8	10.0	29.8 (20–40)	71 (oil) 5 (water)	102 (oil) 32 (water)	1.44 Fold increase one year after frac.
4	Gelled 1% KCl water	8	20.0	27.3 (20–40)	25 (oil) 4 (water)	156 (oil) 173 (water) (30 days)	6.24 Fold increase. (After one year 40 B/D oil and 80 B/D water.)
8	Gelled water	5.0	20.0	300 (20–40)	26 (oil) 6 (water)	113 (oil) 59 (water)	4.35 Fold
Fracture Acidizing Jobs							
5	Viscous emulsion prepad followed by 20% HCl acid	4	10.5	None	90 (oil) 16 (water)	114 (oil) 31 (water)	1.27 Fold increase one year after acidizing.
6	Viscous emulsion prepad followed by 20% HCl with fines suspending agent	6.5	10.0	None	9.1 (oil) 3.9 (water)	35.1 (oil) 3.9 (water)	3.86 Fold increase
7	20% HCl with fines suspending agent	6.5	5.0	None	9.1 (oil) 3.9 (water)	35.1 (oil) 3.9 (water)	Volume small-type treatment questionable. (After 4 mo., 7.2 B/D oil and 10.8 B/D water.)
9	15% HCl with fines suspending agent	1.5	2.5	None	20 (oil) 200 (water)	33 (oil) 400 (water)	1.65 Fold increase after 5 months on old well.

Application of Carbonate Environmental Concepts to Well Completions in Enhanced Recovery Projects[12]

A study of the Monahans Clearfork reservoir in West Texas, reported by Dowling in 1970, clearly shows the value of comprehensive depositional environment studies to both primary and improved recovery projects in carbonate reservoirs.

The Monahans Clearfork reservoir was discovered in 1949 and converted to full waterflood in 1961. By 1969, after injecting 23,700,000 barrels of water into the reservoir, some 20,000,000 barrels of water could not be accounted for. The waterflood was a failure up to that point.

As part of studies designed to plan remedial action, ten wells were drilled and cored. Analyses of slabbed cores and thin sections from these wells, plus all available logs allowed reconstruction of the depositional environment. These studies pin-pointed factors limiting the productive quality of the reservoir, and led to locating additional oil reserves.

Since the porosity of productive dolomites is approximately equal to porosity of non-productive supratidal (formed above mean tide level) dolomites, these two environmental types could not be differentiated with porosity logs. In order to locate productive zones, some type of permeability measurement was required. Drill stem tests, production tests, flowmeters, tracer and temperature surveys, and other production logging techniques were used for this purpose.

Figure 1-39 is a typical cross-section showing net pay by depositional types. Net pay isopachs drawn for each zone and subzone aided in defining the limits, quality, and continuity of each reservoir. These isopachs also provided considerable information on the expected waterflood performance in each productive zone as well as suspected thief zones.

The original poor distribution of injection water is indicated by a 1968 flow profile on Well No. 45 showing 100% of the water going into the G-1 zone, and a 1967 profile on Well No. 65 showing 20% of the injected water going into the G-1 zone and 80% into the G-4 zone.

Use of permeability measurement tools to define pay zones rather than relying on log porosity measurements resulted in the location of additional oil which would not have been recovered. Mapping the vugular porous zones formed by burrowing of marine organisms in calcareous mud near ancient marine channels also indicated additional oil.

Prior to mapping these productive areas near ancient marine channels, zones of lagoon rock type were assumed to be nonproductive. Productive zones were located behind the casing in some wells. Other wells required deepening to penetrate indicated productive zones. New wells were drilled when required.

The geologic reconstruction of the sedimentary environment and post depositional changes as an

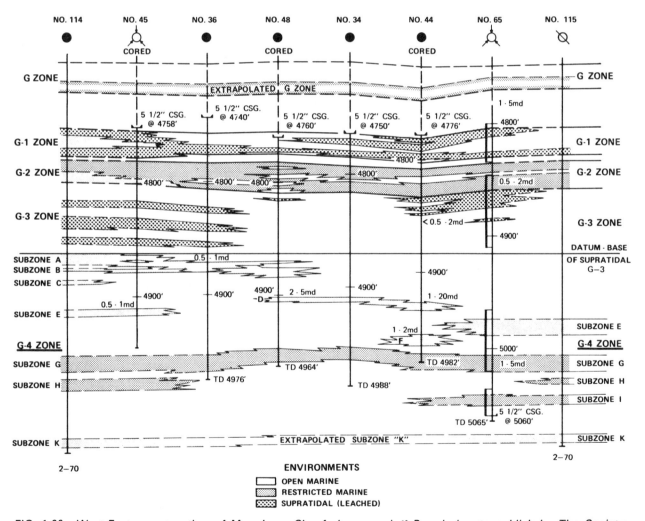

FIG. *1-39—West-East cross section of Monahans Clearfork reservoir.*[12] *Permission to publish by The Society of Petroleum Engineers.*

aid to reservoir definition proved to be very effective in planning redevelopment of the reservoir. The study highlights the fact that much more data is usually required than can be obtained from log analysis to plan well completions for both primary and improved recovery projects. This is especially true of carbonate reservoirs which undergo many changes in the character of porosity after deposition.

Geological-Engineering Team Changes Peripheral Flood to a Pattern Flood to Increase Recovery [20]

The Judy Creek field in central Alberta, producing from a reef in the Beaverhill Lake formation, was discovered in 1959, and put on waterflood in 1962 to arrest the rapidly declining reservoir pressure. By early 1973, it was apparent that the peripheral flood was not satisfactory. A pressure gradient of 1,900 psi existed across the reservoir. Reservoir pressure in the vicinity of injection wells was about 1,000 psi above original pressure, and some internal areas of the pool were below bubble point. It was assumed that barriers within the reef were causing malfunction of the flood.

The first step in remedying the poor performance of the waterflood was to form a study team of reservoir geologists, stratigraphers, log analysts, computer specialists, reservoir engineers, and production engineers.

The study plan was divided into two parts. Phase I included reservoir description and fluid distribution. Phase II covered well workovers, artificial lift equipment, and infill drilling. Phase II also involved the predictive aspects of the study, employing multidimensional, multiphase models based on detailed description and fluid distribution compiled during the initial phase. A work sequence

plan for the study is shown in Figure 1-40.

Figure 1-41 is an east-west cross-section developed by the study group showing porosity distribution of three facies controlled porosity groups, designated as S-3, S-4, and S-5.

S-3 consists of a narrow, peripheral rim of organic reef with an interior lagoon of reef detritus and lime muds. S-4 has a thick organic reef buildup and an extensive interior facies of interbedded, tight and porous limestone. Immediately inside the rim is a zone of coarse detritus. S-5, the uppermost unit, has no organic reef framework. It consists predominantly of coarse organic detritus and lime sand.

A cross plot of K90 md (vertical permeability) vs. percent porosity resulted in subdivision of the environmental facies into three reservoir families, shown in Figure 1-41.

Group I reservoirs occur in the organic-reef and shallow shoal facies and consist of reef framework and associated reef detritus. Average porosity is 12.5% and average vertical permeability (K90) is 170 md.

Group II reservoirs are found in shoals in the fore and backreef where reef detritus and algal laminate rocks were deposited. Porosity averages 9.5% and average permeability is 40 md.

Group III reservoirs occur in deeper water facies of the fore reef and in the interior lagoonal facies. Rock types consist of organic debris and pellets in a lime-mud matrix. Average porosity is 6.5% and average permeability is 3 md.

Pressure maintenance, initiated in 1962, concentrated water injection in the downdip periphery of the reef, principally into the S-3 and S-4 zones. By 1974, the water had advanced far updip in the S-3 unit and has fingered far updip on top of the

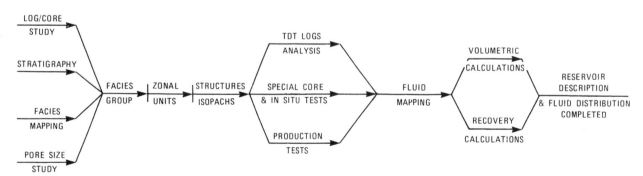

FIG. 1-40—Study sequence of Judy Creek field reservoir description and fluid dynamics. [20] Permission to publish by The Society of Petroleum Engineers.

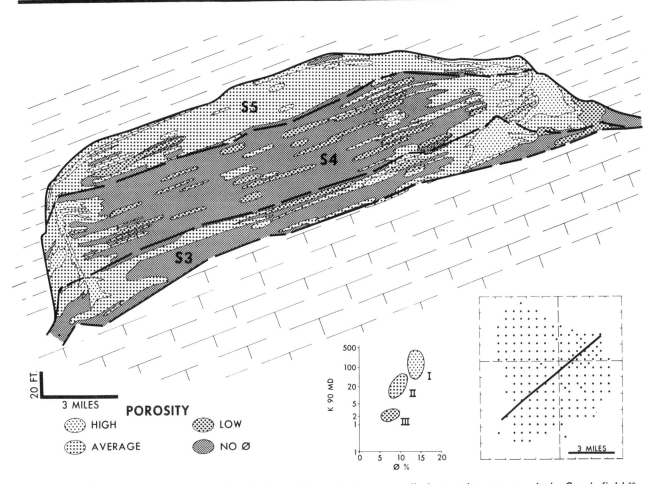

FIG. *1-41—Cross section showing distribution of three facies-controlled porosity groups, Judy Creek field.[20] Permission to publish by The Society of Petroleum Engineers.*

S-5 zone which contains the highest continuous permeability. The discontinuous porosity in S-4 was being bypassed. Figure 1-42 shows the differential advance of water injected through March, 1974.

Figure 1-43 shows pool isobaric map for March, 1974 with the very poor pressure profile from bottom water and peripheral water injection. Also shown is the October, 1975 isobaric map with the much improved pressure profile through the use of the centrally located pattern flood.

The improvement in the flood efficiency resulted from completing pattern injection wells in all porous intervals in S-3, S-4, and S-5 zones. All porous intervals in all producing wells within the pattern waterflood were perforated and high volume pumps were installed to place a high pressure drawdown on all probable oil productive zones. In addition, all wells previously deemed to be flooded out were recompleted and reactivated.

Many reactivated wells produced more than 100 b/d oil, and are expected to recover an additional 10 million bbl of oil. The flood is now under good control with the pattern flood operational along with some water still being injected into peripheral wells. Pressure is being maintained at near the original reservoir pressure of 3500 psi.

The Judy Creek task force study provides these guidelines for future studies:

1. Detailed reservoir description is the key to reservoir development and operation.

2. Because of the usual reservoir heterogeneity in both a lateral and vertical direction in carbonate reservoirs, the facies must be mapped to define detailed geometry and the most likely patterns of fluid movement under various depletion programs.

3. Reservoir description should be updated along with reservoir performance to provide a continu-

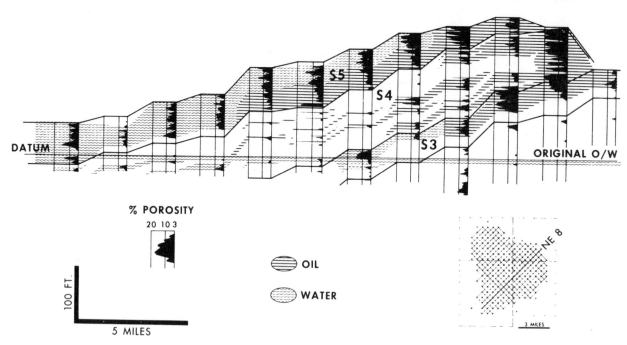

FIG. *1-42—Northeast-southwest structural cross section, Judy Creek, showing the differential advance of water injected downdip.*[20] *Permission to publish by The Society of Petroleum Engineers.*

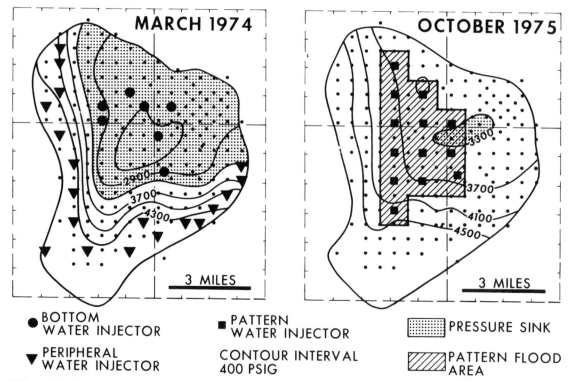

FIG. *1-43—Isobaric maps showing the result of conversion from peripheral to pattern flood, Judy Creek (After Flewitt, 1975).*[20] *Permission to publish by The Society of Petroleum Engineers.*

ously upgraded reservoir model as reservoir depletion continues.

4. A study of this type should be a task force job employing engineers, geologists, log analysts, and other specialists as required. The team effort provides for greater profitability from oil and gas reservoirs, and more efficient use of valuable professional personnel. This type utilization usually provides a bonus in improved morale and enthusiasm of all involved geologists, engineers and other personnel.

Detailed Reservoir Description—Key to Planning Gas-Driven Miscible Flood [25]

A gas-driven LPG bank miscible flood failed to achieve its objectives because of incomplete information on the extent of barriers to vertical sweep in a relatively homogeneous pinnacle reef.

The Golden Spike, Leduc D 3 "A" reef reservoir, located in central Alberta, Canada was discovered in 1949. This pinnacle reef covers 1,385 acres with an average thickness of 480 ft, and originally contained 319 MMstb of oil with no gas cap or water leg. Initial production by dissolved gas drive resulted in rapid pressure decline. Pressure maintenance with gas injection was started in 1953. Analysis of

available geologic data and reservoir performance to 1963, indicated a relatively homogeneous carbonate reef.

Based on this analysis of the reservoir, a miscible flood using hydrocarbon solvent was initiated in 1964. The solvent was injected as a bank across the top of the reef. Computer calculations, based on available reservoir description, indicated the gas-driven solvent bank (7% of hydrocarbon volume) would recover 95% of the original oil in place. This prediction was based on a model study of the apparent reservoir rock characteristics and sweep efficiency of the solvent.

Near the end of the placement of the solvent bank in 1972, productivities of the seven producing oil wells had declined appreciably. Infill drilling was then initiated to increase production and to obtain more geologic information. Observed fluid distribution, Figure 1-44, suggested injected gas was bypassing the solvent bank and underrunning a lateral barrier to vertical permeability.

Because of the continued poor performance of the miscible flood, a major reservoir description and reservoir performance study was carried out with a great deal more core and log data being now available for study.

Figure 1-45 provides detailed environmental

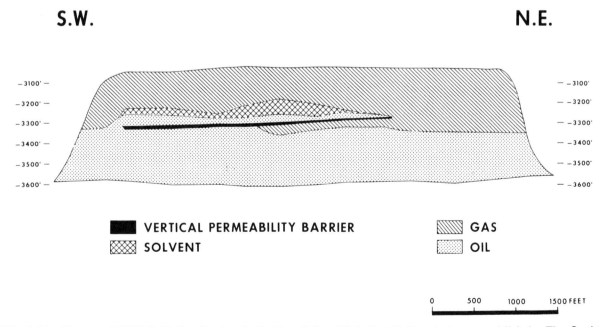

FIG. 1-44—Observed 1973 fluid distribution in Golden Spike D3 A Pool.[25] Permission to publish by The Society of Petroleum Engineers.

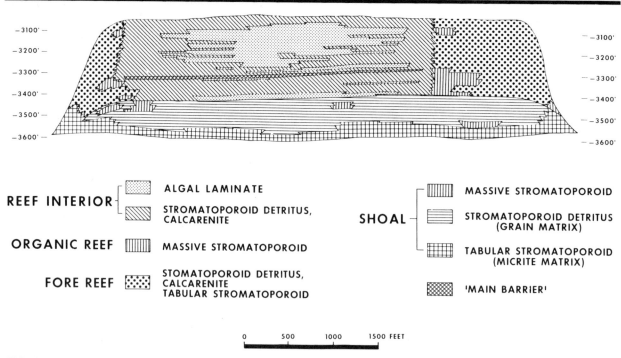

REEF INTERIOR ── ALGAL LAMINATE

STROMATOPOROID DETRITUS, CALCARENITE

ORGANIC REEF ── MASSIVE STROMATOPOROID

FORE REEF ── STOMATOPOROID DETRITUS, CALCARENITE TABULAR STROMATOPOROID

SHOAL ── MASSIVE STROMATOPOROID

STROMATOPOROID DETRITUS (GRAIN MATRIX)

TABULAR STROMATOPOROID (MICRITE MATRIX)

'MAIN BARRIER'

0   500   1000   1500 FEET

FIG. *1-45—Environmental facies of the Golden Spike reservoir.*[25] *Permission to publish by The Society of Petroleum Engineers.*

facies of the Golden Spike reservoir, and Figure 1-46 shows a reinterpreted model of the reservoir showing all barriers to vertical permeability.

Figure 1-47 shows a model of the production

history-matched fluid distribution in 1975, based on this new detailed reservoir description, bottom hole fluid sample data, and reservoir performance data.

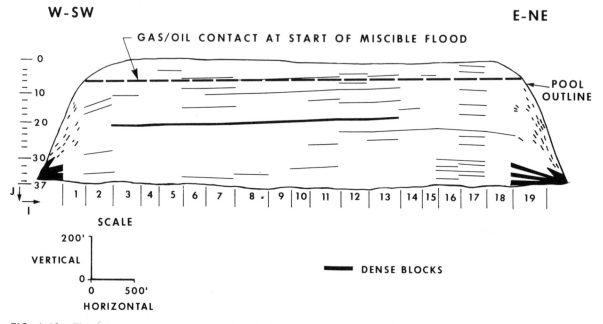

FIG. *1-46—The 2-D cross-sectional model, including vertical permeability barriers, Golden Spike reservoir.*[25] *Permission to publish by The Society of Petroleum Engineers.*

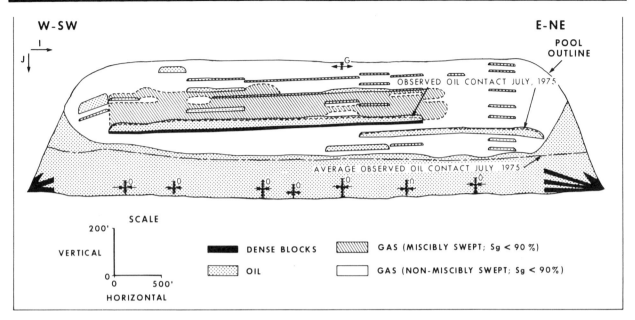

FIG. *1-47—History matched model fluid distribution in 1975, Golden Spike reservoir.*[25] *Permission to publish by The Society of Petroleum Engineers.*

Due to barriers to vertical sweep of the solvent bank, gas had underrun the main barrier and the solvent bank had dispersed into the gas cap. Because the miscible bank was no longer present and effective, the solvent flood was converted to gas drive.

Predicted additional recovery by the miscible bank was reduced to 10 MMstb, compared with the original estimate of 69 MMstb. The original recovery estimate was based on the incorrect assumption that the reef was sufficiently free of barriers to permit a top to bottom sweep of the reef. Current estimate of ultimate recovery is 67%, compared to 95% with an efficient miscible displacement.

The benefits of detailed reservoir description coupled with continued monitoring and prediction of reservoir performance is essential to optimize ultimate recovery and profits. Reservoir operation, including well completions and workovers, cannot be efficiently carried out unless an adequate reservoir description is available along with a continuously updated plan of reservoir operations and control.

REFERENCES

1. Alpay, O. A.: "A Practical Approach to Definition of Reservoir Heterogeneity," J. Pet. Tech., July 1972, p. 841.

2. Black, H. N., Carlile, W. C., Coulter, G. R. and Blalock, S.: "Lithology As a Guide to San Andres Stimulation," Presented at Southwestern Petroleum Short Course, Department of Petroleum Engineering, Texas Tech University, April 17–18, 1975.

3. Black, H. N., and Stubbs, B. A., "A Case History Study—Evaluation of San Andres Stimulation Results," SPE-5649, (Oct. 1, 1975).

4. Busch, D. A.: "Stratigraphic Traps in Sandstones—Exploration Techniques," AAPG Memoir 21 (July, 1974).

5. Busch, D. A.: "Exploration for Sandstone Reservoirs," Training Manual of Oil and Gas Consultants International, (1977).

6. Busch, D. A., Dickey, P. A., Friedman, G. M., Visher, G. S.: Applied Petroleum Geology Training Manual issued by Oil and Gas Consultants International, (1977).

7. Choquette, Phillip W., and Pray, Lloyd C.: "Geologic Nomenclature and Classification of Porosity in Sedimentary Carbonates;" AAPG Reprint Series No. 5; Carbonate Rocks II: Porosity and Classification of Reservoir Rocks.

8. Craig, F. F., Jr., Willcox, P. J., Ballard, J. R., and Nation, W. R.: "Optimized Recovery Through Cooperative Geology and Reservoir Engineering," SPE-6108 (1976).

9. Depositional Environments in Carbonate Rocks, S.E.P.M. No. 14, March, 1969.

10. Dickey, P. A.: "Basic Petroleum Geology," Training Manual of Oil and Gas Consultants International, (1977).

11. Dickey, P. A.: "Development Geology," Training Manual of Oil and Gas Consultants International, (1977).

12. Dowling, Paul L., Jr.: "Application of Carbonate Environmental Concepts to Secondary Recovery Projects, SPE-2987 (1970).

13. Elkins, L. F., and Skov, A. M.: "Some Field Observations

and Heterogeneity of Reservoir Rocks and Its Effect on Oil Displacement Efficiency," SPE-282 (April, 1962).

14. Flewitt, W. E.: "Refined Reservoir Description Maximizes Petroleum Recovery," S.P.W.L.A., Annual Logging Symposium, June, 1975.

15. Friedman, G. M.: "Exploration for Carbonate Reservoirs," Training Manual of Oil and Gas Consultants International, (1977).

16. Groult, J. and Reiss, L. H.: "Reservoir Inhomogeneities Deduced from Outcrop Observations and Production Logging," J. Pet. Tech., July 1966, p. 883.

17. Halbouty, Michel T.: "Needed; More Coordination Between Earth Scientists and Petroleum Engineers," SPE-6107, (Oct. 1976).

18. Harris, D. G.: "The Role of Geology in Reservoir Simulation Studies," J. Pet. Tech., May 1975, pp. 625-632.

19. Hewitt, H. C.: "How Geology Can Help Engineer Your Reservoir," Oil and Gas Journal (11-14-66; pp. 171-178).

20. Jardine, D., Andrews, D. P., Wishart, J. W., Young, J. W.: "Distribution and Continuity of Carbonate Reservoirs," J. Pet. Tech., July 1977, pp. 873-885.

21. LeBlanc, R. J., Sr.: "Distribution and Continuity of Sandstone Reservoirs," SPE-6137, Oct. 1976.

22. LeBlanc, R. J., Sr.: "Geometry of Sandstone Reservoir Bodies," AAPG MEMOIR 18.

23. Morgan, J. T., Cordiner, F. S., Livingston, A. R.: "Tensleep Reservoir Study Oregon Basin Field, Wyoming—Reservoir Characteristics," J. Pet. Tech., July 1977, pp. 886-896.

24. Pittman, G. M.: "Improved Well Completion Through Applied Core Data," API Paper (Oct. 1975).

25. Reitzel, G. A., Callow, G. O.: "Pool Description and Performance Analysis Leads to Understanding Golden Spike's Miscible Flood," J. Pet. Tech., July 1977, pp. 867-872.

26. Robinson, Robert B.: "Classification of Reservoir Rocks by Surface Texture;" AAPG Reprint Series No. 5; Carbonate Rocks II: Porosity and Classification of Reservoir Rocks.

27. Sangree, J. B.: "What You Should Know to Analyze Core Fractures," World Oil, April 1969, pp. 69-72.

28. Sneider, R. M., Richardson, F. H., Paynter, D. D., Eddy, R. E., and Wyant, I. A.: "Predicting Reservoir-Rock Geometry and Continuity in Pennsylvania Reservoirs, Elk City Field, Oklahoma," J. Pet. Tech., July 1977, pp. 851-866.

29. Visher, G. S.: "Stratigraphic Controls for Hydrocarbon Accumulations," Training Manual of Oil and Gas Consultants International, (1977).

30. Wayhan, D. A., and McCaleb: "Elk Basin Madison Heterogeneity—Its Influence on Performance," J. Pet. Tech., Feb. 1969, p. 153.

31. Weber, K. J.: "Sedimentological Aspects of Oil Fields in the Niger Delta," Geologie En Mijnbouw, Vol. 50 (3) pp. 559-576 (1971).

32. Weber, K. J., and Daukoru, E.: "Petroleum Geology of the Niger Delta," World Petroleum Congress, Tokyo 1965.

33. Zeito, George A.: "Interbedding of Shale Breaks and Reservoir Heterogeneities," J. Pet. Tech., Oct. 1965, p. 1,223.

GLOSSARY OF SELECTED GEOLOGIC TERMS USED IN THIS CHAPTER[a]

Abyssal Pertaining to ocean depths below about 6,000 feet.

Anticline Term applies to strata which dip in opposite directions from a common ridge or axis, like a roof of a house.

Calcareous Containing calcium carbonate.

Calcarenite Limestone or dolomite composed of coral or shell sand or of sand derived from erosion of older limestone.

Calcite Calcium Carbonate, $CaCO_3$.

Clastics Rocks composed of fragmental material from pre-existing rocks. Commonest clastics are sandstones and shales.

Conglomerates Water-worn fragments of rock or pebbles cemented together.

Cross-stratification A cross-stratified unit is one with layers deposited at an angle to the original dip of the formation.

Detritus Fragmental material, such as sand, silt and mud, derived from older rocks by disintegration.

Diagenesis Chemical and physical changes in sedimentary rock before consolidation takes place.

Dip The angle at which a stratum is inclined from the horizontal.

Eolian (Aeolian) Deposits Transported and deposited by the wind.

Evaporite Sediments deposited from aqueous solution due to evaporation of the solvent. Salt or anhydrite are examples of evaporites.

Fluviatile Deposits Sedimentary deposits laid down by a river or stream.

Graben Depression produced by subsidence of a strip between normal faults.

Isopach Line or map drawn through points of equal thickness of a designated zone or unit.

Lithology Description of rocks.

Mineralogy The science of the study of minerals.

Neritic or Shelf Area Marine environment extending from low tide to a depth of about 600 feet.

Oolite A spherical to ellipsoidal body, 0.25 to 2 mm in diameter, usually calcareous.

Outcrop That part of a stratum which appears on or near the surface.

Paleoslope The ancient slope, probably the slope at the time of deposition.

Paralic Pertaining to environments of the marine borders, such as lagoonal, or shallow neritic (shelf).

[a] Definitions of terms are from: "Glossary of Geology and Related Sciences with Supplement," Second Edition 1960, by The American Geological Institute.

Pinnacle   A tall, slender, pointed mass of rock.

Progradation   A seaward advance of the shoreline resulting from the nearshore deposition of river-borne deposits.

Sedimentary Basin   A geologically depressed area, with thick sediments in the interior and thinner sediments at the edges.

Silt   An inorganic granular material between .005 mm and .05 mm in diameter or between the size of clay and sand.

Stratum   A single sedimentary bed or layer, regardless of thickness.

Subaerial   Formed or existing on land above water.

Subaqueous   Formed or existing below water

Strike   The direction of a horizontal line in the plane of an inclined stratum, joint, fault, cleavage plane or structure plane and this direction is perpendicular to the dip.

Structural Contour   A contour line drawn through points of equal elevation on a stratum, key bed, or horizon.

Structural Trap   One in which fluid entrapment results from folding, faulting or a combination of both.

Texture   Component particles of a rock, including size, shape and arrangement.

Trench   An elongated but proportionally narrow depression, with steeply sloping sides.

Chapter 2 Reservoir Considerations in Well Completions

Hydrocarbon properties
Components, phases, and molecular behavior
Characteristics of reservoir rocks
Porosity, permeability, and wettability
Fluid distribution
Fluid flow in the reservoir
Pressure distribution near well bore
Radial and linear flow near well bore
Near-well-bore flow restrictions
Reservoir characteristics affecting well completion

INTRODUCTION

Oil and gas wells are expensive faucets that enable production of petroleum reserves or allow injection of fluids into an oil or gas reservoir. A prudently planned initial well completion program is the first and most important step in obtaining satisfactory producing well life to attain maximum recovery with minimum well workover. An optimum initial well completion program must consider not only geologic and fluid conditions occurring in the reservoir at time of discovery, but also changes in fluid saturations adjacent to the well as fluids are produced.

Many times the question of *where* to complete a well is allowed to overshadow the equally important problem of *how* to complete the well. This question of how to complete the well involves effective communication with all desired zones within the completion interval, effective shutoff of undesired zones within or near the completion interval, and solution of mechanical problems such as sand control. Formation damage must be a paramount consideration in any well work.

While it is desirable to minimize future workovers, conditions often occur where workovers are required to correct conditions which, at the time of initial completion, could not be, or were not foretold. In many cases workovers may be forecast as a future requirement in an optimum well completion program.

The purpose of this chapter is to briefly consider the characteristics of reservoir fluids and the flow of those fluids in the area around the wellbore, in order to tie these parameters into well completion, workover and stimulation operations. In this discussion we have borrowed frequently from the work of Norman J. Clark.[1]

HYDROCARBON PROPERTIES OF OIL AND GAS

Crude oil and gas occurring in the earth consist of a large number of petroleum compounds mixed together. These compounds are composed of hydrogen and carbon in various ways and proportions. Petroleum compounds are, therefore, called hydrocarbons, and each compound is made up of different portions of the two elements. Seldom are two crude oils found that are identical and certainly never are two crude oils made up of the same proportions of the various compounds.

Components

Hydrocarbon compounds making up petroleum can be grouped chemically into series. Each series consists of those compounds similar in their molecular makeup and characteristics. Within a given series there is a range of compounds from extremely light to extremely heavy or complex.

The most common hydrocarbon compounds are

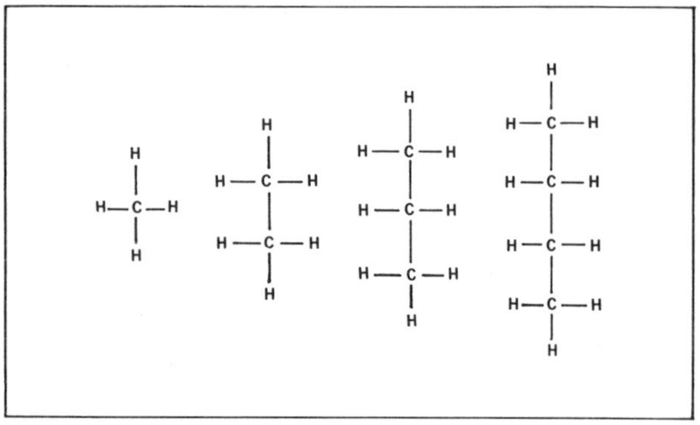

FIG. *2-1—Structural formulas of four lightest paraffin compounds.*

those of the paraffin series which include methane, ethane, propane, butane, etc. See Figure 2-1.

Petroleum deposits include some quantity of nearly all components throughout the entire range of weights and complexities. Gas is not composed entirely of light molecules; the majority of its component molecules are light and simple; whereas, liquid crude oil is made up of a majority of heavier more complex component molecules.

Phases

Generally, all substances can exist as a solid, liquid, or gas. These three forms of existence are termed phases of matter. Whether a substance exists as a solid, liquid, or gas phase is determined by temperature and pressure conditions.

Hydrocarbon compounds, either individually or in a mixture, also change state or phase in response to changing temperature and pressure conditions. This is called "phase behavior" and many times is an important consideration in reservoir and well operations.

Molecular Behavior

The phase behavior of hydrocarbons can be explained by the behavior of the molecules making up the mixtures. Four physical factors are important. As shown in Figure 2-2 these are: (1) pressure, (2) molecular attraction, (3) kinetic energy, and (4) molecular repulsion.

Increased pressure tends to force molecules closer together so that gas will be compressed or possibly changed to a liquid. If pressure is decreased, gas expands and liquid tends to vaporize to gas. These phase changes caused by changes in pressure are termed normal or regular phase behavior.

Pressure and volume are related in that pressure results from molecular bombardment of the walls of the container or a liquid surface. Increased volume tends to reduce pressure by increasing the distance molecules must move to strike the container.

Molecular attraction acts on molecules the same as external pressure. The attraction force between molecules increases as the distance between the

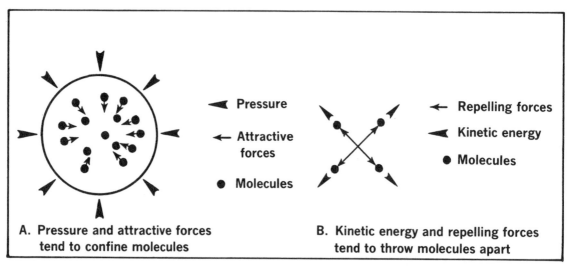

FIG. *2-2—Forces governing hydrocarbon behavior.* Elements of Petroleum Reservoirs. *Permission to publish by The Society of Petroleum Engineers.*

molecules decreases; it also increases as the mass of the molecules increases. With smaller molecules, methane or ethane, there is less attraction between molecules and greater tendency for them to be thrown apart by their kinetic energy into gas; whereas, larger molecules, hexane and heptane, tend to be attracted together into a liquid.

Kinetic energy, or molecular motion, increases with temperature. So, the greater the temperature of a material, the greater the tendency for the material to be thrown apart and thus decrease its density (change from a liquid to a gas or a gas to expand). As temperature decreases, kinetic energy decreases, and all molecules (even the lighter molecules) tend to be attracted together into a liquid state and even frozen into a solid state. This behavior is also called normal phase behavior.

When the molecules get so close together that their electronic fields overlap, a repelling force tends to increase the resistance to further compression.

When hydrocarbon materials appear to be at rest (not expanding, contracting in volume, or changing state), the forces tending to confine the molecules balance the forces tending to throw them apart, and the material is considered to be in "equilibrium."

In petroleum reservoirs, temperature usually remains constant; therefore, only pressure and volume are altered to an appreciable degree in the reservoir during production operations. However, in the well and in surface facilities, temperature, pressure, and volume relations all become important factors.

Pure Hydrocarbons

For a single or pure hydrocarbon such as propane, butane, or pentane, there is a given pressure for every temperature at which the hydrocarbon can exist both as a liquid and a gas, Figure 2-3.

If pressure is increased without a temperature change, the hydrocarbon is condensed to a liquid state. If pressure is decreased without a temperature change, the molecules disperse into a gas. As temperature increases, kinetic energy increases, and

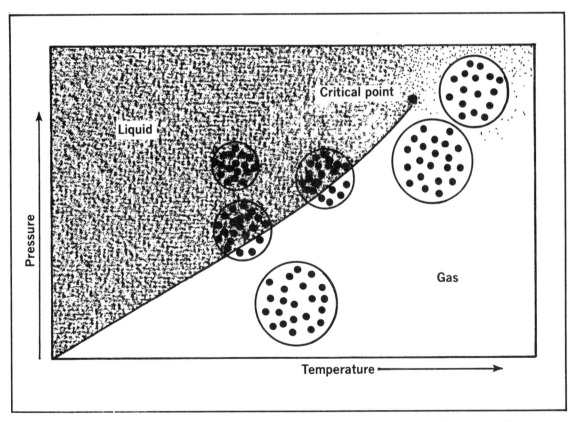

FIG. *2-3—Vapor pressure vs. temperature for a pure hydrocarbon component.* Elements of Petroleum Reservoirs. *Permission to publish by The Society of Petroleum Engineers.*

higher pressure is required for the balanced conditions at which the two phases can exist simultaneously.

The curve Figure 2-3, plotted through the pressure-temperature points where the two phases exist, is called the "vapor-pressure curve." There is a temperature above which the material will not exist in two phases regardless of the pressure. This is called the "critical point," and temperature and pressure at this point are called "critical temperature" and "critical pressure."

Material is commonly considered to be a gas when it exists at temperature and pressure below the vapor-pressure curve and as a liquid above the vapor-pressure curve. However, ranges of temperature and pressure exist in which a material can be classified as either liquid or gas.

In these ranges (shown in the upper right-hand portion of Figure 2-3) the temperature is so great that attractive forces between the molecules are not sufficiently large to permit them to coalesce to a liquid phase. Increased pressure merely causes the molecules to move together uniformly.

Hydrocarbon Mixtures

In a mixture of two components, the system behavior is not so simple. There is a broad region in which two phases (liquid and gas) co-exist.

Figure 2-4 is a diagram of the phase behavior of a 50:50 mixture of two hydrocarbons such as propane and heptane. Superimposed on the correlation are vapor-pressure curves of two components in their pure state.

The two-phase region of the phase diagram is bounded by a "bubble-point" line and a "dew-point" line, with the lines joining at the critical point. At the bubble point, gas begins to leave solution in oil with decreasing pressure. At the dew point, liquid generally begins to condense from gas with increasing pressure; however, above the critical temperature, condensation may occur at some points along the dew point curve with increasing pressure, and at other points with decreasing pressure.

At the critical point, properties of both gas and liquid mixtures are identical. Our previous definition

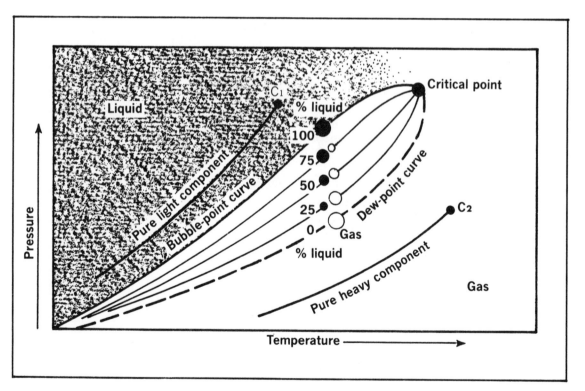

FIG. 2-4—*Vapor pressure curves for two pure components and phase diagram for a 50:50 mixture of the same components.* Elements of Petroleum Reservoirs. *Permission to publish by The Society of Petroleum Engineers.*

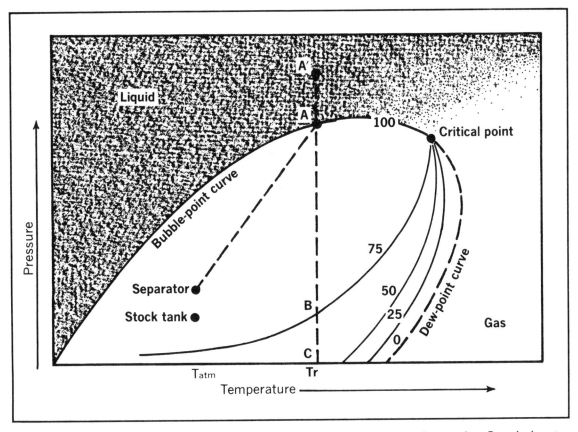

FIG. *2-5—Phase diagram of low shrinkage oil.* Elements of Petroleum Reservoirs. *Permission to publish by The Society of Petroleum Engineers.*

of "critical point" no longer applies because in a multicomponent system both liquid and gas phases exist at temperatures and pressures above the critical point. The variation may be slight in a two-component system, however, with a larger number of components, the pressure and temperature ranges in which two phases exist increase greatly.

An idealized pressure-temperature phase diagram of a crude oil in a reservoir with a temperature Tr is shown in Figure 2-5. Crude oil at its bubble point or saturation pressure is represented by Point A.

The same oil would be "undersaturated" if reservoir pressure were represented by Point A′. Separator and stock-tank temperatures and pressures are also shown in Figure 2-5.

The vertical line, A-B, represents the relative quantities of liquid and gas existing at equilibrium at a particular pressure as reservoir pressure is dropped at constant reservoir temperature. In the production process this is physically represented

by gas coming out of solution in the reservoir, the amount of which is governed by the reduction in reservoir pressure.

Quantities of liquid and gas represented by location of the stock tank point in Figure 2-5, however, do not indicate what would occur in the stock tank because the composition of the original mixture changes at the separator in the production process.

Retrograde Condensate Gas

Some hydrocarbon mixtures exist in the reservoir above their critical temperature as condensate gases. When pressure is decreased on these mixtures, instead of expanding (if a gas) or vaporizing (if a liquid) as might be expected, the intermediate and heavier components tend to condense.

Conversely, when pressure is increased, they vaporize instead of condensing. The process termed "retrograde" is illustrated by temperature condition Tr in Figure 2-6. Condensation of liquids in the

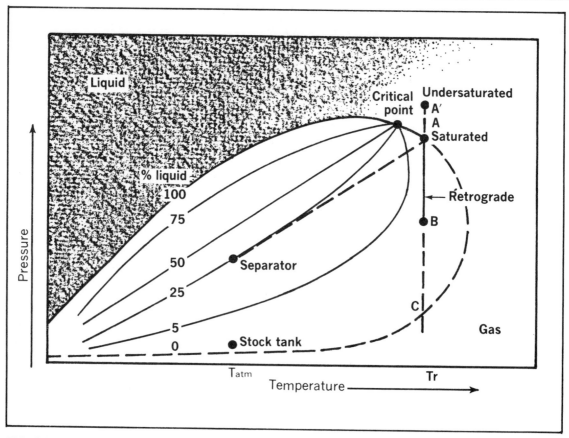

FIG. *2-6—Phase diagram of retrograde condensate gas.* Elements of Petroleum Reservoirs. *Permission to publish by The Society of Petroleum Engineers.*

reservoir alters relative permeability relationships, and usually results in loss of well productivity and also of hydrocarbon recovery.

Gas

Behavior of gases is shown in Figures 2-7 and 2-8. Hydrocarbon in a gas reservoir may be termed as "wet" gas or "dry" gas, depending upon its behavior. Both exist at temperatures above their critical temperature. When the temperature of a wet gas is reduced to stock tank temperature condensation of heavier components results. With a dry gas no condensation of liquids results at stock tank temperature.

Practical Uses of Hydrocarbon Data

The practical approach to the study of reservoir fluid behavior is (1) to anticipate pressure and temperature changes in the reservoir and at the surface during production operations, and (2) to measure by laboratory tests the changes occurring in the reservoir fluid samples. The results of these tests then provide the basic fluid data for estimates of fluid recovery by various methods of reservoir operation, and also for estimates of reservoir parameters through transient pressure testing.

Two general methods are used to obtain samples of reservoir oil for laboratory examination purposes: (1) by means of a subsurface sampler, and (2) by obtaining surface samples of separator liquid and gas. These samples are then recombined in the laboratory in proportions equivalent to gas-oil ratio measured at the separator.

Information concerning the characteristics and behavior of gas needed for work with gas reservoirs depends upon the type of gas and the nature of the problem. If retrograde condensation is involved, needed information may require numerous tests and measurements. If wet gas is involved (with no retrograde condensation) or if dry gas is involved, information is less complex.

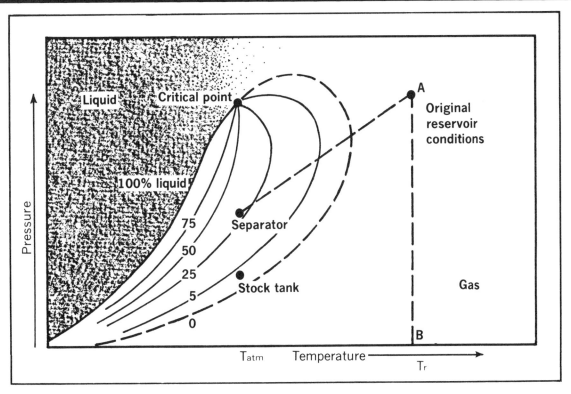

FIG. *2-7—Phase diagram of wet gas.* Elements of Petroleum Reservoirs. *Permission to publish by The Society of Petroleum Engineers.*

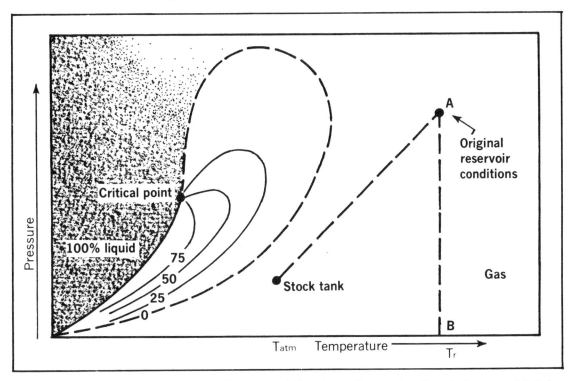

FIG. *2-8—Phase diagram of dry gas.* Elements of Petroleum Reservoirs. *Permission to publish by The Society of Petroleum Engineers.*

EXAMPLE

REQUIRED:

Formation volume at 200°F of a bubble-point liquid having a gas-oil ratio of 350 CFB, a gas gravity of 0.75, and a tank-oil gravity of 30 °API.

PROCEDURE:

Starting at the left side of the chart, proceed horizontally along the 350 CFB line to a gas gravity of 0.75. From this point drop vertically to the 30 °API line. Proceed horizontally from the tank-oil gravity scale to the 200°F line. The required formation volume is found to be 1.22 barrel per barrel of tank-oil.

FORMATION VOLUME OF BUBBLE-POINT LIQUID, BARREL PER BARREL OF TANK-OIL

SOLUTION GAS-OIL RATIO, CUBIC FEET PER BARREL

GAS GRAVITY, AIR = 1

TANK-OIL GRAVITY, °API

TEMPERATURE, °F

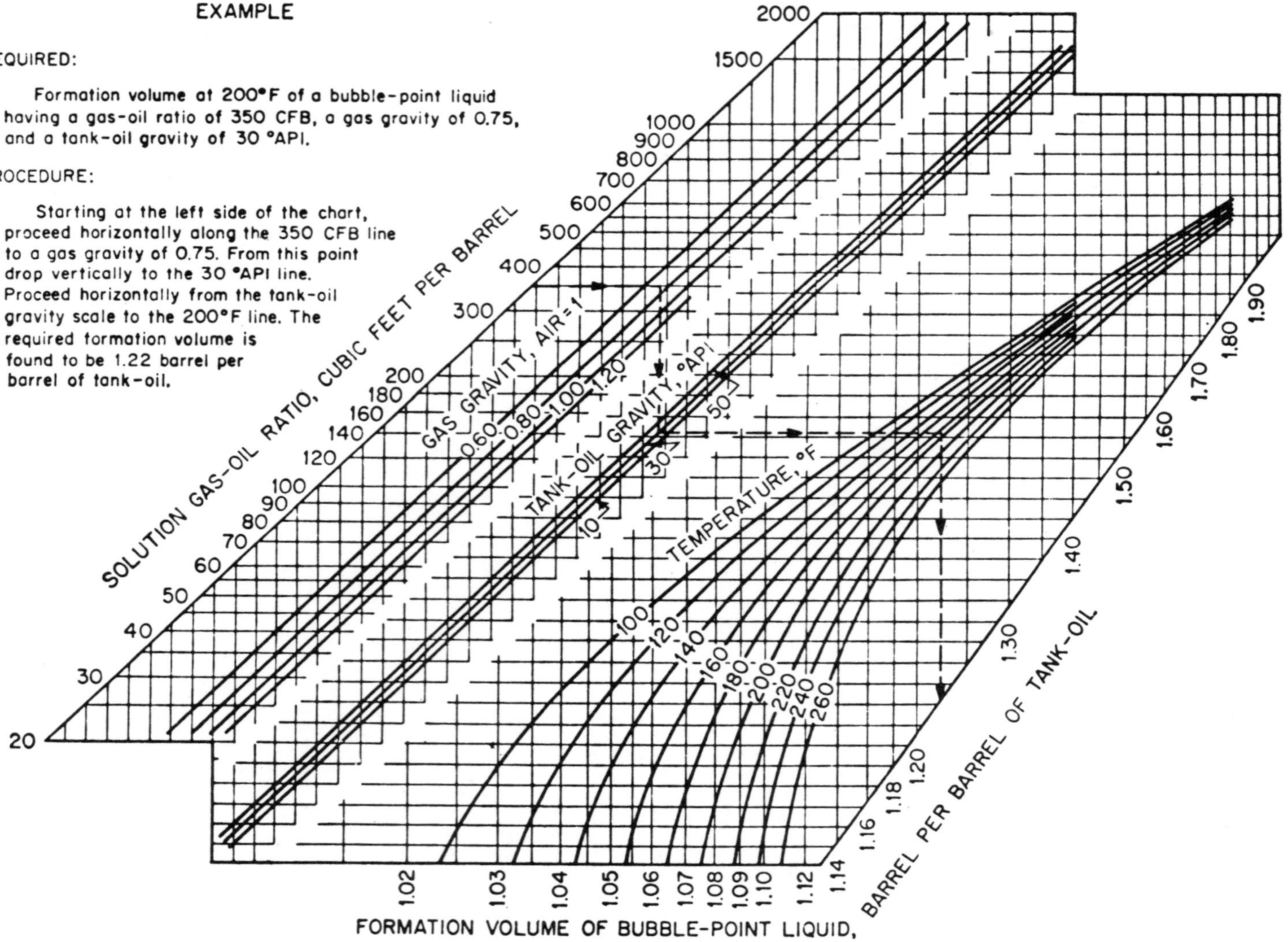

FIG. 2-9—Calculation of oil-formation volume factor by Standing's correlation. Permission to publish by The Society of Petroleum Engineers.

Correlation of Properties of Oils

Several generalizations of oil sample data are available, permitting correlations to be made to minimize the need for oil-reservoir sampling, testing and analysis.

These correlations are valuable for many practical, day-to-day reservoir engineering calculations. Typical of these correlations are those of Standing[2] (GOR vs. formation volume factor, bubble-point pressure and two-phase formation volume factor) and by Beal[3] and Carr et al[4] (viscosities of air, water, natural gas, crude oil, and associated gases). Common correlations are shown in Figures 2-9 and 2-10.

CHARACTERISTICS OF RESERVOIR ROCKS
Porosity

Porosity or pore space in reservoir rock provides the container for the accumulation of oil and gas and gives the rock characteristic ability to absorb and hold fluids. Most commercial oil and gas reservoirs occur in sandstone, limestone, or dolomite rocks; however, some reservoirs even occur in fractured shale. Figures 2-11 and 2-12 show some reservoir rock characteristics.

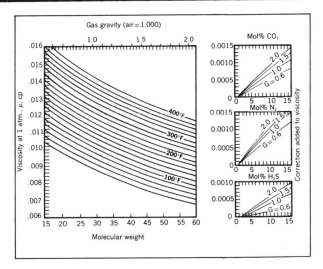

FIG. *2-10B—Viscosity of hydrocarbon gases at one atmosphere and reservoir temperatures, with corrections for nitrogen, carbon dioxide, and hydrogen sulfide.[4] Permission to publish by The Society of Petroleum Engineers.*

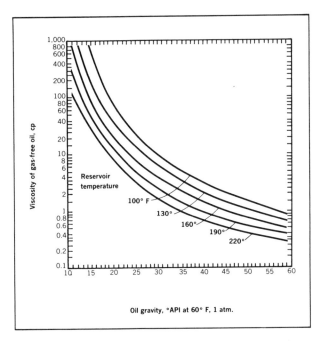

FIG. *2-10A—Viscosity of gas-free crude oil at reservoir temperature.[3] Permission to publish by The Society of Petroleum Engineers.*

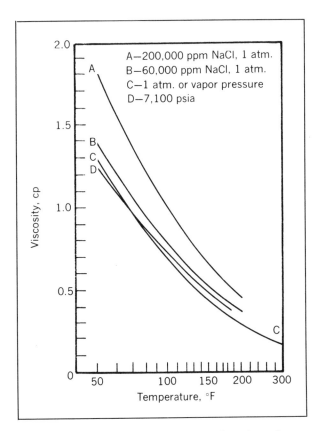

FIG. *2-10C—Viscosity of water as a function of temperature, pressure, and salinity. After Chesnut.*

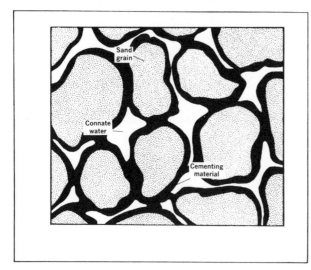

FIG. *2-11—Consolidated sandstones.* Elements of Petroleum Reservoirs. *Permission to publish by The Society of Petroleum Engineers.*

Permeability

Permeability is a measure of the ease with which fluid can move through the inter-connected pore spaces of the rock. Many rocks, such as clays, shales, chalk, anhydrite, and some highly cemented sandstones, are impervious to movement of water, oil, or gas, even though they may actually be quite porous.

In 1856 the French engineer, Henry Darcy, working with water filters, developed a relation which

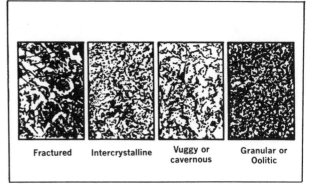

FIG. *2-12—Consolidated limestones.* Elements of Petroleum Reservoirs. *Permission to publish by The Society of Petroleum Engineers.*

describes fluid flow through porous rock. Darcy's Law states that rate of flow through a given rock varies directly with (1) permeability (measure of the continuity of inter-connected pore spaces), and (2) the pressure applied; and varies inversely with the viscosity of the fluid flowing.

In rock having a permeability of 1 darcy, Figure 2-13, 1 cc of a 1-cp viscosity fluid will flow each second through a portion of rock 1 cm in length and having a cross section of 1 cm^2, if the pressure drop across the rock is 1 atmosphere.

$$K = \frac{q \mu L}{A \Delta p} \qquad (1)$$

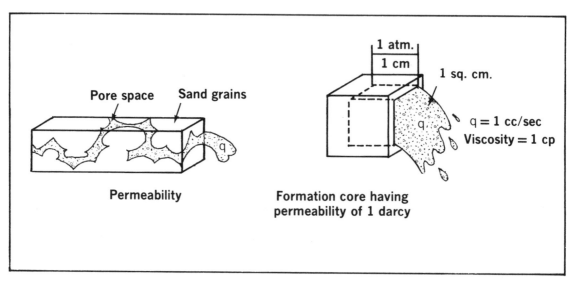

FIG. *2-13—Fluid flow in permeable sand.* Elements of Petroleum Reservoirs. *Permission to publish by The Society of Petroleum Engineers.*

In oil field units the linear form of Darcy's Law for flow of incompressible fluid through a rock filled with only one fluid is as follows:

$$q = 1.127 \times 10^{-3} \frac{kA(p_1 - p_2)}{B\mu L} \qquad (2)$$

where:

q = flow rate, stb/day
k = permeability, md
A = flow area, ft^2
μ = viscosity, cp
L = flow length, ft
p_1, p_2 = inlet and outlet pressures, psi
B = formation volume factor, res. bbl/stb

Relative Permeability

Because two or three fluids—gas, oil, and water—can, and often do, exist in the same pore spaces in a petroleum reservoir, relative permeability relationships must be considered. Relative permeability represents the ease with which one fluid flows through connecting pore spaces in the presence of other fluids, compared to the ease with which one fluid flows when it alone is present.

Consider a rock filled only with oil at high pressure (Figure 2-14A). Gas has not been allowed to come out of solution; therefore, all available space is filled with oil, and only oil is flowing.

If reservoir pressure is allowed to decline (Figure 2-14B), some lighter components of the oil will evolve as gas in the pore spaces. Flow of oil is reduced but gas saturation is too small for gas to flow through the pores.

If pressure continues to decline, gas saturation continues to increase, and at some point (equilibrium gas saturation) gas begins to flow; oil flow rate is further reduced, (Figure 2-14C).

With further increases in gas saturation more and more gas and less and less oil flows through the pores until finally nothing but gas is flowing (Figure 2-14D). Significant amounts of oil may remain in the pore spaces, but cannot be recovered by primary means because relative permeability to oil is now zero.

This same principle governs the flow of oil in the presence of water. The saturation of each fluid present affects the ease of fluid movement or relative permeability.

Figure 2-15 shows typical oil water relative per-

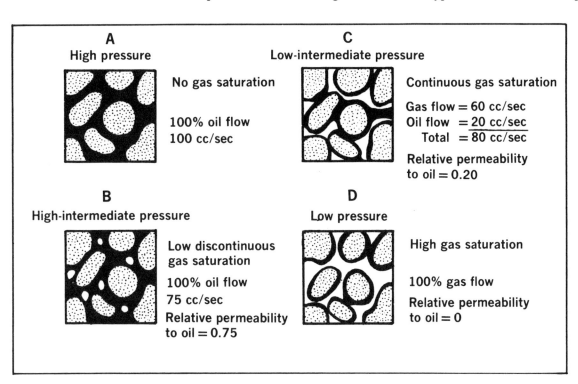

FIG. *2-14—Gas-oil relative permeability concept.* Elements of Petroleum Reservoirs. *Permission to publish by The Society of Petroleum Engineers.*

meability relations for a water wet sandstone. A reservoir represented by Figure 2-15 would have an initial or connate water saturation of about 27%. With an active water drive to maintain pressure, increases in water saturation occurring as water moved in to expel the oil would reduce the relative permeability to oil.

This would result in decreasing oil production and increasing water production. When water saturation reached 75%, relative permeability to oil would be reduced to zero and further oil flow would stop. To reach this point in a practical situation might not be feasible, since very large percentages of water would have to be produced.

The gas-oil or oil-water relative permeability relationships of a particular reservoir rock depend on the configuration of the rock pore spaces, and the wetting characteristics of the fluids and rock surfaces. In an oil-water system the relative permeability to oil is significantly greater when the rock surface is "water wet."

Where two or more fluids are present the "permeability" of equation (2) must represent the permeability of the rock to the desired fluid. This can be done by multiplying the absolute permeability of the rock (permeability to one fluid when completely filled with that fluid) by the relative permeability of the rock to the desired fluid.

$$q_o = 1.127 \times 10^{-3} \frac{k_{abs} k_{ro} A (p_1 - p_2)}{B_o \mu L} \qquad (3)$$

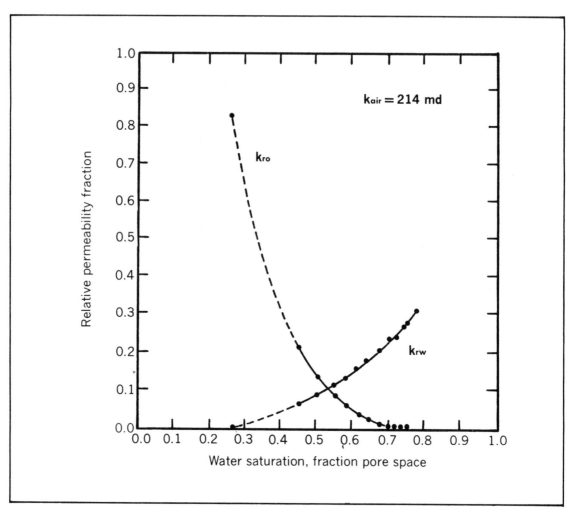

FIG. 2-15—Oil-water relative permeability (water-wet core).[7] Permission to publish by The Society of Petroleum Engineers.

q_o = oil flow rate, stb oil/day
k_{abs} = absolute permeability, md
k_{ro} = relative permeability to oil

For a well producing both water and oil, the "water cut" or the fraction of water in the total flow stream at standard conditions of temperature and pressure can be calculated by this relation:

$$f_w = \cfrac{1}{1 + \cfrac{k_o}{k_w} \times \cfrac{\mu_w}{\mu_o} \times \cfrac{B_w}{B_o}} \quad (4)$$

k_o, k_w = relative permeability
μ_o, μ_w = viscosity
B_o, B_w = formation volume factor

Wettability

Most reservoir rocks were formed or laid down in water, with oil moving in later from adjacent zones to replace a portion of the water. For this reason, most reservoir rocks are considered to be "water wet." The grains of the rock matrix are coated with a film of water, permitting hydrocarbons to fill the center of the pore spaces. Productivity of oil is maximized with this condition.

Actual wettability of a particular reservoir rock is difficult to determine because the process of cutting cores and preparing them for lab tests can, in fact, alter wettability characteristics. Further compounding the problem, most investigators currently believe that there are varying degrees of wettability between strongly water wet and strongly oil wet conditions.

From the standpoint of well completions, stimulation, and workover operations, it is important to realize that the wettability characteristics of the rock near the wellbore can be unfavorably altered by fluids placed in contact with the rock. This is discussed more completely in subsequent chapters on formation damage and surfactants.

Briefly, it is very important to tailor the characteristics of these completion, workover, and stimulation fluids, such that a strongly water wet condition is maintained to maximize relative permeability to oil in an oil-water system, and also, to prevent formation of water-in-oil emulsions in the pore system near the wellbore.

Fluid Distribution

Fluid distribution vertically in the reservoir is important. The relative amounts of oil, water, and gas present at a particular level in the reservoir determine the fluids that will be produced by a well completed at that level, and also influence the relative rates of fluid production.

If oil, water, and gas were placed in a tank, there would be sharp boundaries between the water and oil below, and between the oil and gas above. If the tank were then filled with sand, the contacts between the oil and water and the oil and gas would be quite different, because now the gas, oil, and water exist in capillary spaces. Capillary forces related to wettability and surface tension work against density differences between the fluids to significantly change the previous sharp interfaces between the fluids.

As shown in Figure 2-16, the water saturation in a water wet pore system varies from 100% below the oil zone to progressively lower percentages at points higher in the oil zone. This is because moving higher in the oil zone, the radius of the film between oil and water decreases, due to greater capillary forces. The water fits further back into the crevices between sand grains, and the quantity of water diminishes.

The zone from a point of 100% water (free water level) upward in the sand to some point above which water saturation is fairly constant is called the "transition zone." Relative permeability relations permit both water and oil to flow within the transition zone. Water saturation above the transition zone is termed "irreducible water saturation" or more commonly the "connate" water saturation. Above the transition zone only oil may flow in an oil-water system.

Connate water saturation is related to permeability. Pore channels in lower permeability rocks are generally smaller. For a given height above the free water level, capillary pressure will be the same in two pores of different sizes. Therefore, the film between the water and oil will have the same curvature, and the amount of water occurring in the crevice will be about the same. As shown in Figure 2-17, more oil is contained in the large pore space, however, and the percent of water in the small pore will be greater.

The nature and thickness of the transition zones between water and oil, oil and gas, and water and

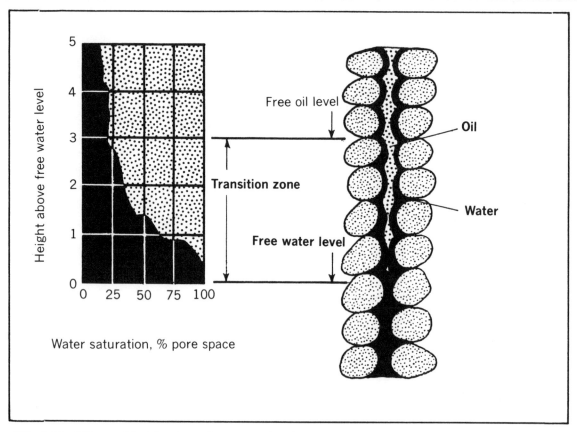

FIG. *2-16—Effects of height above free water level on connate-water content in oil sand.* Elements of Petroleum Reservoirs. *Permission to publish by The Society of Petroleum Engineers.*

gas are influenced by several factors, among which are uniformity, permeability, and wettability of the rock, and the surface tension and density differences between the fluids involved. Generally these statements can be made concerning fluid distribution:

—The lower the permeability of a given sand, the higher will be the connate water saturation.

—In lower permeability sands, the transition zones will be thicker than in higher permeability sands.

—Due to the greater density difference between gas and oil as compared to oil and water, the transition zone between oil and gas is not as thick as the transition zone between oil and water.

A well completed in the oil-water transition zone will be expected to produce both oil and water, depending on the saturations of each fluid present at the completion level. Figure 2-18 summarizes oil, water, and gas saturation in a typical homogeneous rock situation.

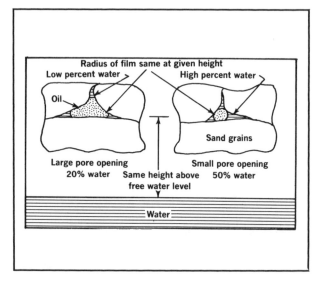

FIG. *2-17—Effect of pore size and shape on connate-water content.* Elements of Petroleum Reservoirs. *Permission to publish by The Society of Petroleum Engineers.*

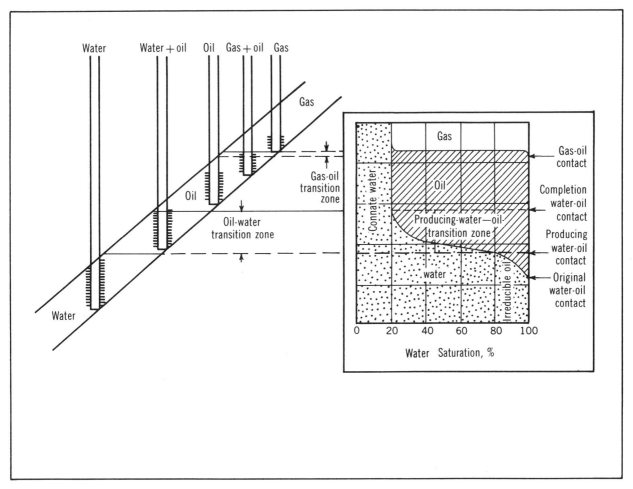

FIG. *2-18—Fluid distribution in a uniform-sand reservoir containing connate water, oil, and a gas cap.*

FLUID FLOW IN THE RESERVOIR

Oil has little natural ability to produce itself into a well bore. It is produced principally by pressure inherent in gas dissolved in oil, in associated free gas caps, or in associated aquifers.

Pressure Distribution Around the Well Bore

Pressure distribution in the reservoir and factors which influence it are of great significance in interpreting well production trends caused by pressure charateristics.

Figure 2-19 shows pressure distribution around a producing oil well completed in a homogeneous zone. Some distance away from the well, pressure is assumed to be 3,000 psi. Moving nearer the well, pressure gradually declines to about 2,700 psi. From nere to a point within the well bore opposite the completion interval, pressure sharply declines to 2,000 psi. At the wellhead, influenced by hydrostatic and also frictional effects in the tubing, pressure is down to about 600 psi.

In a radial flow situation where fluids move toward the well from all directions, most of the pressure drop in the reservoir occurs fairly close to the wellbore. As shown in Figure 2-20, in a uniform sand, the pressure drop across the last 15 ft of the formation surrounding the wellbore is about one-half of the total pressure drop from the well to a point 500 ft away in the reservoir.

Obviously flow velocities increase tremendously as fluid approaches the wellbore. This area around the wellbore is the "critical area." To maximize well productivity everything possible must be done to prevent flow restriction in this critical area.

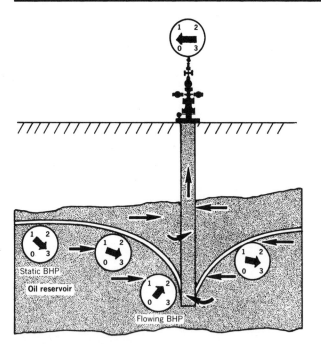

FIG. *2-19—Pressure conditions around a flowing well.*
Elements of Petroleum Reservoirs. *Permission to publish by The Society of Petroleum Engineers.*

Radial Flow Around the Wellbore

Steady state radial flow of incompressible fluid is described by Darcy's Law in the oil field units of Figure 2-21:

$$q = \frac{.00708\, kh\,(p_e - p_w)}{B\mu \ln (r_e / r_w)} \tag{5}$$

Corrections are required to account for flow of compressible fluids, and for turbulent flow velocities.

For non-homogeneous zones (the usual case) permeabilities must be averaged for flow through parallel layers of differing permeabilities (Figure 2-22):

$$\bar{k} = \frac{k_1 h_1 + k_2 h_2 + k_3 h_3}{h_1 + h_2 + h_3} \tag{6}$$

Varying permeabilities in series as shown in Figure 2-23 can be averaged as follows:

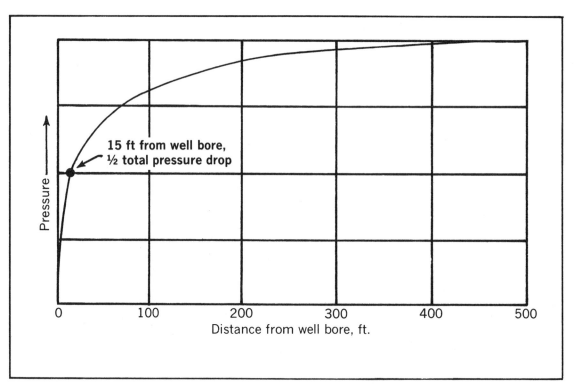

FIG. *2-20—Pressure distribution near the well in radial flow.* Elements of Petroleum Reservoirs. *Permission to publish by The Society of Petroleum Engineers.*

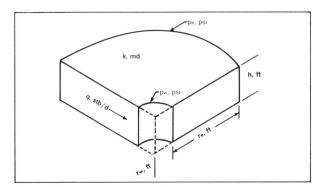

FIG. *2-21—Units for Darcy's Law equation.*

$$\bar{k} = \frac{\ln(r_e/r_w)}{\dfrac{\ln(r_1/r_w)}{k_1} + \dfrac{\ln(r_2/r_1)}{k_2} + \dfrac{\ln(r_3/r_2)}{k_3}} \quad (7)$$

Linear Flow Through Perforations

Ideally (but usually not true) the "perforation tunnel" through the casing and cement sheath is thought to be a void space completely open to flow of fluid. In the ideal case, the perforation tunnel does not offer much restriction to flow.

In sand problem wells, highly permeable gravel used to hold the formation sand in place must fill the perforation tunnel to prevent movement of sand into the tunnel. In this case, the flow restriction of the sand or gravel-filled perforation becomes important.

Flow through the "perforation tunnel" takes on a linear, rather than radial configuration. The linear form of Darcy's Law must be corrected for the fact that turbulent flow usually exists.

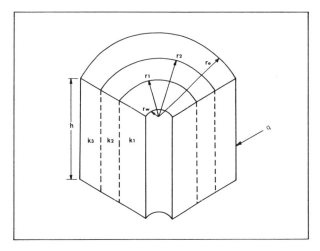

FIG. *2-23—Radial flow, series combination of beds.*

Experimental measurements of pressure drop through gravel-filled perforations, compared with uncorrected linear flow Darcy Law calculations, are shown in Figure 2-24.

Curve A indicates that plugging by even high-permeability (one darcy) sand gives large pressure drop. Actual test data with very high-permeability sand, Curve B, proves turbulent flow results in higher pressure drop than Darcy's Law calculations, Curve C, predict.

Saucier[8], as well as other investigators, have provided turbulence correction factors, which can be applied to the Darcy equation, to permit calculation of pressure drop through the perforation tunnel.

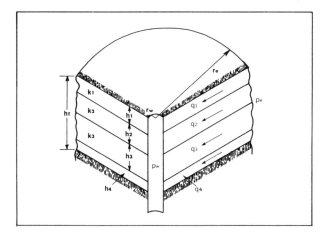

FIG. *2-22—Radial flow, parallel combination of beds.*

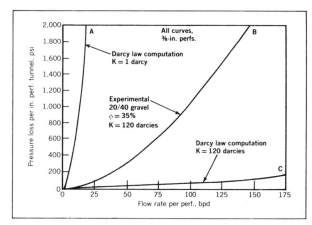

FIG. *2-24—Pressure drop vs. flow rate through perforation (R. H. Torrest, reported by Bruist).[9] Permission to publish by The Society of Petroleum Engineers.*

Causes of Low Flowing Bottom-Hole Pressure

In a reservoir with uniform sand and fluid conditions and no artificial restrictions at the well bore, two factors may cause low flowing bottom-hole pressure in a well. These are permeability and producing rate as shown in Figure 2-25.

With low permeability or excessive rate of production, pressure drawdown will be appreciably higher, thus reducing flowing bottom-hole pressures and possibly requiring that a well be put on artificial lift if high rates of production are required.

Low flowing bottom-hole pressure many times occurs through damage to permeability adjacent to the well bore, caused by drilling or completion operations. This is particularly unfortunate because at this point in the reservoir, restriction is greatly magnified in effect.

Figure 2-26 shows a normal pressure sink compared to a pressure sink in a well where the formation has been damaged. Formation damage may result from any of a number of causes as discussed in detail in other sections of this manual.

The existence of a zone of reduced permeability near the wellbore can be determined through well testing and calculation techniques. Generally, average permeability of the drainage area away from the wellbore (determined by pressure buildup analysis), is compared with permeability, which is a combination of the permeability near the wellbore and that in the drainage area (determined by a productivity test).

Hurst and van Everdingen[5,6] introduced the term Skin or Skin Effect to describe the abnormal pressure drop through the damage zone. This abnormal pressure drop is in addition to the normal radial flow pressure drop, and can be calculated by:

$$\Delta p_s = \frac{141.2\, qB\mu}{kh} \times s \qquad (8)$$

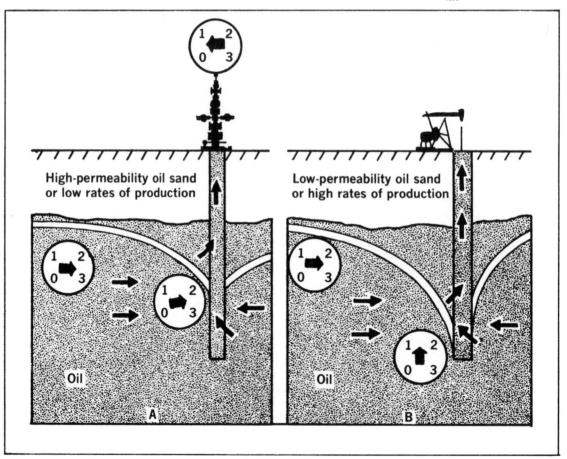

High-permeability oil sand or low rates of production

Low-permeability oil sand or high rates of production

Oil

Oil

A

B

FIG. *2-25—Effects of permeability and production rates on bottom-hole and well-head pressures.* Elements of Petroleum Reservoirs. *Permission to publish by The Society of Petroleum Engineers.*

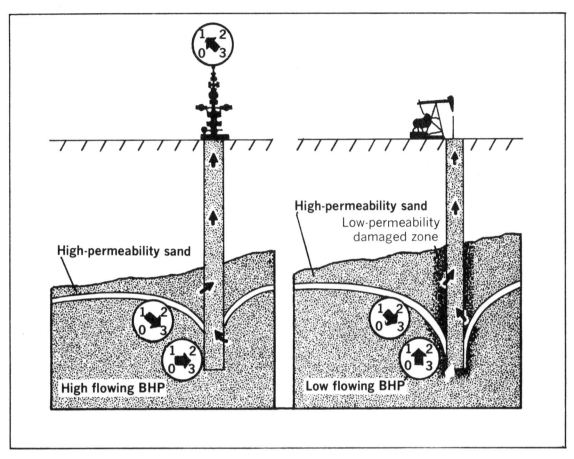

FIG. *2-26—Influence of skin effect on pressure around a well bore.* Elements of Petroleum Reservoirs. *Permission to publish by The Society of Petroleum Engineers.*

Other terms used to quantify formation damage are Damage Ratio and Flow Efficiency.

Damage Ratio:

$$DR = \frac{q_t}{q_a} =$$

$$\frac{\text{Theoretical flow rate without damage}}{\text{Actual flow rate observed}} \quad (9)$$

Also:

$$DR = \frac{J_{ideal}}{J_{actual}} = \frac{\bar{p} - p_{wf}}{\bar{p} - p_{wf} - \Delta p_s} \quad (10)$$

Flow Efficiency:

$$FE = \frac{J_{actual}}{J_{ideal}} = \frac{\bar{p} - p_{wf} - \Delta p_s}{\bar{p} - p_{wf}} \quad (11)$$

In multizone completion intervals, where transient pressure-testing techniques may give questionable results concerning formation damage, production logging techniques may be helpful. Flow profiling may point out zones in an otherwise productive interval, which are not contributing to the total flowstream. The noncontributor zones are likely damaged.

EFFECTS OF RESERVOIR CHARACTERISTICS ON WELL COMPLETIONS

Reservoir Drive Mechanisms

In an oil reservoir, primary production results from the utilization of existing pressure. Basically, there are three drive mechanisms: dissolved gas, gas cap, and water drive; however, as a practical matter most reservoirs produce through some combination of each mechanism.

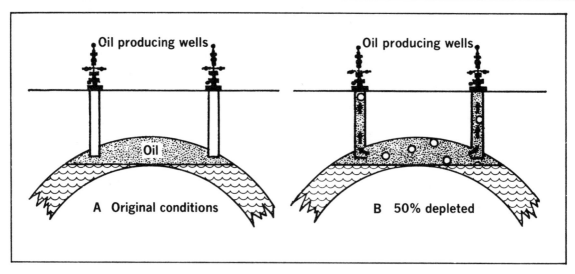

FIG. *2-27—Dissolved-gas-drive reservoir.* Elements of Petroleum Reservoirs. *Permission to publish by The Society of Petroleum Engineers.*

In a dissolved gas drive reservoir, the source of pressure is principally the liberation and expansion of gas from the oil as pressure is reduced, Figure 2-27.

A gas-drive reservoir uses principally the expansion of a cap of free gas over the oil zone, Figure 2-28.

A water drive uses principally expansion or influx of water from outside and below the reservoir, Figure 2-29.

The effect of the reservoir drive mechanism on producing well characteristics must be taken into account in making well completions initially, and later in recompleting wells to systematically recover reservoir hydrocarbons. Figures 2-30 and 2-31 show typical reservoir pressure vs. production and gas-oil ratio vs. production for the three basic drive mechanisms.

In a dissolved gas drive reservoir, (with no attempt to maintain pressure by fluid injection) pressure declines rapidly; gas-oil ratio peaks rapidly, and then declines rapidly; and primary oil recovery is relatively low. Recompletions could not be expected to reduce gas-oil ratio.

In a gas-cap-drive reservoir, pressure declines less rapidly. Gas-oil ratios increase as the gas cap expands into the up-structure well completion intervals. But recompletion or shutting in of up-structure wells provide possibilities for overall gas-oil ratio control. In a water drive reservoir, pressure remains relatively high. Gas-oil ratios are low; but down-structure wells soon begin to produce water. This must be controlled by recompletion or shutting in of these wells. Eventually even up-structure wells must produce significant amounts of water in order

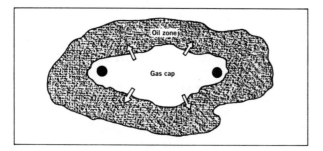

FIG. *2-28—Gas-cap-drive reservoir.* Elements of Petroleum Reservoirs. *Permission to publish by The Society of Petroleum Engineers.*

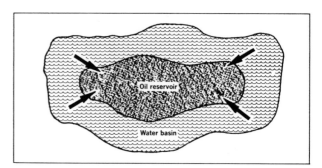

FIG. *2-29—Water-drive reservoir.* Elements of Petroleum Reservoirs. *Permission to publish by The Society of Petroleum Engineers.*

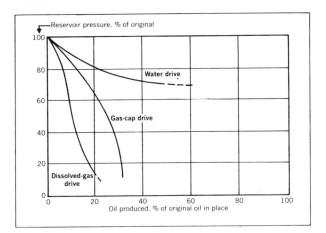

FIG. *2-30—Reservoir-pressure trends for various drive mechanisms.* Elements of Petroleum Reservoirs. *Permission to publish by The Society of Petroleum Engineers.*

to maximize oil recovery.

Obviously many factors must be considered in developing a reservoir, however, the main factors deal with the reservoir itself and the procedures used in exploitation. Well spacing, or better, well location, is one important factor. Money, time, labor, and materials consumed in drilling wells are largely non-recoverable. Therefore, if development drilling proceeds on close spacing before the drive mechanism is correctly identified the investment will already have been made when the recovery mechanism is finally determined.

This does not present an impossible problem,

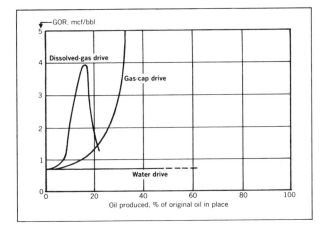

FIG. *2-31—Gas-oil ratio trends for various drive mechanisms.* Elements of Petroleum Reservoirs. *Permission to publish by The Society of Petroleum Engineers.*

even when the predominant drive cannot be determined early in the development. A certain number of wells must be drilled in any event if the field is of appreciable size. Enough wells are needed to define the reservoir—that is, to establish the detailed geologic picture regarding zone continuity and to locate oil-water and gas-oil contacts. Beyond this minimum, the number of infill wells and the well spacing can be varied in many instances.

The development program should be based on reservoir considerations and conditions, rather than on surface conditions or on some arbitrary grid pattern. The development program can be outlined schematically with subsurface stratigraphic cross-sections and a surface plan for well locations on the structure map. Detailed knowledge of the geology of the reservoir and its depositional environment is the key to an effective development plan, as is pointed out in Chapter 1, Geologic Considerations.

Many case histories are available to show the problems resulting from reservoir development without sufficient consideration of the stratigraphy of the reservoir.

With regard strictly to the effect of reservoir drive mechanism, general statements can be made as to well development patterns as outlined in the next topics.

Dissolved-Gas-Drive Reservoirs—Well completions in a dissolved-gas drive reservoir with low structural relief can be made in a regularly spaced pattern throughout the reservoir, and provided the rock is not stratified, can be made low in the reservoir bed, Figure 2-32.

A regular spacing pattern could also be used for a dissolved-gas drive reservoir with a high angle of dip, Figure 2-33.

Again the completion intervals should be structurally low because of the angle of structural dip, and exact subsurface location would vary with well location on the structure. Here it is expected that the oil will drain down-structure in time so that higher-than-usual oil recovery will be realized with minimum investment in wells. The operator must recognize the reservoir situation soon enough to eliminate drilling the structurally high wells.

Due to low recovery by the primary mechanism, some means of secondary recovery will almost certainly be required at some point in the life of the reservoir. Initial well completions need to be designed with this in mind.

Limits of oil production

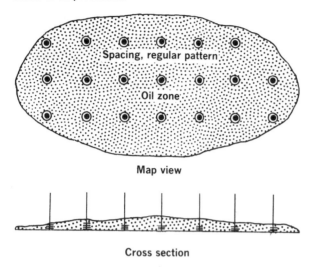

Map view

Cross section

FIG. *2-32—Dissolved-gas-drive reservoir, low angle of dip.* Elements of Petroleum Reservoirs. *Permission to publish by The Society of Petroleum Engineers.*

Gas-Cap-Drive Reservoirs—Wells may be spaced on a regular pattern in a gas-cap drive reservoir where sand is thick, dip angle is low, and the gas-cap is completely underlain by oil, Figure 2-34.

Again, completions should be made low in the section to permit the gas cap to expand and drive oil down to the completion intervals for maximum recovery with minimum gas production.

A gas-cap drive reservoir in a thin sand with a high angle of dip is likely to be more efficiently controlled by having completion spaced irregularly but low on the structure to conform to the shape of the reservoir, Figure 2-35.

Because of the high angle of dip, a regular spacing pattern may cause many completions to be located

Limits of oil production

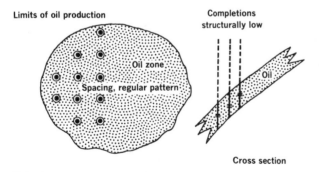

Cross section

FIG. *2-33—Dissolved-gas-drive reservoir, high angle of dip.* Elements of Petroleum Reservoirs. *Permission to publish by The Society of Petroleum Engineers.*

Limits of oil production

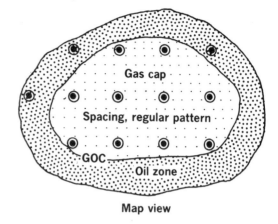

Map view

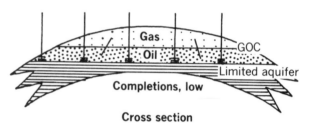

Cross section

FIG. *2-34—Gas-cap-drive reservoir, low angle of dip.* Elements of Petroleum Reservoirs. *Permission to publish by The Society of Petroleum Engineers.*

too near the gas-oil contact. Such an oil reservoir is common where multiple thin sands are found on a single structure and the oil column is only a fraction of the total productive relief.

Water-Drive Reservoirs—Wells may be spaced on a regular pattern in a water drive reservoir having a thick sand and low angle of dip, Figure 2-36.

Completion intervals should be selected high on the structure to permit long producing life while oil is displaced up to the completion intervals by invading water from below.

A water-drive reservoir in a thin sand with high angle of dip may best be developed with irregular well-spacing because of the structural characteristics, Figure 2-37.

The completions, however, should be made high on the structure to delay encroachment of water into the producing wells. Spotting the wells on a regular spacing pattern not only may cause a number of wells to produce water early in the life of the reservoir and result in their early abandonment, but also may reduce the effectiveness of the water

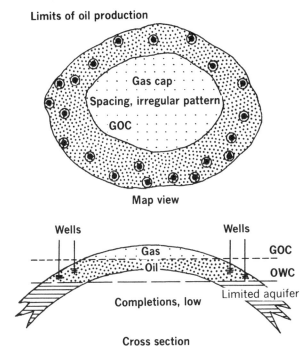

Limits of oil production

Map view

Cross section

FIG. *2-35—Gas-cap-drive reservoir, high angle of dip.* Elements of Petroleum Reservoirs. *Permission to publish by The Society of Petroleum Engineers.*

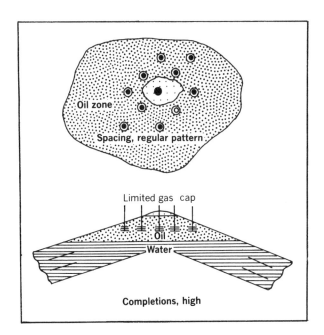

FIG. *2-36—Water-drive reservoir, low angle of dip, thick sand.* Elements of Petroleum Reservoirs. *Permission to publish by The Society of Petroleum Engineers.*

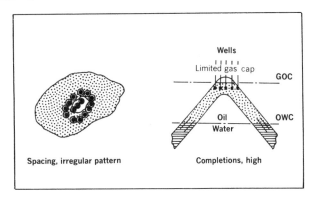

FIG. *2-37—Water-drive reservoir, high angle of dip, thin sand.* Elements of Petroleum Reservoirs. *Permission to publish by The Society of Petroleum Engineers.*

drive through excessive early water production. Fewer wells would then remain to produce the remainder of the oil, thus lengthening unnecessarily the length of time required to deplete the reservoir.

Significant amounts of water must be produced in the later life of the field in order to maximize recovery.

Reservoir Homogeneity

The general procedure, as previously described and shown by the illustrations, is to complete high for water drive and low for dissolved gas and gas-cap drive reservoirs to have an adequate number but

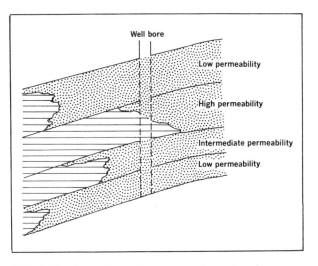

FIG. *2-38—Irregular water encroachment and premature water breakthrough in high-permeability layers of reservoir rock.* Elements of Petroleum Reservoirs. *Permission to publish by The Society of Petroleum Engineers.*

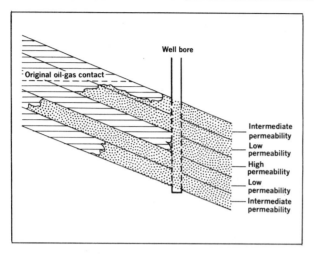

FIG. *2-39—High gas-oil ratio production caused by early encroachment of gas through high-permeability zones of stratified reservoir.* Elements of Petroleum Reservoirs. *Permission to publish by The Society of Petroleum Engineers.*

not too many wells. It would be practical, however, to make such completions only if the reservoir were quite uniform.

Most sandstone formations were originally laid down as stratified layers of varying porosity and permeability. Similar statements can be made regarding carbonate, and even reef-type reservoirs. Thus, the normal sedimentary process results in reservoirs of a highly stratified nature. Fluids flow through alternate layers with different degrees of ease, and many times impermeable zones separate the permeable beds so that no fluid can move from bed to bed. This is shown in Figures 2-38 and 2-39.

In thin beds or highly stratified beds "fingering"

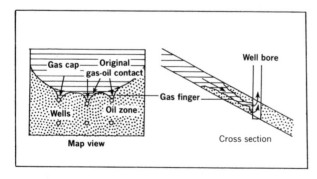

FIG. *2-40—Fingering of free gas into well along bedding planes.* Elements of Petroleum Reservoirs. *Permission to publish by The Society of Petroleum Engineers.*

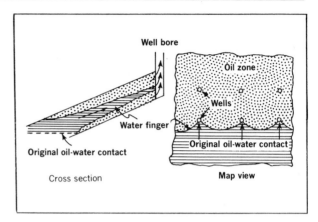

FIG. *2-41—Fingering of water into well along bedding planes.* Elements of Petroleum Reservoirs. *Permission to publish by The Society of Petroleum Engineers.*

(movement of fluid parallel to bedding planes) of free gas down from a gas cap, or water up from a water basin is always a possibility when short completion intervals combined with high rates of production are encountered.

Figures 2-40 and 2-41 illustrate these two problems. If the reservoir is stratified either by shale breaks or by variations in permeability, it probably will be necessary to stagger the completion intervals in various members of the reservoir to be sure that each member is drained. Some vertical staggering of completion intervals can be effected during development to secure proportionate withdrawals from the various strata.

Additional distribution of completions between the various members of this pay may then be made during later workovers, on the basis of experience and operational conditions.

For maximum recovery from reservoir completions, intervals should be limited to one identifiable zone wherever practical. Single-zone completions are preferred to facilitate thorough flushing for higher recoveries and to obtain flexibility in recompletion work for control of reservoir performance.

Completions comprising more than one reservoir are termed multi-zone completions. Such completions may be needed for low permeability reservoirs which require long completion intervals for obtaining economic quantities of production.

REFERENCES

1. Clark, Norman J.: "Elements of Petroleum Reservoirs," Henry L. Doherty Series, AIME, Dallas, 1960. Revised 1969.

2. Standing, M. B.: "Volumetric and Phase Behavior of Oil Field Hydrocarbon Systems," Reinhold Publishing Corp., New York, 1952.

3. Beal, Carlton: "The Viscosity of Air, Water, Natural Gas, Crude Oil and its Associated Gases at Oil Field Temperatures and Pressures," Trans. AIME, 1946.

4. Carr, Norman L.; Kobayashi, Riki; and Burrows, David B.: "Viscosity of Hydrocarbon Gases Under Pressure," Trans. AIME, 1954.

5. van Everdingen, A. F.: "The Skin Effect and its Influence on the Productive Capacity of a Well," Trans. AIME, 1953.

6. Hurst, William: "Establishment of Skin Effect and its Impediment to Fluid Flow Into a Well Bore," Petroleum Engineer, Oct., 1953.

7. Morgan, J. T.; and Gordon, D. T.: "Influence of Pore Geometry on Water-Oil Relative Permeability," J. Pet. Tech., Oct. 1970, p. 1,199.

8. Saucier, R. J.: "Gravel Pack Design Considerations,"

J. Pet. Tech., Feb. 1974, p. 205.

9. Bruist, E. H.: "Better Performance of Gulf Coast Wells," SPE No. 4777, New Orleans, 1974.

10. Dowling, Paul L., Jr.: "Application of Carbonate Environmental Concepts to Secondary Recovery Projects," SPE 2987, 1970.

11. Weber, K. J.: "Sedimentological Aspects of Oil Fields in the Niger Delta," Geologie En Mijnbouw, Vo. 50 (3), 1971.

12. Jardine, D.; Andrews, D. P.; Wishart, J. W.; and Young, J. W.: "Distribution and Continuity of Carbonate Reservoirs," SPE 6139, New Orleans, Oct. 1976.

13. Reitzel, Gordon A., and Callow, George O.: "Pool Description and Performance Analysis Leads to Understanding Golden Spike's Miscible Flood," SPE 6140, New Orleans, October 1976.

14. Earlougher, Robert C., Jr.: "Advances in Well Test Analysis," Monograph Volume 5, Henry L. Doherty Series, AIME, 1977.

Chapter 3   Well Testing

Well production testing
Periodic production tests
Productivity or deliverability tests
Transient pressure tests
Basics of buildup, drawdown, injection, and multiple well tests
Drill stem testing
Basics of DST tools, and procedures
Pressure buildup analysis
Procedures for good test data
Eyeball interpretation of charts

Well Production Testing

INTRODUCTION

The objectives of Well Production Testing vary from a simple determination of the amount and type of fluids produced to sophisticated transient pressure determinations of reservoir parameters and heterogeneities. Briefly, Well Testing procedures are a set of tools which properly used can provide valuable clues as to the condition of production or injection wells. The Well Completion or Production Engineer needs to be able to design and conduct the simplier testing procedures, and to be familiar with the possibilities and limitations of the more sophisticated procedures.

The purpose of this section is to present a brief discussion of the basics of production well testing. A somewhat more complete discussion of Drill Stem Testing follows. References at the end of this section should be consulted for detailed explanation and example calculations using the many well testing methods.

Generally Oil or Gas Well Production Tests may be classified as:

—Periodic Production Tests
—Productivity or Deliverability Tests
—Transient Pressure Tests

Periodic production tests have as their purpose, determination of the relative quantities of oil, gas and water produced under normal producing conditions. They serve as an aid in well and reservoir operation and also in meeting legal and regulatory requirements.

Productivity or deliverability tests are usually performed on initial completion or recompletion to determine the capability of the well under various degrees of pressure drawdown. Results may set production allowables, aid in selections of well completion methods, and design of artificial lift systems and production facilities.

Transient pressure tests require a higher degree of sophistification and are used to determine formation damage or stimulation related to an individual well, or reservoir parameters such as permeability, pressure, volume and heterogeneities.

PERIODIC PRODUCTION TESTS

Production Tests are run routinely to physically measure oil, gas and water produced by a particular well under normal producing conditions. Test results may then be used to allocate total field or lease production between wells where individual well production is not monitored continuously. They provide the basis for periodic reports to legal or regulatory groups.

From the standpoint of well and reservoir operation, they provide periodic physical evidence of well conditions. Unexpected changes, such as extraneous water or gas production may signal well or reservoir problems. Abnormal production declines may mean artificial lift problems, sand fillup in the casing, scale buildup in the perforations, etc.

For oil wells results are usually reported as oil production rate, barrels of oil per day; gas-oil ratio (GOR), cubic feet per barrel; and water-oil ratio

(WOR), percentage of water in the total liquid stream. Test equipment can consist of nothing more than a gas-oil separator and a stock tank, with appropriate measuring devices such as an orifice meter for gas and a hand tape for oil and water. The current trend is toward more sophisticated systems, the ultimate providing recording meter measurement of water, and oil, as well as gas, along with automatic switching and unattended operation.

Accuracy of measurements, and careful recording of the conditions under which the test was run are of obvious importance. Choke size, tubing pressures, casing pressures, details of the artificial lift system operation, in short everything affecting the ability of the well to produce should be recorded. Problems such as measurement of emulsion, power oil fluid with hydraulic pumping systems, input gas with gas lift systems, must be recognized and properly handled. Stabilized producing conditions are of obvious importance where short test periods are used. Tests should be run at the "normal" production rate, since changes of rate often influence the relative quantities of oil, gas and water.

For gas wells routine production tests per se are less common, since gas production is usually metered continuously from individual wells. Gas production is reported in Mcfd (thousands of standard cubic feet per day) or MMcfd (millions of standard cubic feet per day). Hydrocarbon liquids or water are reported in barrels condensate per million cubic feet (BCPMM) or barrels water per million cubic feet. Stabilized flow condition, careful metering and reporting of volumes and pressures are again of obvious importance.

PRODUCTIVITY OR DELIVERABILITY TESTS

Productivity or Deliverability Tests represent the second degree of sophistication in oil or gas well production testing. They involve a physical or empirical determination of produced fluid flow versus bottom hole pressure drawdown. With a limited number of measurements they permit prediction of what the well should produce at other pressure drawdowns. They do not rely on a mathematical description of the flow process. They are successfully applied to non-darcy, below-the-bubble point flow conditions, even though fluid properties and relative permeabilities are not constant around the wellbore.

They do not permit calculation of formation permeability or the degree of abnormal flow restriction (formation damage) near the wellbore. They do, however, include the effects of formation damage. Thus, they can be used as an indicator of well flow conditions, or as a basis for a simple comparison of completion effectiveness among wells in a particular reservoir.

Deliverability tests represent stabilized producing conditions. They involve the measurement of bottom hole static and flowing pressure, as well as fluid rates produced to the surface.

Commonly used deliverability tests for oil wells may be classified as:

—Productivity Index
—Inflow Performance
—Flow after Flow
—Isochronal

Gas-well-deliverability tests are designed to establish the "absolute open flow potential," or the production rate if flowing bottom hole pressure could be reduced to zero. Termed multipoint back-pressure tests, or simply back pressure tests, they can be classified according to test procedure as:

—Flow after Flow
—Isochronal

Oil Wells

Productivity Index—The Productivity Index test is the simplest form of Deliverability Test. It involves the measurement of shut-in bottom hole pressure; and, at one stabilized producing condition, measurement of the flowing bottom pressure and the corresponding rate of liquids produced to the surface. Productivity Index is then defined as:

$$\text{PI} = J = \frac{q}{p_i - p_{wf}} \qquad (1)$$

q = Total liquids stb/d
p_i = Shut-in bottom hole pressure, psi
p_{wf} = Flowing bottom hole pressure, psi
$p_i - p_{wf}$ = Pressure drawdown, psi

Specific PI accounts for the length of the producing section:

$$\text{Specific PI} = \frac{\text{Productivity Index}}{\text{Length of producing zone}} \qquad (2)$$

With a well producing above the bubble point, the PI may be constant over a wide range of pressure drawdowns. However, with flow below the bubble point, and gas occupying a portion of the pore system, PI falls off with increased drawdown.

Productivity Index also declines during the life of a well due to many factors, among which are changes in reservoir pressure, composition and properties of produced fluids, and flow restriction or formation damage near the wellbore. Productivity Index does however give the Well Completion Engineer a useful index of well and wellbore conditions, and with recognition of the limitation involved, a yardstick for comparison between wells.

Inflow Performance Test—The simple concept of Productivity Index attempts to represent the inflow performance relation of a well as a straight line function, (Figure 3-1, Well A). The true inflow performance relation or IPR usually declines at greater drawdowns as shown in Figure 3-1, Wells B and C.

An inflow-performance test should, in effect, consist of PI tests at several production rates in order to provide a better representation of the true inflow performance relation of the well. Vogel[11], based on a computer simulation of dissolved gas drive reservoirs, wherein he calculated IPR's using a wide range of reservoir and fluid parameters, proposed the general IPR curve of Figure 3-2. Often

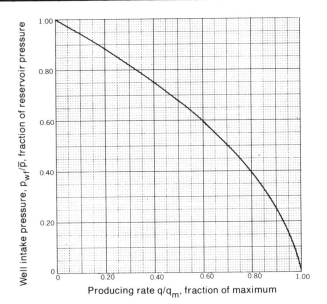

FIG. *3-2—Inflow performance relationship. (Vogel). Permission to publish by The Society of Petroleum Engineers.*

this same Vogel relation is successfully applied to other types of reservoir drive systems.

A primary advantage of the Vogel relation is that, with a value for static reservoir pressure, one well test, recording production rate and corresponding bottom hole pressure, is all that is needed to provide a reasonable value for production rate at any other flowing bottom hole pressure. In conducting the test a key point is to insure that stabilized flow conditions are established. Standing[14] extended the Vogel relation such that, if the "Flow Efficiency" could be estimated, then the effect of flow restriction or damage near the wellbore could be included in a simple graphical analysis to predict the IPR with the restriction, and also the IPR with the restriction removed. (See chapter on well-completion design.

Again, it should be pointed out that the inflow performance relation is an empirical representation, and that it changes during the life of the well. Methods for predicting the IPR at future times were suggested by Standing involving current versus future mobility ratio and formation volume factors.[14]

Flow After Flow—Back-pressure tests have been used for empirical determination of Gas Well capability for many years.[7] Essentially a plot of flow rate, versus "squared drawdown pressure," on a log-log paper provides a straight line which can be extended to predict flow rate for any drawdown.

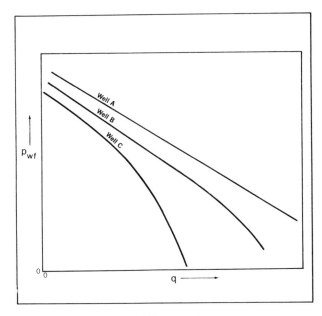

FIG. *3-1—Typical inflow performance curves.*

Oil Well deliverability can be represented in the same manner.[15] The method again is particularly useful for reservoirs producing below the bubble point where mathematical description is impractical. Oil rate is related to pressure drawdown empirically as follows:

$$q_o = J' (p_i^2 - p_{wf}^2)^n \qquad (3)$$

J' = productivity coefficient
n = empirically determined exponent: $0.5 < n < 1.0$

Figure 3-3 shows ideal flow rate and bottom-hole flowing pressure versus time for a properly run flow after flow test.

A log-log plot of flow rate, q, vs. $(p_i^2 - p_{wf}^2)$ should define a straight line with slope "n." This plot can then be used to predict flow rate for any possible drawdown pressure, Figure 3-4. At least four rates should be run, and each rate should continue until the well reaches a stabilized flowing pressure condition. This requirement for many wells means that long testing periods are required—and often limits the usefulness of the flow after flow method.

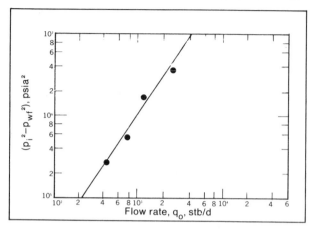

FIG. 3-4—Isochronal deliverability curve.[18] Permission to publish by The Society of Petroleum Engineers.

Modified Isochronal—To reduce testing time the Gas Well Isochronal Test procedure has also been adapted to Oil Wells.[15] The test is run as shown ideally in Figure 3-5, with a series of relatively short equal-time flow and shut-in periods, and one final flow period where flowing pressure is permitted to stabilize. Four to six hours is sufficient duration for the equal flow or shut-in period.

In making the flow rate versus squared drawdown pressure plot, each flowing period is divided into time increments; i.e. 1/4 hr., 1 hr., and 4 hrs. after the start of the flow period. A plot of q vs. $(\bar{p}^2 - p_{wf}^2)$ is made for each time increment as shown in Figure 3-6, where $\bar{p}$ = average reservoir pressure. Shut-in pressure for each calculation must be the shut-in pressure just prior to the start of that particular flow period.

The slope of the resulting plots determines the proper slope for the "stabilized" curve which is drawn through the one data point where flow conditions were actually allowed to stabilize. The stabilized curve is then used for predictive purposes.

Gas Wells

Gas-well deliverability testing was formalized through work done by the U.S. Bureau of Mines and the Railroad Commission of Texas.[6,7] The original empirical procedure was called the Multipoint Back Pressure Test. The technique requires careful measurement of flow rates and surface pressures at four stabilized flow conditions, and a surface shut-in pressure. Surface pressures

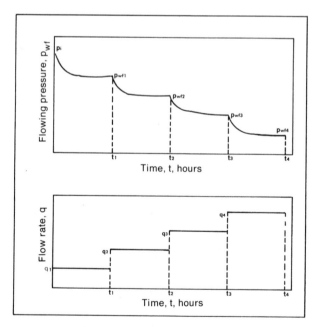

FIG. 3-3—Pressure-rate history for a flow after flow test.[18] Permission to publish by The Society of Petroleum Engineers.

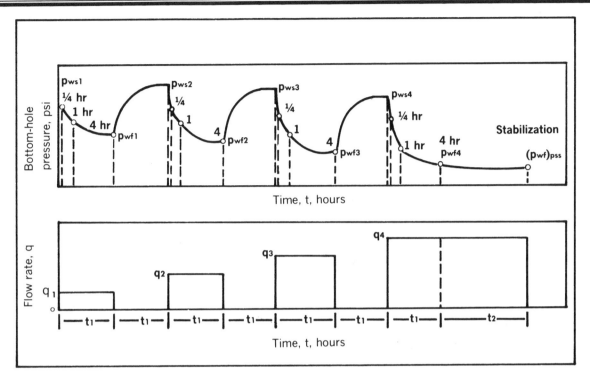

FIG. *3-5—Pressure-rate history for a modified-isochronal-flow test.* [18] *Permission to publish by The Society of Petroleum Engineers.*

are then converted to bottomhole pressures by calculation procedures.

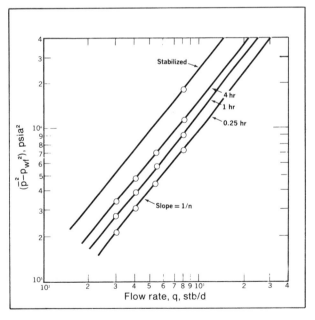

FIG. *3-6—Modified-isochronal-test data plot.* [18] *Permission to publish by The Society of Petroleum Engineers.*

A log-log plot of flow rate vs. $(p_i^2 - p_{wf}^2)$ should define a straight line which can be extended to determine the flow rate at zero bottom hole pressure (or maximum drawdown) which is termed the absolute open-flow potential of the well. The slope of the line usually varies between 0.5 and 1.0. In some cases, (for example DST's) one flowing pressure is used to estimate the AOF range, assuming that the slope is 1.0 for maximum AOF, or 0.5 for minimum AOF.

Properly carried out to stabilized conditions, the original multipoint back-pressure test procedure is a flow-after-flow test. As a practical matter, stabilization for low permeability formation may require many hours, thus, becomes impractical.

The isochronal test for gas wells was first suggested in 1955,[8] subsequently modified, is currently used as standard test procedure for many conditions. This procedure described previously for oil wells reduces the test time requirements to a practical range and apparently provides results comparable to the original flow-after-flow procedure.

Some areas where permeabilities are very low and stabilization impractical have standardized short term test for gas wells. One such test requires that the well be shut in then flowed for twenty minutes at which time the measured flow rate indicates well capability.

TRANSIENT PRESSURE TEST
Basis for Transient Pressure Analysis

Assume that the only well completed in a reservoir is shut in until a completely stable situation is reached. If this well is then put on production, and pressure is reduced in the wellbore, a wave of reduced pressure gradually moves outward into the reservoir establishing a pressure gradient or sink toward the well. With continued fluid withdrawals from the wellbore, the pressure wave moves further outward. Each point passed by the wave experiences a continuing pressure decline. At a particular time, the maximum distance the wave has traveled is called the drainage radius of the well.

When the wave front reaches a closed boundary, pressure at each point within the boundary continues to decline, but at a more rapid rate. If the wave front encounters a boundary, which supplies fluid at a rate sufficient to maintain a constant pressure at the boundary, pressure at any point within the drainage radius will continue to decline, but at a slower rate. With either the closed boundary, or the constant pressure boundary, the pressure gradient toward the well tends to stabilize after a sufficient time. Pressure level at a particular point may continue to decline.

For the constant pressure boundary a "steady state" will be approached where both pressure gradient and pressure level do not change with time. For the closed boundary a "pseudo steady state" condition is reached where pressure gradient is constant, but pressure declines linearly with time at each point within the drainage radius.

Changes in production rate, or production from other wells will cause additional pressure wave movements, which affect pressure decline and pressure gradients at every point within the drainage radius of the first well.

The basis for Transient Pressure analysis is the observation of these pressure changes, and the fluid withdrawal or injection rates which caused them; along with mathematical descriptions of the flow process, involving properties of the rock through which the movement occurred, and the characteristics of the fluids moving within.

The diffusivity equation, describing fluid flow through reservoir rock, assumes horizontal flow, negligible gravity effects, a homogeneous and isotropic rock, and a single slightly compressible fluid.[10] Also Darcy's Law must apply, and porosity, permeability, viscosity and compressibility must be independent of pressure. With these limitations the diffusivity equation can be easily solved. Where minor variances occur, approximation methods result in reasonable solutions. Where major variances occur, numerical reservoir simulation techniques must be used in an attempt to model these variations.

Limitations and Application

While assumptions and complications of analysis techniques appear to greatly limit application of transient pressure testing for the Well Completion Engineer, under favorable circumstances—and with experience—it can be a useful tool in solving well problems. It should not be oversold—but it does provide a chance to recognize near-wellbore problems; i.e.—formation damage—which might not be detected in the composite picture shown by Deliverability Tests.

For the reservoir analyst the chance to determine average reservoir permeability and pressure, reservoir volume, and perhaps something about reservoir heterogeneities is intriguing.

Pressure transients are subject to all of the normal well completion problems of communication behind casing, partial or ineffective perforations, or partial penetration of the producing zone. Proper analysis must consider all available clues. Production tests, details of perforating and well completion or workover operations, flow profiles or other production logging data may be helpful.

Wellbore Damage or Stimulation—From the standpoint of the Well Completion Engineer wellbore damage or stimulation indicators are of practical importance. There are several ways to quantify damage or improvement. One method uses the idea of "skin" or "*skin effect*". Shown graphically in Figure 3-7.

Pressure drop across the infinitesimally thin skin, Δp_s, is added to the transient pressure drop in the reservoir to represent the wellbore pressure. The pressure drop across the skin can be calculated as follows:

$$\Delta p_s = \frac{141.2 \, qB\mu}{kh} s \qquad (4)$$

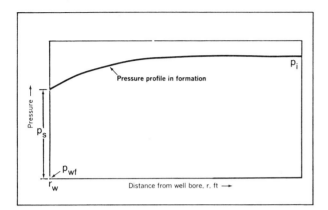

FIG. 3-7—*Pressure distribution around a well with a positive skin factor.*[18] *Permission to publish by The Society of Petroleum Engineers.*

Δp_s = pressure drop across skin, psi
B = formation volume factor, reservoir bbl/stb
μ = viscosity, cp
s = skin factor, dimensionless
k = permeability, md
h = height, ft

The value of the skin factors can vary from about -5 for a hydraulically fractured well, to ∞ for a completely plugged well. One problem with the concept of skin effect is that the numerical value of the skin "s" does not directly show the degree of damage.

Flow Efficiency (or Condition Ratio) describes the well's actual flow capacity as a fraction of its capacity with no damage.

$$FE = \frac{J_{actual}}{J_{ideal}} = \frac{\bar{p} - p_{wf} - \Delta p_s}{\bar{p} - p_{wf}} \qquad (5)$$

Damage Ratio is the inverse of flow efficiency. Skin factor, flow efficiency, or damage ratio can be determined from most of the transient pressure techniques that follow.

Wells completed with only a part of the producing zone open—through ineffective perforating or the fact that the well was not drilled completely through the zone will appear to be damaged even if there is no physical flow restriction.

Deviated holes penetrating the reservoir at an angle with no other problems will appear stimulated.

Reservoir Heterogeneities—Reservoir heteroge- neities affect transient pressure behavior. Vari- ation in rock properties due to depositional or post depositional changes—vugs—fractures—di- rectional or layered permeability; changes in fluid properties—gas-oil, oil-water contacts—or changes due to pressure variation; changes due to stimulation effects or injection fluids—all affect pressure tran- sients.

Multiple well tests are more severely affected by heterogeneities, and as a result can be used to define heterogeneities in simple cases. A primary difficulty is that many different conditions can give the same transient pressure response. Without other evi- dence, geologic, seismic, fluid flow or well per- formance, transient pressure tests should not be used to infer heterogeneous reservoir properties. If individual layers of a multiple layered reservoir are not isolated, meaningful individual layer in- formation on permeability, skin factor or average reservoir pressure cannot be estimated with current technology.

Description of Transient Pressure Tests

Almost every conceivable method of creating and observing changing wellbore pressures has been employed for reservoir analysis. A reasonable classification of transient pressure tests by type is:

—Pressure buildup
—Pressure drawdown
—Multiple rate
—Injection buildup or fall-off
—Multiple well interference
—Drill-stem tests

Each type presents certain advantages and limita- tions, and certain factors which are particularly important for reasonable results. The following description is intended to give the Completion Engineer an acquaintance with each type. More details are provided concerning pressure buildup tests. Drill-stem testing, being more in the province of completion engineering, is handled as a separate section.

For an excellent and practical treatment of tran- sient pressure analysis and calculation procedures for oil wells, S.P.E. Monograph No. 5, "Advances in Well Test Analysis", by Robert C. Earlougher, Jr.[18] is recommended.

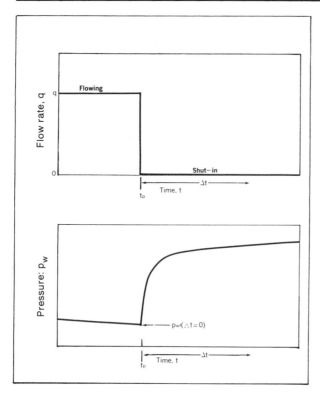

FIG. *3-8—Idealized rate and pressure history for a pressure buildup test.*[18] *Permission to publish by The Society of Petroleum Engineers.*

Pressure-Buildup Testing—Pressure-buildup testing is probably the most familiar transient well testing technique. Essentially the well is produced at constant rate long enough to establish a stabilized pressure distribution, then is shut in as shown in Figure 3-8. Stabilization is important, otherwise erroneous data will be calculated. If stabilization is impossible, other techniques, such as variable rate testing must be used.

Referring to Figure 3-8, t_p is production time, and Δt is shut in time. Pressure is measured immediately before shut in, and is recorded as a function of time during the shut-in period. The resulting pressure buildup curve is then analyzed for reservoir properties and wellbore condition.

Horner Method—After wellbore storage effects have diminished, wellbore pressure during shut in is:

$$p_{ws} = p_i - \frac{162.6\, qB\mu}{kh} \log\left(\frac{t_p + \Delta t}{\Delta t}\right) \quad (6)$$

p_{ws} plotted versus $\log \dfrac{t_p + \Delta t}{\Delta t}$ is a straight line with slope $(-m)$, and intercept (p_i). The value of t_p (hrs) should be estimated as follows: cumulative production from last pressure equalization, divided by the stabilized production rate just before shut in, times 24—or $t_p = \dfrac{24\, V_p}{q}$.

This plot, Figure 3-9 is called the Horner Plot. The straight line portion of the curve may be extrapolated to infinite time to obtain p_i. The value of p_i is accurate for short production periods, but is somewhat too high for long production periods.

The measured slope value (m) can be used to determine reservoir permeability:

$$k = \frac{162.6\, qB\mu}{mh} \quad (7)$$

Skin factor (s) does not appear in the Horner Equation—but does affect the early-time shape of the curve—as do wellbore storage effects. Skin factor "s" can be calculated from the following equation:

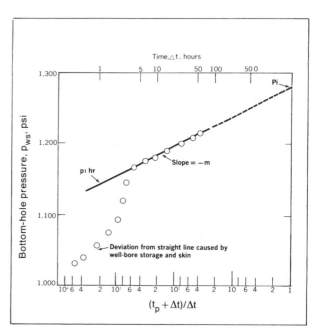

FIG. *3-9—Horner plot of pressure buildup data.*[18] *Permission to publish by The Society of Petroleum Engineers.*

$$s = 1.151 \left[\frac{p_{1\,hr} - p_{wf}(\Delta t = 0)}{m} \right.$$

$$\left. - \log \left(\frac{k}{\phi\,\mu\,c_t\,r_w^2} \right) + 3.23 \right] \qquad (8)$$

The value for $p_{1\,hr}$ must be taken from the straight line extrapolation.

If wellbore storage effects last so long that a semi-log straight line does not develop, "type curve" techniques presented by McKinley, Ramey, and Earlougher and Kersch, may be helpful in salvaging estimates from otherwise unusable data. Type curve methods should be considered a last resort, however.

Miller-Dyes-Hutchinson Plot—The Horner build-up plot may be simplified if the producing time is much greater than the shut-in time. The Miller-Dyes-Hutchinson (MDH) buildup equation is:

$$p_{ws} = p_{1\,hr} + m \log \Delta t \qquad (9)$$

$$m = \frac{162.6\,q\,B\,\mu}{kh} \qquad (10)$$

A plot of p_{ws} vs. log Δt should be a straight line—with a typical example shown in Figure 3-10. Extrapolated shut-in pressure p_i is again somewhat too high.

Since the *MDH* method is somewhat easier to use it is usually employed—except for the case

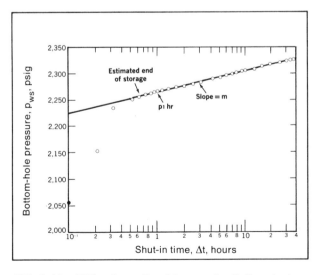

FIG. *3-10—Miller-Dyes-Hutchinson plot.*[18] *Permission to publish by The Society of Petroleum Engineers.*

of Drill Stem Tests where flow time is about the same magnitude as shut-in time. Here the Horner method must be used.

Frequently pressure buildup tests are not as simple as discussed. Many factors can affect the shape of the pressure buildup plot. These include wellbore storage effects, formation damage, partial penetration, hydraulic fractures, unstabilized flow conditions, fluid and rock interfaces, water-oil or gas-oil contacts, and rock heterogeneities. Practical problems such as a leaking pump or lubricator or a bottom-hole pressure gauge or reading device in poor condition, may render otherwise good data unusable. Or if used, it may provide erroneous results.

A major cost consideration of the Pressure Build-up Test is the fact that income is deferred while the test is being run.

Pressure Drawdown Testing—Pressure draw-down tests have two advantages over pressure buildup tests. First, production continues during the test period. Second, in addition to formation permeability and formation damage information, an estimate can be made of the reservoir volume in communication with the wellbore. Thus the "Reservoir Limits" test can be used to estimate if there is sufficient oil—or gas (if a dry gas reservoir)—in place to justify additional wells in a new reservoir.

Multiple Rate Testing—Pressure buildup and drawdown tests require a constant flow rate, which is sometimes impossible or impractical to maintain for a sufficiently long period. Multiple rate analysis can be applied to several well flow situations: for example, uncontrolled variable rates; a series of constant rates; or constant bottom hole pressure with continually changing flow rate.

Multiple rate tests have the advantage of providing transient test data without the requirement of well shut in. They minimize wellbore storage effects and phase segregation effects; thus, sometimes provide good results where buildup or drawdown tests would not. This is perhaps their greatest advantage.

Accurate flow rate and pressure measurements are essential. Rate measurements are much more critical than in constant rate tests. Rate changes must be sufficient to significantly effect the transient pressure behavior. The analysis procedure is direct and simple but computations needed to make the plot are more bothersome, and are sometimes left to the computer.

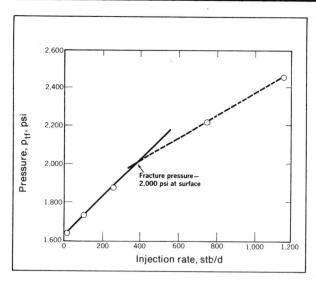

FIG. *3-11—Step-rate injectivity plot.*[18] *Permission to publish by The Society of Petroleum Engineers.*

Injection-Well Testing—Injection well transient pressure testing is basically simple as long as the mobility ratio between the injected and in-situ fluids is about unity. Injection is analogous to production—thus an Injectivity test parallels a Drawdown test, and a Pressure Falloff test parallels a Pressure Buildup test. Calculation of reservoir characteristics is similar.

A Step Rate Injectivity test, Figure 3-11, is normally used to estimate fracture pressure in an injection well. Fracturing may not be important in a water injection well, but could be critical in a tertiary flood to avoid injecting an expensive fluid through uncontrolled fractures. Fluid is injected at a series of increasing rates. Each time step should be about the same length—usually one hour with formation permeability less than 5 md—or 30 minutes with permeability more than 10 md. Six to eight rates are desirable to bracket the estimated fracture pressure. Bottom hole pressures should be plotted, but if friction loss in the flow string is small, surface pressure can be used.

Interference Tests (multiple well testing)—In an Interference test, a long-duration rate change in one well creates a pressure change in an observation well that can be related to reservoir characteristics. A Pulse test is an interference test that provides similar data by changing production rate in a cyclic manner to produce short term pressure pulses. The pressure response to the pulses is measured at one or more observation wells. Pressure responses are

very small—perhaps 0.01 *psi*—thus special measuring equipment is required.

If enough observation wells are used, interference-test data can sometimes be analyzed by computer methods to give a description of the variation of reservoir properties with location. Vertical pulse testing in one well (under ideal well conditions) may indicate vertical formation continuity. Orientation and length of vertical fractures may be estimated through pulse testing and reservoir simulation techniques.

REFERENCES

1. Miller, C. C., Dyes, A. B., and Hutchinson, C. A., Jr.: "The Estimation of Permeability and Reservoir Pressure from Bottomhole Pressure Buildup Characteristics", Trans AIME 1950.

2. Horner, D. R.: "Pressure Buildup in Wells", Proceedings Third World Petroleum Congress, The Hague 1951.

3. van Everdingen, A. F.: "The Skin Effect and Its Influence on the Productive Capacity of a Well", Trans. AIME 1953.

4. Hurst, William: "Establishment of the Skin Effect and Its Impediment to Fluid Flow Into a Wellbore", Petr. Engr. Oct. 1953.

5. Gladfelter, R. E., Tracy, G. W., and Wilsey, L. E.: "Selecting Wells which will Respond to Production Stimulation Treatment", Drill. and Prod. Prac., API 1955.

6. Rawlins, E. L., and Schellhardt, M. A.: "Back Pressure Data on Natural Gas Wells and Their Application to Production Practices, Monograph 7 USBM 1936.

7. Railroad Commission of Texas, "Back Pressure Testing of Natural Gas Wells".

8. Cullender, M. H.: "The Isochronal Performance Method of Determining the Flow Characteristics of Gas Wells", Trans AIME 1955.

9. Brons, F., and Martin V. E.: "The Effect of Restricted Fluid Entry on Well Productivity", J. Pet. Tech., August 1961, p. 172.

10. Matthews, C. S., and Russel, D. G.: "Pressure Buildup and Flow Tests in Wells", Monograph Series, Society of Petroleum Engineers of AIME, Dallas 1967.

11. Vogel J. V.: "Inflow Performance Relationships for Solution Gas Drive Wells", J. Pet. Tech., January 1968, p. 83.

12. Ramey, H. J., Jr.: "Short-Time Well Test Data Interpretation in the Presence of Skin Effect and Wellbore Storage", J. Pet. Tech., January 1970, p. 97.

13. McKinley, R. M.: "Wellbore Transmissibility From After Flow-Dominated Pressure Buildup Data, "J. Pet. Tech., July 1971, p. 863.

14. Standing, M. B.: "Concerning the Calculation of Inflow Performance of Wells Producing Solution Gas Drive Reservoirs, J. Pet. Tech., September 1971, 1, 141.

15. Fetkovich, M. J.: "The Isochronal Testing of Oil Wells", SPE paper 4529, Las Vegas, September 1973.

16. Earlougher, Robert C., Jr., and Kersch, Keith M.: "Analysis of Short-Time Transient Test Data by Type-Curve Matching, J. Pet. Tech., July 1974, p. 793.

17. Ramey, H. J.: "Practical Use of Modern Well Test Analysis," SPE 5858, October 1976.

18. Earlougher, Robert C., Jr.: "Advances in Well Test Analysis," SPE Monograph Series No. 5, Dallas, 1977.

Drill Stem Testing

BACKGROUND
Objective

Drill stem testing provides a method of temporarily completing a well to determine the productive characteristics of a specific zone. As originally conceived, a Drill Stem Test provided primarily an indication of formation content. The pressure chart was available, but served mainly to evaluate tool operation.

Currently, analysis of pressure data in a properly planned and executed DST can provide, at reasonable cost, good data to help evaluate the productivity of the zone, the completion practices, the extent of formation damage, and perhaps the need for stimulation.

Many times actual well producing rates can be accurately predicted from DST data. The DST shows what the well will produce against gradually increasing back-pressure. From this a Productivity Index (P.I.), or Inflow Performance Relationship (IPR) can be established, and if the flowing pressure gradient in the tubing can be estimated, then actual producing rate can be determined.

Reservoir Characteristics

Reservoir characteristics that may be estimated from DST analysis include:

—*Average Effective Permeability*—This may be better than core permeability since much greater volume is averaged. Also, effective permeability rather than absolute permeability is obtained.

—*Reservoir Pressure*—Measured, if shut-in time is sufficient, or calculated, if not.

—*Wellbore Damage*—Damage ratio method permits estimation of what the well should make without damage.

—*Barriers—Permeability Changes—Fluid Contacts*—These reservoir anomalies affect the slope of the pressure buildup plot. They usually require substantiating data to differentiate one from the other.

—*Radius of Investigation*—An estimate of how far away from the wellbore the DST can "see."

—*Depletion*—Can be detected if reservoir is small and test is properly run.

In summary, the DST, properly applied, has become a very useful tool for the Well Completion Engineer.

Basics of DST Operations

Simply a Drill Stem Test is made by running in the hole on drill pipe a bottom assembly consisting of a packer and a surface operated valve. The DST valve is closed while the drill string is run, thus pressure inside the drill pipe is very low compared to hydrostatic mud column pressure. Once on bottom, the packer is set to isolate the desired formation zone from the mud column, and the control valve is opened to allow formation fluids to enter the drill pipe.

After a suitable period, the valve is closed, and a pressure buildup occurs below the valve as formation fluids repressure the area around the wellbore. After a suitable buildup time, the control valve usually is opened again, and the flowing and shut-in periods repeated, (several times if desired), to obtain additional data and verification. Figure 3-12 shows schematically a simple single-flow-period operational sequence.

Pressure vs Time Plot

The entire sequence of events is recorded on a pressure vs time plot shown schematically in Figure 3-13. As the tools are run in the hole, the pressure bomb records the increase in hydrostatic mud column pressure. The pressure at Point A is termed the initial hydrostatic mud pressure.

Initial Flowing and Shut-in Periods—One objective of the DST is to determine the static or shut-in reservoir pressure of the zone. To measure true static reservoir pressure, any over-pressured condition near the wellbore, due to drilling fluid filtration or fluid compression in setting the DST packer, must first be relieved by a short flowing period.

The control valve is then closed and pressure builds up toward static reservoir pressure. Depend-

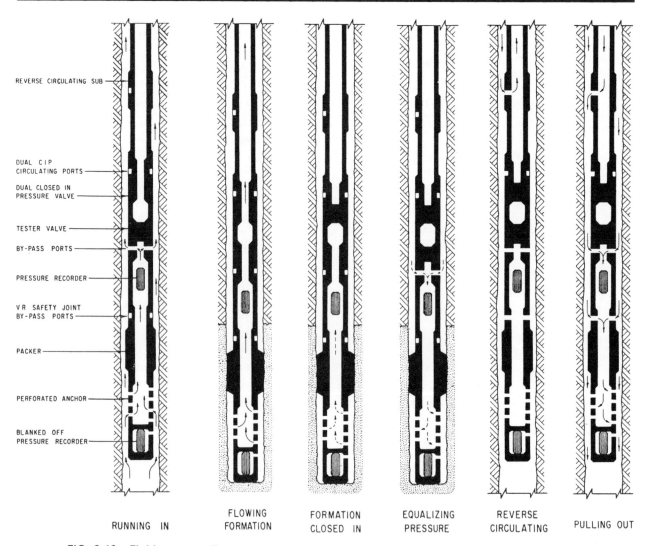

REVERSE CIRCULATING SUB

DUAL C I P
CIRCULATING PORTS

DUAL CLOSED IN
PRESSURE VALVE

TESTER VALVE

BY-PASS PORTS

PRESSURE RECORDER

V R SAFETY JOINT
BY-PASS PORTS

PACKER

PERFORATED ANCHOR

BLANKED OFF
PRESSURE RECORDER

| RUNNING IN | FLOWING FORMATION | FORMATION CLOSED IN | EQUALIZING PRESSURE | REVERSE CIRCULATING | PULLING OUT |

FIG. *3-12—Fluid-passage diagram, open-hole drill stem test. Courtesy of Halliburton Services.*

ing on the length of the flowing period, the length of the shut-in period, and certain reservoir parame-

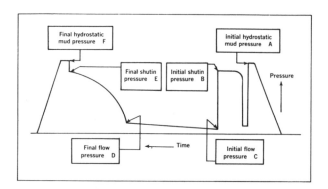

FIG. *3-13—Sequence of events in a DST.*

ters, the pressure may or may not level out at the true static reservoir pressure. Thus Point B, initial shut-in pressure, may be somewhat less than true static reservoir pressure.

Second Flowing and Shut-in Periods—The objective of the second flowing and shut-in periods is to permit the calculation of reservoir parameters, as well as to determine fluid production. As the control valve is opened, pressure falls to Point C (initial flowing pressure). As fluids move into the drill pipe above the recorder, pressure increases, reflecting the increase in hydrostatic head of liquid and in back pressure, due to flow restrictions in the tools.

After a suitable period of time, depending on test objectives, hole conditions and the cost of rig time, the tool is shut in. The pressure at the moment of shut in, Point D, is termed the final flowing pressure. Recorded pressure then rises toward Point E, the final shut-in pressure. Usually the final shut-in pressure is significantly less than the static reservoir pressure, again due to limitations of time relative to reservoir parameters.

This second pressure buildup portion of the DST, following a reasonably long flowing period, usually provides data for transient pressure analysis. Sometimes DST's are designed having two reasonably long flowing periods and subsequent shut-in periods, Figure 3-14, to provide verification of calculations, and to permit more sophisticated well testing techniques.

Field-Recorded DST Pressure Chart

Figure 3-14 is typical of actual multiple-flow-period charts with key points described therein.

THEORY OF PRESSURE BUILDUP ANALYSIS

Horner Equation—Transient pressure analysis of a DST is based on the Horner pressure buildup equation. This equation describes the repressuring of the wellbore area during the shut-in period, as formation fluid moves into the "pressure sink" created by the flowing portion of the DST:

$$p_{ws} = p_i - \frac{162.6 \, q \mu B}{kh} \log_{10} \left(\frac{t_p' - \Delta t'}{\Delta t'} \right) \quad (1)$$

Where

p_{ws} = measured pressure in the wellbore during buildup, psig
t_p' = flowing time, minutes
$\Delta t'$ = shut-in time, minutes
p_i = shut-in reservoir pressure, psig
q = rate of flow, stb/day
μ = fluid viscosity, cp
B = formation volume factor, reservoir bbl/stock tank bbl
k = formation permeability, md
h = formation thickness, ft

Conditions which must be assumed during the buildup period for Equation 1 to be strictly correct are: radial flow, homogenous formation; steady-state conditions; infinite reservoir; single-phase flow. Most of these conditions are met on a typical DST. Steady state flow is perhaps the condition causing the primary concern, particularly at early shut-in time.

Horner Buildup Plot—Assuming these conditions are met, then a plot of p_{ws} vs $\log_{10} \left(\dfrac{t_p' + \Delta t'}{\Delta t'} \right)$ should yield a straight line, and the slope (m) of the straight line should be:

$$m = \frac{162.6 \, q \mu B}{kh} \quad (2)$$

The constant m is representative of a given fluid having physical properties μB flowing at a rate q through a formation having physical properties kh.

Figure 3-15 is an idealized Horner Plot. The DST pressure chart of Figure 3-15 shows very simply how flowing time t_p', and formation pressure p_{ws} at various shut-in times $\Delta t'$, are picked from the chart and related to the Horner Plot. Usually p_{ws} is determined at 5-minute intervals along the shut-in pressure curve.

In a multiple flow period DST, selecting a value for t_p' creates some problem mathematically. However, very little error is caused by assuming that t_p' is the time of the flowing period immediately preceding the particular shut-in period. With equal flowing periods on a multiple flow period DST, this is usually done. With a very short initial flowing period, compared to a longer second flowing period, t_p' can be assumed to be the total of the flowing times with very little error.

In Figure 3-15 the slope m of the "straight line" is numerically the difference between the p_f' pressure value at $\log_{10} \dfrac{t_p' + \Delta t'}{\Delta t'} = 0$, and at $\log_{10} \dfrac{t_p' + \Delta t'}{\Delta t'} = 1.0$. If the points are plotted on semilog paper, m is the change in pressure over one log cycle.

The ideal situation of Figure 3-15, where all points line up as a straight line, is seldom seen in actual DST's, since "after flow" or wellbore storage effects cause a deviation from the straight line during the early times. As a rule of thumb, four points are needed to determine the straight line.

The length of shut-in time required to approach a steady state, or straight line condition fluctuates

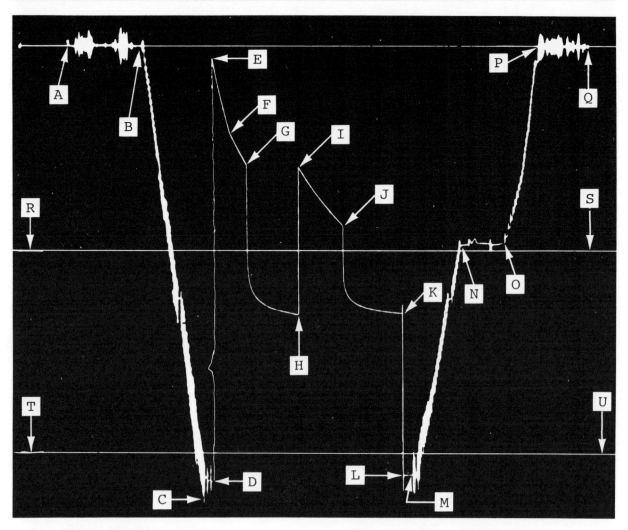

A-Q	Baseline drawn by recorder.
A	Recorder clock started.
A-B	Tools being made up.
B-C	Running in hole.
C-D	On bottom—completing surface hookup—initial hydrostatic mud pressure.
D	Tester valve opens.
E	Beginning 1st flow period (initial flow pressure).
E-F	Fluid filling small ID drill collars.
F-G	Fluid filling larger ID drill pipe.
G	End 1st flow period (final flow pressure).
G-H	First closed-in pressure period.

H	Final closed-in pressure of 1st period.
I	Begin 2nd flow period (initial flow pressure).
J	End 2nd flow period (final flow pressure).
K	End 2nd shut-in period (final closed-in)
K-L	Equalizing hydrostatic pressure across packer.
M-N	Pulling out of hole.
N-O	Reached top of fluid fillup in drill pipe—reversing out fluid to surface.
O-P	Continued trip out of hole.
P-Q	Breaking down tools.
R-S	1,000-psi lines drawn by
T-U	chart interpreter.

FIG. 3-14—Typical chart for multiple-flow-period DST. Courtesy of Halliburton Services.

with reservoir and fluid characteristics and flow conditions. Experience has formulated certain rules of thumb to help determine what is sufficient shut-in time. Generally the shut-in pressure must reach at least 65% of static pressure in order to produce

a straight line on the Horner Plot.

Prior to work by McKinley, Ramey and others, unless the straight line portion of the buildup plot could be identified, no further DST analysis was possible. By use of type-curve methods, however,

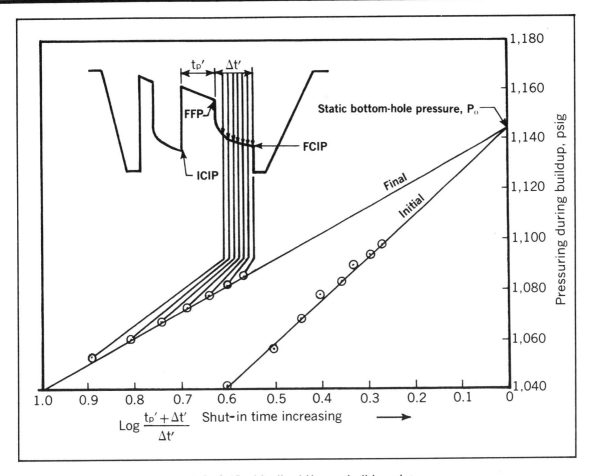

FIG. *3-15—Idealized Horner buildup plot.*

it is sometimes possible to make reasonable estimates of formation parameters even though the DST tool was not shut in long enough to indicate the Horner straight line.

Reservoir Parameters Obtained by Buildup Analysis

Permeability (k)—Assuming the Horner buildup plot produces a straight line such that m can be determined; then the average effective permeability k can be calculated as follows:

$$k = \frac{162.6\,q\mu B}{mh} \qquad (3)$$

Parameters μ (viscosity) and B (formation volume factor) can be estimated from available correlations, if the API gravity of the crude oil and the gas-oil ratio can be determined through wellsite measurements after the test.

Formation thickness h must be the net thickness of productive zone, which should be determined from log analysis. If net thickness h is unavailable, kh or formation capacity is determined:

$$kh = \frac{162.6\,q\mu B}{m} \qquad (4)$$

If all the reservoir parameters are unknown, transmissibility $\dfrac{kh}{\mu B}$ is determined:

$$\frac{kh}{\mu B} = \frac{162.6\,q}{m} \qquad (5)$$

Static Reservoir Pressure (p$_i$)—Static, or shut-in reservoir pressure, is obtained by extrapolating the Horner straight line to an "infinite" shut-in time.

At infinite shut-in time, $\dfrac{t_p' + \Delta t'}{\Delta t'} = 1.0$; or as shown

in Figure 3-15, $\log_{10} \dfrac{t_p' + \Delta t'}{\Delta t'} = 0$.

In Figure 3-15, both the 1st buildup plot and the 2nd buildup plot extrapolate to the same static pressure. This lends confidence to the analysis. If the 2nd buildup static pressure is lower than the 1st, then depletion of the reservoir is a possibility.

Depending on the length of the initial flowing period and reservoir parameters, the initial shut-in buildup may provide a stabilized value of static reservoir pressure that can be read directly from the pressure gauge. Pressure may be assumed to be stabilized if it holds the same value for 15 minutes or longer.

Wellbore Damage—Many times DST results are affected by formation damage. Thus, to be meaningful, the effect of flow restriction caused by the damaged zone must be accounted for in analyzing a specific DST.

Hurst and van Everdinger presented the following empirical equation for a dimensionless value "s" denoting "skin factor:"

$$ s = 1.151 \left[\frac{p_i - p_{ff}}{m} - \log \frac{k t_p'}{\phi \mu c r_w^2} + 2.85 \right] \quad (6) $$

The skin factor is useful in comparing damage between wells, however, cannot be readily applied to a specific formation to show what that zone should make if damage were removed.

Griffin and Zak carried Equation 6 one step further introducing the concept of Damage Ratio (DR), which compares flow rate observed on a DST (q_a) to the theoretical flow rate without damage (q_t):

$$ DR = \frac{q_t}{q_a} \quad (7) $$

An equation for calculation of DR based on the skin factor relation of Hurst and van Everdingen is:

$$ DR = \frac{(p_i - p_{ff})}{m \left(\log \dfrac{k t_p'}{\phi \mu c r_w^2} - 2.85 \right)} \quad (8) $$

Where:

p_i = shut-in reservoir pressure, psi

p_{ff} = formation pressure at flow time T, psi (final flowing pressure)

c = fluid compressibility, vol/vol/psi

ϕ = formation porosity, fraction

μ = viscosity of reservoir fluid, cp

r_w = well bore radius, inches

k = effective permeability, md

t_p' = flowing time, minutes

DR substantially greater than 1.0 indicates damage. Griffin and Zak simplified this equation by assigning average values to the formation parameters k, ϕ, c, μ, and r_w. This produced an equation for Estimated Damage Ratio:

$$ EDR = \frac{(p_i - p_{ff})}{m (\log t_p' + 2.65)} \quad (9) $$

Reservoir and Fluid Anomaly Indications

Many times the assumptions of the Horner buildup equation, homogeneous formation, single phase flow and infinite reservoir, do not hold in an actual case. If changes occur within the radius of investigation of the DST, they can be detected by a change in slope of the Horner buildup plot.

Permeability or Viscosity—Examining the Horner slope equation, it is seen that if rate of flow q remains constant, then permeability k, or fluid viscosity μ, are likely suspects for change as the wave of increasing pressure travels toward the wellbore.

$$ m = \frac{162.6 \, q \mu B}{kh} \quad (10) $$

Permeability may change due to natural lensing or due to formation damage, Figure 3-16A. It is doubtful, however, that formation damage would affect sufficient volume of formation to be detected as a change of slope on the buildup plot. Fluid viscosity could change due to a change in fluid phase or type (i.e., gas to oil). "Seeing" the gas-liquid contact from the upstructure well of Figure 3-16B would be difficult, due to the normally short radius of investigation through a gas column. Seeing the gas-liquid contact from the downstructure well is a much more likely possibility.

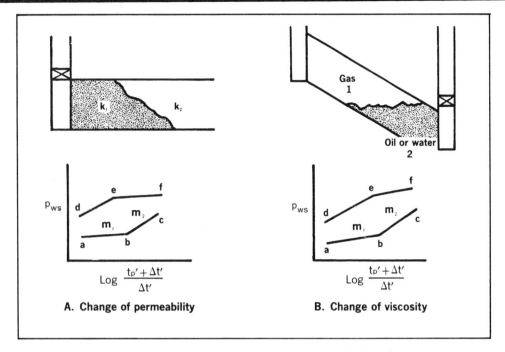

FIG. *3-16—Effect of changing permeability and viscosity.*

Barrier—A sealing barrier such as a fault or permeability pinchout can cause a change of slope m. If the barrier is a straight line as A-A' in Figure 3-17A, then the buildup slope will change by a factor of 2, Figure 3-17B.

In summary, a change in permeability, or viscosity, or existence of a barrier, all can cause a change in the slope of the Horner plot. Thus, the fact that a change in slope appears on the buildup plot leaves open the question of what caused the anomaly. This must be resolved through other geologic or reservoir information.

Distance to Anomaly—The distance to the anomaly, (r_a) whether it be a barrier, or change of permeability, or a fluid contact, can be calculated using Horner's equation:

$$-E_i\left(\frac{-3793\, r_a^2 \phi \mu c}{k\, t_p}\right) = 2.303\, ln\left(\frac{t_p + \Delta t_a}{\Delta t_a}\right) \tag{11}$$

r_a = distance to anomaly in feet
t_p = flow time in hours
Δt_a = shut-in time at the point of slope change, hours
$-E_i$ = exponential integral value—See Figure 3-18.

Radius of Investigation

The following equation from Van Poollen may be used to estimate the radius of investigation of a particular DST in an infinite radial flow system:

$$r_i = \sqrt{\frac{k\, t_p'}{5.76 \times 10^4 \phi \mu\, c}} \tag{12}$$

r_i = radius of investigation, ft
t_p' = flow time in minutes
Other units as previously noted

Obviously the longer the flowing time, the further back away from the wellbore we are looking with our DST.

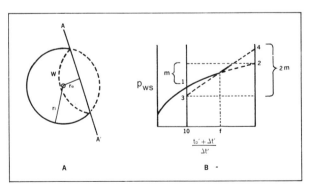

FIG. *3-17—Effect of a fault.*

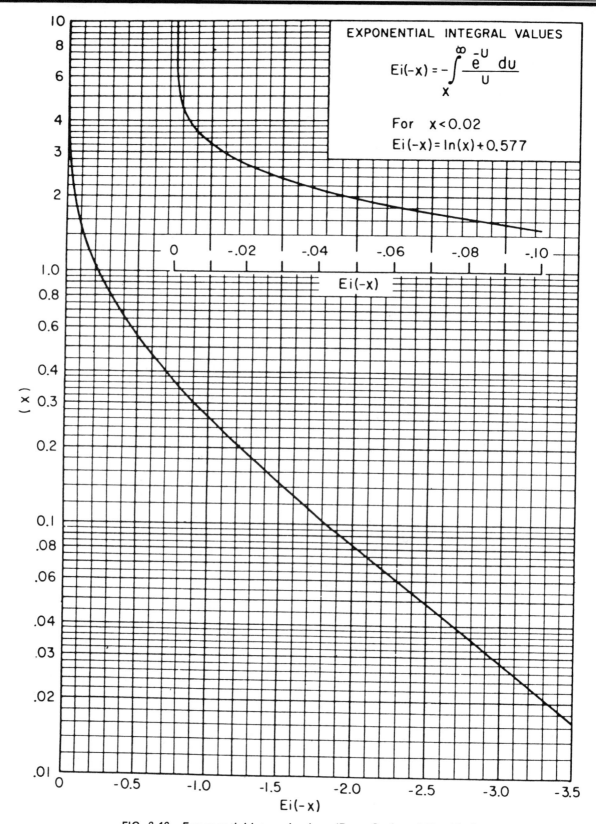

FIG. 3-18—Exponential integral values (From Craft and Hawkins).

Depletion

If the extrapolated or stabilized shut-in pressure from the second buildup is lower than the initial shut-in pressure, a depleting reservoir may be the cause. Obviously a reservoir must be extremely small for depletion to occur on a DST, but many field examples are available to prove that it can occur.

Another possibility, however, is that the recorded initial shut-in pressure may be higher than true shut-in reservoir pressure. This effect is called supercharge. Supercharge could be due to leak off of drilling fluid filtrate overpressuring the formation around the wellbore, or to compression of well fluid below the DST packer as it was set.

In some formations a short initial flowing period (1 to 3 minutes) is not sufficient to relieve the overpressured condition. Where this condition is suspected to exist, longer initial flowing periods (perhaps 20 minutes) should be used.

The important point is that the question of supercharge must be resolved before depletion can be diagnosed. A second DST is sometimes required to define depletion.

Reservoir Parameters (Gaseous System)

In DST's of gas zones, flow rate is usually reported in terms of cubic feet per day (or more conveniently, Mcfd) rather than barrels per day. This requires accounting for deviation of the reservoir gas from the Perfect Gas Law using the gas deviation factor (Z) and the absolute formation temperature (°R).

For the Horner buildup plot, the square of the formation pressure (p_{ws}) during the buildup is plotted versus $\left(\dfrac{t'_p + \Delta t'}{\Delta t'}\right)$ as shown in Figure 3-19.

Values of Z and μ for gas may be obtained from the literature knowing the specific gravity of the gas. Equations assume that compressibility and viscosity of gas remain reasonably constant over the range of temperature and pressure changes occurring during the flow period. Large pressure drawdown between the wellbore and the external boundary of the reservoir such as may occur in a low permeability zone may render this assumption invalid. If this is true, a bottom-hole choke should be used to reduce drawdown.

Equations for permeability, estimated wellbore

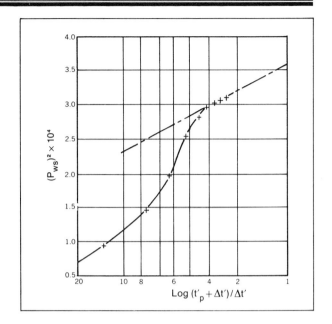

FIG. 3-19—Horner plot for gas well.

damage and absolute open flow potential for a gas zone are:

Permeability

$$k = \frac{1637\, q_g T_f \mu Z}{m_g h} \qquad (13)$$

Z = gas deviation factor
q_g = rate of flow mcf/d
T_f = formation temperature, °R = (°F + 460)
m_g = Horner buildup slope for gas well

Wellbore Damage (EDR)

$$EDR = \frac{p_i^2 - p_{ff}^2}{m_g\,(\log t'_p + 2.65)} \qquad (14)$$

Absolute Open Flow Potential (MCF/Day)

Using the single-point back-pressure test method:

$$AOF = \frac{q_g\,(p_i^2)^n}{(p_i^2 - p_{ff}^2)^n} \qquad (15)$$

Where:

n is an exponent varying between 0.5 and 1.0.

If $n = 1.0$:   Max. AOF $= \dfrac{q_g p_i^2}{p_i^2 - p_{ff}^2}$ (16)

If $n = 0.5$:   Min. AOF $= \dfrac{q_g p_i}{\sqrt{p_i^2 - p_{ff}^2}}$ (17)

Analysis of DST Data Using Type-Curve Methods

Several "type-curve methods" are available for analyzing early time data from pressure transient tests. Although these methods have perhaps greatest application in longer term tests, they can sometimes be used in DST analysis to salvage at least some information from a test where sufficient data is not available to obtain a "straight line" on the Horner plot. Ramey[11], McKinley[12], and Earlougher-Kersch[14] methods have application, with McKinley reportedly easier to use, but Earlougher-Kersch, and Ramey perhaps more accurate.[15]

The important point is that where the Horner plot can be made it should be used. In this case, type-curve methods may be helpful in picking the correct straight line on the Horner plot by indicating when wellbore storage or afterflow effects have ended.

RECOMMENDATIONS FOR OBTAINING GOOD TEST DATA

The key to DST evaluation is obtaining and recording good data. The DST must be planned to fit the specific situation.

Recording surface events, both character and time, is important to chart analysis. For example: What types and amounts of fluids were recovered? What were the characteristics of each fluid (salinity and perhaps resistivity of water, was it water cushion, mud filtrate or formation water, how much of each; API gravity and gas-oil ratio of recovered oil, etc.)? What size chokes were used? When were they changed? When did fluid come to the surface?

Time intervals allotted to each of the basic DST operations should ideally be adjusted during the test based on surface observations. The "closed chamber" method of analyzing the initial flowing period (discussed later) provides very early indications of formation fluid types and rates for use by the on-site supervisor in running the DST. Experience is a good teacher—past experience in the area should be studied in planning subsequent tests.

The Initial Flowing Period—This must be sufficient to relieve the effect of supercharge or over-

pressure in the formation immediately surrounding the wellbore. Normally 5 to 20 minutes is sufficient; however, longer times may be desirable in low productivity reservoirs in order to positively differentiate supercharge from depletion.

Some DST analysts prefer that the initial flowing period be extended to a time equal to the final flowing period, in order that the initial build-up curve can be analyzed the same as the final build-up curve. Thus, two sets of calculations are available for comparison. In this case, the static reservoir pressure probably cannot be actually measured by the pressure gauge, but can be obtained by extrapolation of the Horner plot.

Initial Shut-in Period—With no previous experience, the length of Initial Shut-In Period may be based on statistical studies as follows:

—With 30 minute shut-in, only 50% of tests reached static reservoir pressure.

—With 45 minute shut-in, 75% of tests reached static reservoir pressure.

—With 60 minute shut-in, 92% of tests reached static reservoir pressure.

Final Flow Period—This should be at least one hour of good to strong blow. The longer the flow period, the deeper the radius of investigation. If fluid reaches surface, additional time is desirable to obtain accurate volume gauges and gas-oil ratios. If blow quits, nothing is gained by continuing flow test.

Final Shut-In Period—This is the most important portion of the DST as far as formation evaluation is concerned. The length of shut-in should be based on events during the flow period.

—If formation fluid surfaces, FSI period should be one-half the flowing time (but never less than 30 minutes).

—With good to strong blow, FSI period should equal the flowing time (but never less than 45 minutes).

—With poor blow FSI period should be twice the flowing time (120 minutes if possible).

Closed Chamber DST Technique

This technique, in effect, permits a more scientific look at the bubbles in the "bubble bucket." It gives a good indication, in the first minutes of the initial flowing period, what fluid (gas, water, or perhaps oil) is entering the drill pipe, and an estimate of the fluid production rate.

To use the technique, a conventional downhole DST assembly is run. Conventional surface equipment is also used; however, we must be able to shut in the drill pipe at the surface, and we must be able to measure surface pressure with a reasonably accurate gauge. Figure 3-20 shows a suitable surface hook up.

In conducting the closed chamber test, the DST tool is opened conventionally, but as soon as bubbles appear in the bubble bucket, giving positive indications that the tool is open, the bubble hose is shut in. As formation fluids move into the drill pipe, surface pressure rises. This rate of pressure buildup is recorded for 5 to 10 minutes and provides the clues needed to estimate fluid type and entry rate. With gas entry, even at low rates, surface pressure buildup is quite rapid.

Figure 3-21 shows gas flow rate versus rate of surface pressure increase for various drill pipe internal capacities. With water entry, even at high rates, surface pressure buildup is quite slow.

Figure 3-22 compares surface pressure rise with water and gas entry at various rates. Entry of crude oil, due to gas breakout, usually shows a relatively slow initial pressure rise which increases more rapidly as gas breakout continues. Experience permits differentiation between oil and water.

With a properly run closed chamber technique,

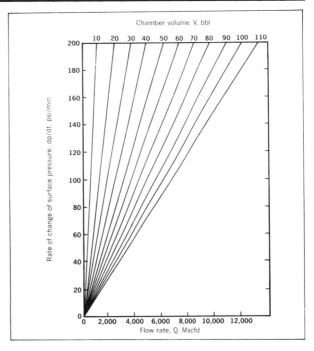

FIG. *3-21—Rate of change of surface pressure vs flow rate for various chamber volumes.*[16] *Permission to publish by The Society of Petroleum Engineers.*

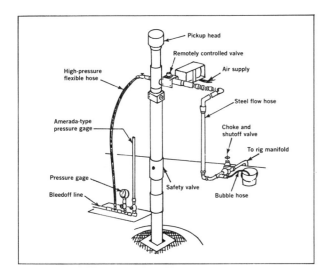

FIG. *3-20—Surface equipment for closed-chamber testing.*[16] *Permission to publish by The Society of Petroleum Engineers.*

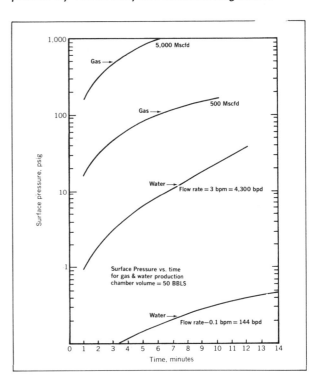

FIG. *3-22—Surface pressure rise with water and gas entry at various rates.*[16] *Permission to publish by The Society of Petroleum Engineers.*

the well site supervisor should reasonably know fluid types and rates in the first 5 to 10 minutes of the DST, such that he can perhaps terminate the DST at that point—or adjust subsequent flow and shut-in periods to maximize the usefulness of the data obtained.

DST's in High Capacity Wells

Meaningful interpretation of formation characteristics in high capacity zones is difficult with conventional DST tools, since flow restrictions within the tools prevent flowing the zones at a high enough rate to provide sufficient drawdown.

Where this is a problem, DST assemblies are available, which provide straight through flow patterns with a minimum flow diameter of $2\frac{1}{4}$ inches. Where desirable, perforating guns can be run through the DST assembly. Stimulation can also be performed below these DST tools.

DST's from Floating Vessels

The incentive for effective formation evaluation of offshore exploratory wells is obvious and DST's

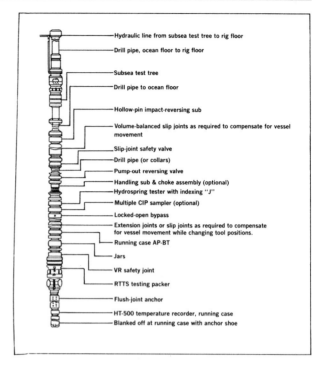

FIG. *3-24—Floating vessel test string operated by drill-pipe reciprocation.*

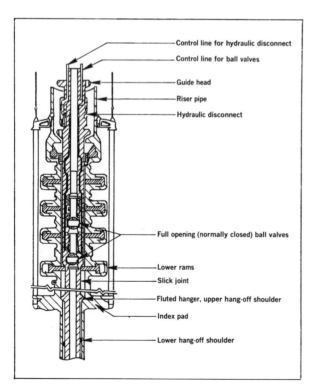

FIG. *3-23—Subsea test tree.*

from floating vessels are becoming more and more common.

Safety considerations are paramount in conducting DST's from a floating vessel. The subsurface test tree (Figure 3-23) was the first step in this regard. Typically, it provides a fail-safe means of shutting in the well at the ocean floor, and a hydraulic disconnect to allow removal of the section of drill pipe above the ocean floor BOP, if necessary to move the vessel with the DST tool on bottom.

Two systems of DST tool operation have been developed. In the first system, operation of the tools is strictly by reciprocation of the drill pipe, Figure 3-24. Tools are similar to those of a land operation, except that slip joints (also serving as a shut-in valve when closed) are required to compensate for vessel motion. Drill collars usually provide the string weight below the slip joints to keep the DST packer set.

In the second, opening and closing of the tester valve is controlled by annulus pressure (annular pressure responsive). Thus, all pipe movement is eliminated during the test, from the time the packer is set and the subsea test tree is spaced out, until the packer is pulled loose after the test is finished.

Wireline Formation Tester

The wireline formation tester is essentially a logging type device, which permits confirmation of formation fluid, indications of productivity and formation pressure. A recent tool design essentially consists of a packer, which can be forced against the wall of the borehole to isolate the mud column, a hydraulic piston used to create pressure drawdown, and two sample chambers (2¾ gal capacity is typically used) to collect formation fluid samples.

On one trip in the hole, any number of formation pressure measurements can be made in different zones, using the piston device to create drawdown, and two fluid samples can be obtained in promising zones. Thus, a rapid method of "looking" at multiple zones is available to aid and confirm log analysis.

The Schlumberger RFT tool is shown schematically in Figure 3-25. The pretest chambers provide an indication of whether or not a packer seal has been obtained and if so, an estimate of flow rate from the zone. Pressures are recorded during flow and buildup.

In a high permeability zone, buildup occurs very rapidly as shown in Figure 3-26.

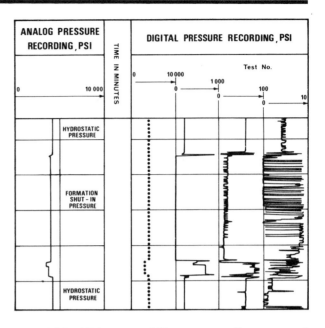

FIG. *3-26—High-permeability zone. Courtesy of Schlumberger.*

Figure 3-27 shows analog and digital pressure recording of a test in a low permeability carbonate zone. When the pretest chamber was opened, pressure dropped from hydrostatic of 4426 psi to a flow

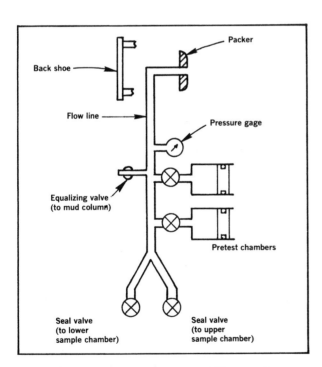

FIG. *3-25—Schematic diagram of RFT sampling system. Courtesy of Schlumberger.*

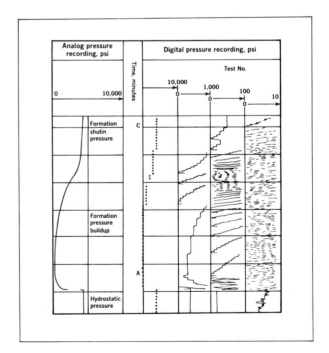

FIG. *3-27—Low-permeability zone. Courtesy of Schlumberger.*

pressure of 290 psi (point A). Buildup began immediately and reached a nearly stabilized maximum of 4060 psi (point C) 5½ minutes later. Total time required to test this zone was 8 minutes.

"EYEBALL" INTERPRETATION OF PRESSURE CHARTS

Present day DST tools provide very accurate pressure vs time data for use in evaluating a formation with the Horner calculation procedures. The first and the most important step before calculations, however, is a careful examination of the pressure charts. Preliminary checks should be made to compare pressures and chart indications with known data:

—Flowing pressures should be indicative of drill pipe fillup. Final flow pressure should equal the hydrostatic head of recovery in the drill pipe.

—Initial flow pressure should reflect the use or the absence of a water cushion.

—Possibility of mechanical malfunctions should be investigated by careful study of chart indications. Tie down the cause of all the important "wiggles."

—Compare indications of top and bottom charts to show clock and tool plugging problems.

Even where calculation procedures are not possible, there may be significant information which can be interpreted by "Eyeball" methods.

The actual DST charts on the following pages show examples of DST problems[8] which restrict calculation possibilities. Included also are example situations which can be reasonably interpreted by "eyeball" methods.

Base Line Drawn Incorrectly—The base line is the basis for all pressure measurements on a formation test chart. Figure 3-28 shows the base line inconsistent with the initial pressures of the test.

Leaking Drill Pipe—At points of delay, while going into the hole or at total depth, a decrease in recorded pressure is indicative of a loss in the hydrostatic head. There are two possibilities:

—The hole may be taking fluid.

—The drill pipe may be leaking, Figure 3-29.

Stair-Stepping Gauge—A "stair-stepping" appearance in a buildup curve Figure 3-30, may be caused by:

—The chart drum lugs and/or inner case runners may be dirty.

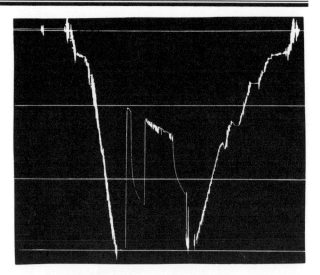

FIG. *3-28—Base line drawn incorrectly.*[8]

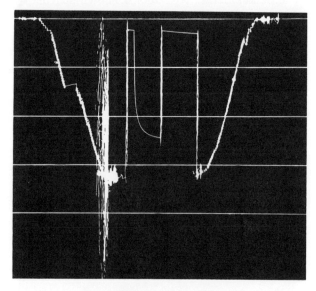

FIG. *3-29—Leak in drill pipe.*[8]

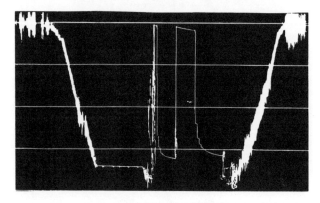

FIG. *3-30—Stair-stepping.*[8]

—The lead screw may not be straight.

—The inner case or cover may be crooked or rough.

Clock Failure—This problem is characterized by time discontinuance, as shown in Figure 3-31.

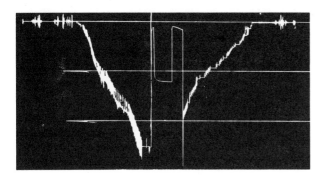

FIG. *3-31—Clock failure.*[8]

Clock Running Away—A clutch spring malfunction may cause the clock to "run away;" this occurrence is usually caused by excessive rough treatment and does not damage the clock, Figure 3-32.

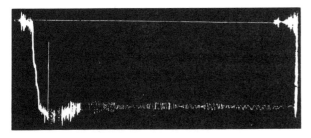

FIG. *3-32—Clock running away.*[8]

Leaking Closed-In Pressure Valve—Closed-in pressure equipment that leaks will cause erratic buildup curves. Leakage is characterized by a decrease in the buildup pressure and normally a subsequent rise, Figure 3-33.

False Buildup—A common point of great concern in a dual closed-in operation of formation testing is the frequent difference in the initial and final buildup pressures. This may be attributed to a short first flow period resulting in a higher buildup. Supercharge is the most common condition resulting in an abnormally high initial buildup of pressure. This is shown in Figures 3-34 and 3-35.

Limited Reservoir—A depleting reservoir will also exhibit a high initial buildup curve compared to the final buildup curve. Flow period will usually indicate a decreasing rate of flow. The difference

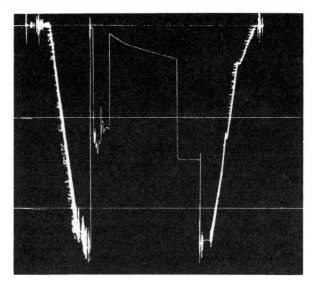

FIG. *3-33—Leaking closed-in pressure valve.*[8]

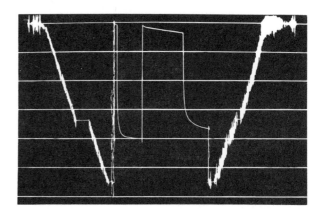

FIG. *3-34—Chart indicating supercharge.*[8]

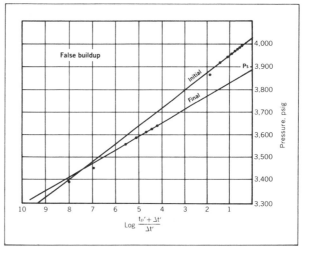

FIG. *3-35—False buildup, due to supercharging.*[8]

in the extrapolated static reservoir pressures of the initial and final buildup curves indicates the loss in reservoir pressure during the second flow period. These conditions are indicated in Figures 3-36 and 3-37 for a depleting liquid and gas reservoir.

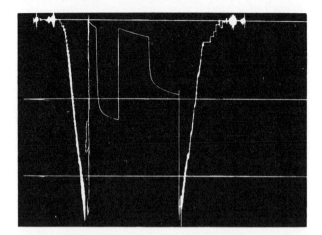

FIG. 3-36—Depleting liquid reservoir.[8]

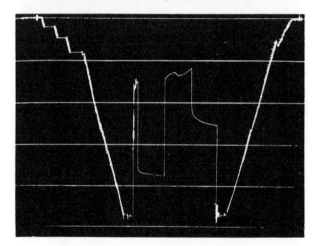

FIG. 3-37—Depleting gas reservoir.[8]

Barrier—A barrier may be detected in the plot of the buildup curve of a formation test. The final buildup curve may have a peculiar appearance as compared to the initial as shown in Figures 3-38 and 3-39.

Formation Damage (Liquid Zone)—Formation damage in an oil zone is usually indicated by:

—The very sharp rise after shut-in.

—A short radius curve.

—A reasonably flat slope.

—A high differential pressure between closed-in and final flow pressure.

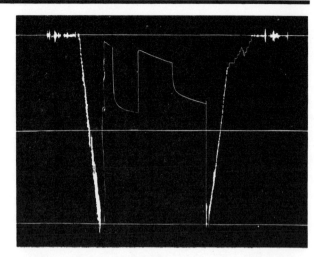

FIG. 3-38—Barrier within radius of investigation.[8]

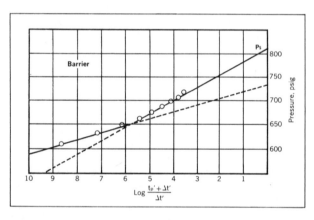

FIG. 3-39—Extrapolation of a buildup curve, indicating a barrier.[8]

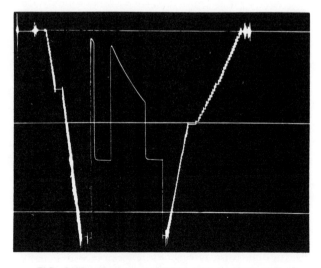

FIG. 3-40—High productivity, high damage.[8]

Figure 3-40 shows high productivity, high damage.
Formation Damage (Gas Zone)—High-capacity gas wells exhibit damage characteristics due to the back-pressure through the bottomhole choke, therefore, damage may be present in a high or low production formation, as shown in Figure 3-41.

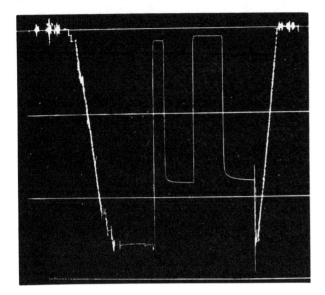

FIG. *3-41—Low productivity, high damage.*[8]

Cleanup of Formation Damage—The initial build-up may exhibit damage; however, damage may no longer be present after the final flow period. This indicates that the damage was removed during the flow period, as shown in Figure 3-42.

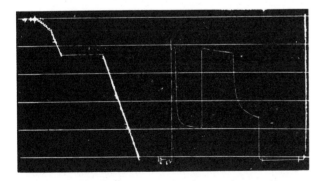

FIG. *3-42—Formation damage indicated on initial buildup; no damage indicated on final buildup.*[8]

Buildup Closure Time—Approximately 75% closure must be attained for an accurate extrapolation plot of a buildup curve. This figure will vary with well conditions. If a single phase is produced, less closure will be required while multiple phases usually require more closure. Figures 3-43 and 3-44

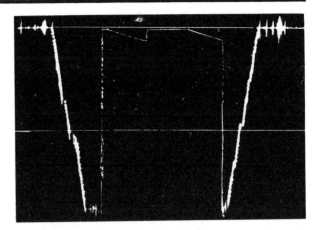

FIG. *3-43—Low-permeability formation with a low reservoir pressure.*[8]

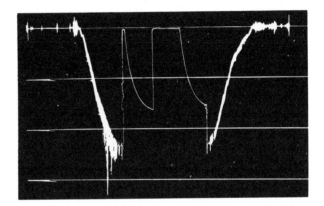

FIG. *3-44—Low-permeability formation with a high reservoir pressure.*[8]

show low permeability and varying bottom hole pressure.
Plugging in Test Tool—Plugging, one of the most common mechanical problems in formation testing, is usually characterized by sharp pressure fluctuations, if the plugging alternately plugs and frees itself. However, when the plugging is sustained momentarily the action will appear as small buildup curves, as shown in Figure 3-45.

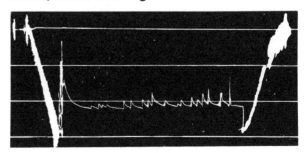

FIG. *3-45—Plugging-in tool.*[8]

Plugging in the Flow Perforation—Plugging flow perforations are evident when Figure 3-46 indicates little or no change in flowing pressures, while the blanked-off chart, Figure 3-47, indicates an increased pressure.

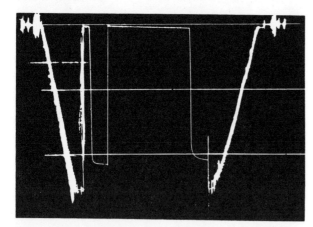

FIG. *3-46—Top gauge—plugging in the flow perforations.*[8]

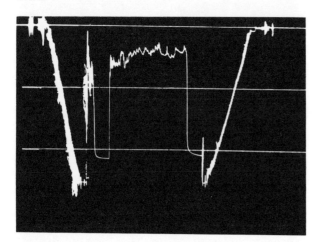

FIG. *3-47—Bottom gauge—plugging in the flow perforations.*[8]

Changes in Flow Rate—An abrupt change of rate in the flow period can usually be attributed to a change in the ID of the pipe. Ordinarily this change to a slower rate will indicate when the fill-up has cleared the drill collars, Figure 3-48.

Swabbing—Some wells are swabbed during a formation test. This action will cause a drawdown in the flow period outlined by a decreasing sequence of small "inverted fishhooks" as shown in Figure 3-49. A subsequent flow and swabbing action may very often be noted.

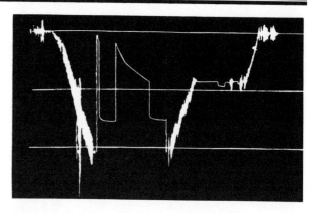

FIG. *3-48—Fillup transition from drill collar to drill pipe.*[8]

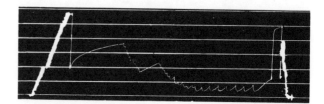

FIG. *3-49—Swabbing during a formation test.*[8]

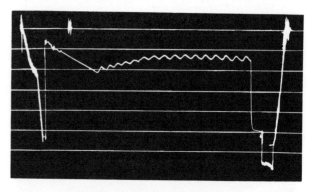

FIG. *3-50—Flowing in heads.*[8]

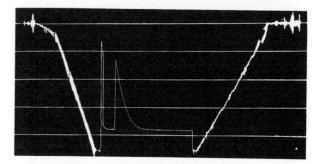

FIG. *3-51—Equalized flow.*[8]

Flowing In Heads—The flow pressures of a single phase flow are usually very uniform; however, a combination of phases will result in an irregular pattern. If gas is breaking through the fluid, this pattern may be very "lumpy" in appearance. A uniform "rippling" appearance occurs when a well is flowing in heads, as shown in Figure 3-50.

Equalized Flow—The second flow period should be of such duration as to establish a uniform rate of flow. A longer flow period can spoil the buildup curve by allowing the well to equalize before attempting a closed-in pressure, Figure 3-51.

REFERENCES

1. Horner, D. R.: "Pressure Buildup in Wells," Proc. Third World Pet. Congr., Sect. II, E. J. Brill, Leiden, Holland (1951).

2. van Everdingen, A. F.: "The Skin Effect and Its Influence on the Productive Capacity of a Well," Trans. AIME (1953) 198, 171.

3. Miller, Dyes & Hutchinson: "The Estimation of Permeability and Reservoir Pressure From Bottom Hole Pressure Buildup Characteristics," Trans. AIME, Vol. 189, 1950, 91-104.

4. Zak, A. J., Jr. and Griffin, P., III: "Here's a Method for Evaluation DST Data," Oil and Gas Journal (April 1957).

5. van Poollen, H. K.: "Radius-of-Drainage and Stabilization-time Equations," Oil and Gas Journal, Vol. 62, No. 37, Sept. 14, 1964.

6. Maier, L. F.: "Recent Developments in the Interpretation and Application of DST Data," J. Pet. Tech., Nov. 1962, p. 1,213.

7. McAlister, J. A., Nutter, B. P. and Lebourg, M.: "Multi-Flow Evaluator Better Control and Interpretation of Drill Stem Testing," J. Pet. Tech., Feb. 1965, p. 207.

8. Murphy, W. C.: "The Interpretation and Calculation of Formation Characteristics from Formation Test Data," Halliburton Services.

9. Johnston Testers, Review of Basic Formation Evaluation.

10. Matthews, C. S., Russell, D. G.: "Pressure Buildup and Flow Tests in Wells, AIME Monograph No. 1, H. L. Doherty Series.

11. Ramey, H. J., Jr.: "Short-Time Well Test Data Interpretation in the Presence of Skin Effect and Wellbore Storage," J. Pet. Tech., Jan. 1970, p. 97.

12. Milner, E. E., and Warren, D. A., Jr.: "Drill Stem Test Analysis Utilizing McKinley System of After-Flow Dominated Pressure Buildup," SPE-4123, October 1972 Annual Meeting, San Antonio, Texas.

13. Edwards, A. G., and Shryock, S. H.: "A Summary of Modern Tools and Techniques Used in Drill Stem Testing," Halliburton Services, Sept. 1973.

14. Earlougher, R. C., Jr., and Kersch, K. M.: "Analysis of Short-Time Transient Test Data by Type-Curve Matching," J. Pet. Tech., July, 1974, 793.

15. Sinha, B. K., Sigmon, J. E., and Montgomery, J. M.: "Comprehensive Analysis of Drill-Stem Test Data With the Aid of Type Curves," SPE-6054, Oct. 1976, Fall Meeting, New Orleans.

16. Alexander, L. G.: "Theory and Practice of the Closed-Chamber Drill-Stem Test Method," SPE-6024, October, 1976, Fall Meeting, New Orleans.

Chapter 4 Primary Cementing

Cementing materials
Slurry characteristics and design
Adjustment of slurry properties—additives
Factors affecting "bonding"
Slurry flow properties
Displacement mechanics—the key
Cost considerations
Special problems, new developments
Primary cementing practices

INTRODUCTION

Probably the single most important factor in the well completion operation is obtaining a satisfactory primary cementing job. An effective primary cement job is the necessary starting point for all subsequent operations. With a defective primary cement job all remaining operations are adversely affected. While primary cementing is often the responsibility of the drilling group, it is the completion, production, and workover groups who are most affected by, and perhaps should be most interested in primary cementing.

Fig. 4-1 shows the usual procedure for placing cement and details much of the equipment needed to facilitate a typical cementing operation. This section is primarily concerned with cements and cementing practices, and less concerned with casing and cementing equipment.

CEMENTING MATERIALS
Function of Cement in Oil Wells

In well completion operations cements are almost universally used to fill the annular space between casing and open hole. Two principal functions of primary cement are:

—To restrict fluid movement between formations.
—To support the casing.

Cement materials properly placed around the casing, having permeabilities less than 0.1 md, and compressive strengths greater than 100 to 300 psi should be satisfactory for these functions.

The key to success is proper placement of the cement completely around the casing.

Cementing compositions can be broadly classified as neat or tailored. Properties of neat cement are relatively inflexible and tailored cements are almost always used both to reduce the cost of the cementing operation, and to improve the capability of cementing materials as oil well requirements become more rigorous.

Manufacture, Composition, and Characteristics of Cement

Portland cements are a finely-ground mixture of calcium compounds. They are made from limestone (or other high calcium carbonate materials) and clay or shale. Some iron and aluminum oxides may be added if necessary. These materials are finely ground and mixed, then heated to 2600°–2800°F. in a rotary kiln. The resulting clinker is ground with a controlled amount of gypsum to form cement.

Typical cement particle size distribution is such that 85% passes a 325-mesh screen (44 microns), 90% passes a 200-mesh screen (74 microns), and 100% passes a 150-mesh screen (100 microns).

Principal compounds formed in the burning process and their functions are:

Tricalcium silicate (C_3S) is the major compound in most cement and is the principal strength-

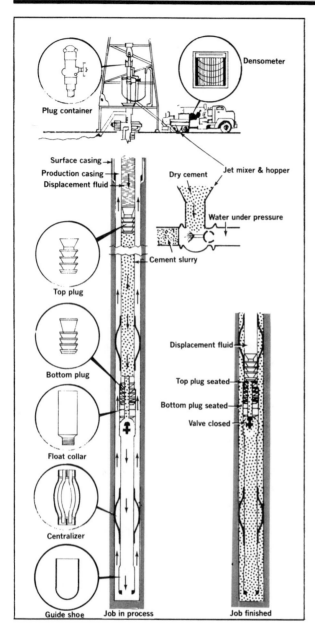

FIG. 4-1—Schematic diagram of how cement placement is carried out.

producing material. It is responsible for early strength (1 to 28 days).

Dicalcium silicate (C_2S) is the slow hydrating compound and accounts for the gradual gain in strength which occurs over an extended period.

Tricalcium aluminate (C_3A) promotes rapid hydration and controls the initial set and thicken-

ing time. It affects the susceptibility of cement to sulfate attack; high sulfate resistant cement must have 3% or less C_3A.

Tetracalcium aluminoferrite (C_4AF) is the low-heat-of hydration compound in cement. It gives color to the cement. An excess of iron oxide will increase the amount of C_4AF and decrease the amount of C_3A in the cement.

All API classes of Portland cements are manufactured in essentially the same way from the same ingredients, but proportions and particle size are adjusted to give the desired properties. The water requirement of each type of cement varies with the fineness of grind or surface area. High early strength cements have high surface area (fine grind). Retarded cements have low surface area—for higher temperatures organic retarders are also added to provide greater retardation. Table 4-1 compares typical compositions and grinds of API classes of Portland cement.

Cement sets through a crystal growth process. Figure 4-2 shows a typical arrangement of C_2S, C_3S, C_3A and C_4AF crystals in set cement. During the setting process disturbances such as gas bubbles percolating through a cement column may leave "worm holes" for subsequent communication of low-viscosity fluids.

Once the cement is in place around the casing, it is important that this crystal growth process proceed as quickly as possible to reduce exposure time to "disturbance mechanisms."

Pozmix cement combines Portland cement with pozzolan. Pozmix cement consists of Portland cement, a pozzolanic material and about 2% bentonite. By definition a pozzolan is a siliceous material which reacts with lime and water to form calcium silicates having cementitious properties. Since Portland cement releases about 15% free lime when it reacts with water, the addition of pozzolan reacts with this free lime to form a more durable mass of calcium silicates. The pozmix composition is less expensive than other basic cementing materials because more mix water is used per weight of material.

Calcium aluminate cements or Refractory cements (trade names: Cement Fondu, Luminite Cement) are manufactured by heating bauxite and limestone until liquefied, then cooling and grinding. High-alumina cements are used to cement casing through the hot zone in in-situ combustion wells,

TABLE 4-1
Typical Composition of Portland Cement

API Class	Compounds					Fineness, Sq cm/gram
	C_3S	C_2S	C_3A	C_4AF	$CaSO_4$	
A	53	24	8	8	3.5	1600–1900
B	47	32	3	12	2.9	1500–1900
C	58	16	8	8	4.1	2000–2400
D & E	26	54	2	12	3.0	1200–1500
G	52	32	3	12	3.2	1400–1600
H	52	32	3	12	3.3	1200–1400

Several other cementacious compositions are used to meet special requirements.

where temperatures may range from 750° to 2000°F. as the fire front passes. High-alumina cements resist attack by sulfates—and their quick setting characteristics sometimes recommend their use where formation temperatures are low. Limitations include high cost, and questionable long term strength.

Gypsum cements (Cal-Seal), usually a hemihydrate form of gypsum ($CaSO_4 \cdot 1/2H_2O$), set very rapidly, expand significantly (0.3%) on setting, but tend to deteriorate in contact with water. They are not used very often, except in connection with Portland cement.

Permafrost cements, a blend of gypsum cement with Portland cement, have low heat of hydration, and will set at 15°F before freezing. They are used in cementing through frozen formations in Arctic areas.

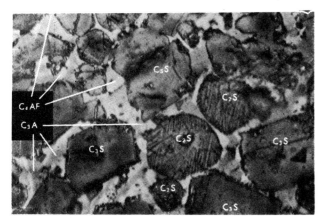

FIG. *4-2—Typical arrangement of crystals in set cement.*

Selection of Cement For Specific Well Application

The problem of selecting a cementing material for a specific well application is one of designing an economical slurry that:

1. Can be placed effectively with the equipment available.

2. Will achieve satisfactory compressive strength soon after placement.

3. Will thereafter retain the properties necessary to isolate zones and to support and protect the casing.

Most API cements have been replaced by Class G (Basic) cement. API classification of cements for various depth and temperature conditions, with recommended mixing water quantity and resulting slurry weight, are shown in Table 4-2:

API Class A and B cements are more economical than premium types. Class B is resistant to sulfate attack.

API Class C (High Early) cement has higher compressive strength than Class A in the first 30 hours. However, Class A with calcium chloride will give better early strengths than Class C without accelerators. Class C is available only in limited areas.

API Class D-E-F (Retarded) cements are delayed by organic compounds and/or a coarser grind. The added cost of these cements should be justified by special well conditions. Availability of these cements is limited.

API Class G & H (Basic) cements are similar to Class B, but are manufactured to more rigorous chemical and physical specifications resulting in a more uniform product. They contain no accelerators, retarders, or viscosity control agents. Class

TABLE 4-2
API Cement Classification

API Class	Mixing water gal/sk	Slurry weight, lb/gal	Well depth, ft	Static temperature, °F.
A	5.2	15.6	0–6,000	80–170
B	5.2	15.6	0–6,000	80–170
C	6.3	14.8	0–6,000	80–170
D	4.3	16.3	6–12,000	170–260
E	4.3	16.3	6–14,000	170–290
F	4.3	16.3	10–16,000	230–320
G	5.0	15.8	0–8,000	80–200
H	4.3	16.3	0–8,000	80–200

H is similar to Class G except it has slightly coarser grind providing slightly greater retarding effect for deeper, hotter conditions. Class G or H cements are compatible with accelerators or retarders for use over the complete range of API conditions; thus, with proper additives, Class G or H can be used in any cementing situation.

Worldwide, Class G with additives (if needed) is replacing most other API cements.

Mix water quantity affects slurry. Mixing water quantities shown in Table 4-2 are selected to provide pumpable viscosity and minimum free water (less than 1%). The effects of reducing mixing water quantities are:

—Increased slurry density and compressive strength.
—Increased slurry viscosity.
—Decreased pumpability time.
—Decreased slurry volume per sack of cement.

Increasing mixing water quantities has just the opposite effect. Well conditions sometimes dictate variation in mix water quantities to obtain desired properties. If mix water proportions are increased above recommendations, bentonite (2 to 4%) should be used to "tie up" resulting free water.

Early compressive strength is affected by curing conditions. Compressive strength of neat cement increases over a period of several weeks. The compressive strength at any particular time during the curing process depends heavily on mixing and curing conditions. Typical values after 24 hours are shown in Table 4-3.

The pumpability time, or thickening time of a cement slurry, is an important factor in slurry design. Generally these statements can be made regarding thickening time:

—Higher the temperature—faster the set. This is a primary factor affecting thickening time.
—Higher the pressure—faster the set. This effect is more pronounced below 5000 psi, but continues up through limits of lab test equipment.
—Loss of water from slurry accelerates set.
—Shutdown during placement results in cement gellation which will shorten pumpability and accelerate set and strength development.

TABLE 4-3
Typical Compressive Strength—psi @ 24 Hours

Curing Conditions temp.	press.	Class A	Class C	Class D	Class G	Class H
60°F.	0 psi	615	780	(a)	440	325
80°F.	0 psi	1,470	1,870	(a)	1,185	1,065
95°F	800 psi	2,085	2,015	(a)	2,540	2,110
110°F	1,600 psi	2,925	2,705	(a)	2,915	2,525
140°F	3,000 psi	5,050	3,560	3,045	4,200	3,160
170°F	3,000 psi	5,920	3,710	4,150	4,830	4,485

(a)—Not recommended at this temperature.

TABLE 4-4
High-Pressure Thickening Time (Hours:Minutes)

Circulating temperature	Class A	Class C	Class D	Class G	Class H
91°F	4:00+	4:00+	—	3:00+	3:57
103°F	3:36	3:10	4:00+	2:30	3:20
113°F	2:25	2:06	4:00+	2:10	1:57
125°F	1:40	1:37	4:00+	1:44	1:40

The effect of temperature on thickening time of various cement types is shown typically in Table 4-4.

The effect of extreme pressure on the thickening time of a particular cement slurry is shown in Table 4-5.

Lab tests must simulate placement conditions. In the laboratory, thickening time (elapsed time between application of pressure and temperature and the attainment of a slurry viscosity of 100 B_c) is measured using the AMOCO Thickening Time Tester according to procedures of API Testing Code RP 10B.

With no shut-down or dehydration, thickening time of a particular slurry depends on formation temperature gradient, depth, displacement rate, and hydrostatic and circulating pressures: i.e., the temperature-pressure history of the cement slurry as it is placed. API Code RP 10B Well Simulation Schedules standardize temperature-pressure history vs. depth and type of job (casing, squeeze, liner).

These API Schedules represent typical well conditions and average temperature-pressure-depth relations along the U.S. Gulf Coast, and may not correspond to conditions in other areas. Schedule 8, 14,000 feet Casing-Cementing Well-Stimulation Test is shown in Table 4-6.

Thickening time values reported by Service Companies are often based on modified test schedules to fit individual well conditions rather than on API Code RP 10B Schedules.

For critical cementing situations (static temperature above 260–275°F.) special care should be taken to insure that lab test schedules fit actual pressure-temperature-time conditions, and that cementing materials and mix-water samples are similar to those that will actually be used on the job.

Required Properties and Characteristics of Oil Well Cements

Viscosity—Should be low for better flow properties and mud removal. Cement is a non-Newtonian

TABLE 4-5
Effect of Pressure on Thickening Time
(Class H Cement with Retarder)

Depth, ft	Temperature °F. static	circulating	Pressure, psi	Thickening time, hours: minutes
10,000	230	144	5,000	2:10
			10,000	1:34
			15,000	1:18
14,000	290	200	10,000	5:50
			15,000	4:30
			20,000	3:20
16,000	320	248	10,000	4:11
			15,000	3:39
			20,000	2:30
			25,000	2:08

TABLE 4-6

SCHEDULE 8
14,000-ft (4270 m) Casing Cementing
Well-Simulation Test

Field Conditions Assumed

Surface temperature:	80 F(27°C)
Surface pressure:	1750 psi(12100 kPa)
Mud density:	16 lb per gal
	(1.9 kg/litre)
Bottom-hole temperature:	206 F(97°C)
Bottom-hole pressure:	13390 psi(92300 kPa)
Time to reach bottom:	52 min

1	2		3	
Time, min	Pressure,		Temperature,	
	psi	kPa	F	°C
0	1750	12100	80	27
2	2200	15200	85	29
4	2600	18000	90	32
6	3100	21400	95	35
8	3500	24100	99	37
10	4000	27600	104	40
12	4400	30300	109	43
14	4900	33800	114	46
16	5300	36500	119	48
18	5800	40000	124	51
20	6200	42700	128	53
22	6700	46200	133	56
24	7100	49000	138	59
26	7600	52400	143	62
28	8000	55200	148	64
30	8500	58600	153	67
32	8900	61400	158	70
34	9400	64800	162	72
36	9800	67600	167	75
38	10300	71000	172	78
40	10700	73800	177	81
42	11200	77200	182	83
44	11600	80000	187	86
46	12000	82700	191	88
48	12500	86200	196	91
50	12900	88900	201	94
52	13390	92300	206	97

Final temperature and pressure should be held constant to completion of test, within ±2 F(±1 °C) and ±100 psi(±700 kPa), respectively.

fluid; thus viscosity is a function of shear rate. Fann Viscometer must be used to determine viscosity characteristics.

Thickening Time—Two and one-half to 3 hours generally provides necessary placement time plus safety factor. Pressure-Temperature Thickening Time Tester should be used with samples of actual cement and mix water where placement conditions are critical.

Water Separation—Set volume of cement should be equal to slurry volume. Free water should be less than 1%. Settling tests are the basis for measurement.

Strength Development—Compressive strength of 500 psi is more than sufficient to support pipe. Compressive strength is the basis for most WOC regulations, and with proper cement and accelerators WOC time can be as short as 3 to 6 hours. Laboratory specimens should be pressure-temperature cured to simulate down-hole conditions.

Sulfate Resistance—Low C_3A (Class B) cement best. Basis for laboratory measurement is to cure cements in 5% sodium sulfate solution and observe cracking and disintegration.

Pozzolans help sulfate resistance of Class A cement. Dense slurry gives better resistance. High water ratio slurry has poor resistance.

Sulfate resistance of cement is primarily dependent upon the tricalcium aluminate (C_3A) content of the cement. C_3A reacts with sulfates in mix water or formation water to form calcium sulfoaluminate. Resulting enlargement of crystals causes cracks to form providing paths for further attack. Cement should have three percent or less C_3A to be highly sulfate resistant. Some cements have no C_3A (Maryneal Incor, El Toro 35, Dykerhoff Class B).

Maximum sulfate reaction takes place at low temperature (80°–100°F.). Above 180°F. sulfate action is nil. Sulfate concentration in formation water less than 1,500 ppm should not harm set cement.

Density—Most slurries range from 11 to 18.5 lb/gal. Pozzolans or bentonite are effective means for reducing density—any water increasing material will reduce density. For accuracy cement slurries should be weighed under pressure to negate effect of air entrainment.

Storage Stability—Dry stored cements will remain good for long periods of time. However, slight changes can occur in humid conditions which will affect thickening time on critical cementing situations.

Mixing Water—Fresh water is preferable, any drinkable water is satisfactory for cement. Inorganic compounds usually accelerate set of cement. Organic compounds; such as, mud thinners, fluid loss agents, corrosion inhibitors, and bacteriacides, retard set. Few waters cause "flash setting." Sea water is satisfactory but thickening time should be checked. Carbonates and bicarbonates have an unpredictable effect on thickening time; thus, waters containing high carbonates or bicarbonates (greater than 2,000 ppm) should be avoided.

Effect of Acid on Cements—HCl reacts with

cement to form silica gel which resists further attack. Communication after an acid job could be due to a mud channel since acid could shrink mud particles and open channel. HF acid may affect cement to a limited extent.

Heat of Hydration—Heat developed by cement setting in a 2-inch annulus will increase temperature 30°–40°F. above formation temperature. In washouts, more cement means higher temperature. In cementing through ice lenses in permafrost regions, heat of hydration becomes an important factor since heat liberated by cement in setting tends to melt the ice lenses and prevent bonding. Gypsum-blend cement (Permafrost cement) has a low heat of hydration (18–20 BTU/lb slurry) compared to high alumina cement (60–90 BTU/lb slurry).

Expansion—This property of cement is increasingly being considered. Measurement criteria are not yet defined. Salt cement exhibits expansive characteristics.

Fluid Loss—Desirable for squeeze cementing. Measurement is made on filter press at 1,000 psi using 325-mesh screen. Neat cement has a fluid loss greater than 1,000 cc in 30 minutes at 1,000 psi. For primary cementing fluid loss value of 150 to 400 cc desirable.

Where gas communication through a cemented annulus is a concern, as in gas storage wells or high pressure gas wells, use of very low fluid loss (20 cc in 30 minutes at 1,000 psi on 325 mesh screen) appears to reduce the problem. For squeeze cementing fluid loss control is a primary concern, 50 to 200 cc is a desirable range.

Permeability—Most set cement slurries have very low values—less than any producing formations. If, however, cement is allowed to freeze when setting, or if strength retrogression has occurred due to a high temperature environment, permeability will be much higher (5–10 darcies). Disturbance of cement in the setting process by gas percolation may also provide communication for low viscosity fluids.

CEMENT ADDITIVES
Functions of Cement Additives

Oil well cement slurries usually contain additives to in some way modify basic properties. Additives can:

—Vary cement density.

—Increase or decrease strength.

—Accelerate or retard setting time.

—Control filtration rate.

—Reduce slurry viscosity.

—Bridge for lost circulation control.

—Improve economics.

Many additives affect more than one property of the cement slurry; thus, type and concentration must be chosen with care based on laboratory tests simulating actual well conditions.

Additives are usually free flowing powders which can be blended in bulk plants to obtain uniform distribution. Most additives can be used in mix water if bulk blending facilities are lacking. Text material shows Halliburton trade names. The Appendix lists corresponding products for other Service Companies. Quantity of additives are usually specified in percent additive by weight of cement (where lb/sk is used the weight of a sack is assumed to be 94 lb).

Cement Accelerators

Cement having a compressive strength of 50 psi or greater adequately supports casing. Many operators, however, base WOC time on the period required for the cement to reach a compressive strength of 500 psi. Where cementing temperatures range from 40° to 100°F. WOC time becomes significant. For example, at 40°F., Class A cement requires 48 hours to reach a compressive strength of 500 psi.

Several accelerators added to bulk cement or mixing water are available i.e.: calcium chloride, sodium chloride, and seawater. Calcium chloride is most frequently used. Densified cement slurries (using a friction reducer to permit lower water ratios) provide faster strength buildup. Cal-Seal blended with Class A Portland cement will produce a flash set in about 20 minutes at 75°F.

Calcium chloride is available in regular 77% grade, or as anhydrous 96% grade. Normally 2.0% regular grade or 1.6% anhydrous is used. Above 3.0% little additional acceleration advantage is gained. See Table 4-7.

Sodium chloride in low concentrations accelerates cement setting, but acts as a retarder in high concentrations. 2.0 to 2.5 wt% NaCl in cement produces optimum acceleration. Somewhat more should be used with higher water ratio bentonite

TABLE 4-7
Effect of CaCl₂ on Thickening Time and Strength
(Class A Cement)

Thickening Time, Hours:Minutes

Calcium Chloride Percent	Simulated Well Depths — Feet							
	API Casing — Cementing				API Squeeze—Cementing			
	1000′	2000′	4000′	6000′	1000′	2000′	4000′	6000′
0.0	4:40	4:12	2:30	2:25	3:30	3:29	1:52	0:58
2.0	1:55	1:43	1:26	1:10	1:30	1:20	0:54	0:30
4.0	0:50	0:52	0:50	0:58	0:48	0:53	0:37	0:23

Compressive Strength, psi

Curing Time Hours	Calcium Chloride Percent	Atmospheric Pressure			API Curing Pressure	
		40°F	60°F	80°F	95°F 800 psi	110°F 1600 psi
6	0	N. S.	20	75	235	860
12	0	N. S.	70	405	1065	1525
18	0	5	620	1430	2210	2750
24	0	30	940	1930	2710	3680
36	0	185	1500	2490	3640	4925
48	0	505	2110	3920	4820	5280
6	2	N. S.	460	850	1170	1700
12	2	65	785	1540	2360	2850
18	2	170	1810	3080	3250	4300
24	2	415	2290	3980	4450	5025
36	2	945	2900	4810	5770	6000
48	2	1460	4205	6210	6190	5680

cement. The Table 4-8 shows thickening time and compressive strength:

Seawater contains small amounts of sodium, magnesium, and calcium chlorides (20,000 to 40,000 ppm); thus, provides some acceleration, but not nearly as much as 4.0% CaCl₂.

Densified cement slurries (low water ratio) have

TABLE 4-8

SODIUM CHLORIDE CEMENT ACCELERATOR
Portland Cement — API Class A
Thickening Time — Hours:Minutes
Pan American Pressure Thickening Time Tester
Water Ratio — 5.2 Gals/Sk   Slurry Weight — 15.6 Lbs/Gal

Sodium Chloride Percent	API Casing — Cementing Simulated Well Depths — Feet			
	1000′	2000′	4000′	6000′
0.0	4.40	4.12	2.30	2.25
2.0	3.05	2.27	1.52	1.13
4.0	3.05	2.35	1.35	1.20

COMPRESSIVE STRENGTH — PSI

Curing Time Hours	Sodium Chloride Percent	Atmospheric Pressure		API Curing Pressure	
		60°F	80°F	95°F 800 psi	110°F 1600 psi
12	0	70	405	1065	1525
24	0	940	1930	2710	3680
48	0	2110	3920	4820	5280
12	2	290	960	1590	2600
24	2	1230	2260	3200	3420
48	2	3540	3250	3900	4350
12	4	280	1145	1530	2575
24	4	1390	2330	3150	3400
48	4	3325	3500	3825	4125

particular application for plugs where short pumping time and fast strength buildup are desired. At higher formation temperatures a retarder may be needed to provide sufficient pumping time. Also 2% CaCl₂ can be used with low water ratios to obtain even faster set and strength buildup. Table 4-9 shows the effect of various water ratios.

Cal-Seal is a high-strength controlled setting gypsum cement. It is sometimes used by itself, but more frequently, in 50-50 ratio with Portland cement to increase durability. Due to its fast set it is usually placed with a dump bailer.

Cement Retarders

Increasing well depths and higher formation temperatures have led to the development of cement retarders to extend pumpability time. Most retarders are compatible with all API Class A, B, D, E, G or H cements. The current trend, however, is toward use of a Class G basic cement with the addition of retarder to fit individual well conditions.

The primary factor governing the use of retarders is well temperature. Additives which require higher water ratios also require additional retarder, and the increased water dilutes the retarder.

Above static bottomhole temperatures of 260–275°F., thickening time tests should be run with the actual materials to be used on the job.

Halliburton retarders include HR-4, HR-7, HR-12, HR-20, and HR-13L:

HR-4 is a lignin-type (calcium lignosulfonate) retarder for use with Class A, B, G or H Portland or Pozmix cement, up to static temperatures of 260 to 290°F. HR-4 should not be used with high bentonite (12–25%) cement since it does not act as a dispersant. See Figure 4-3.

HR-12 and HR-20 are blends of HR-4 and an organic acid. HR-12 is recommended for use in API Class D, E, G or H cements where static well temperatures are greater than 260°F. HR-12 is not as effective as HR-4 in API Class A Cement. HR-20 is similar to HR-12 but contains more organic acid. See Figure 4-4.

HR-13L is a liquid retarder (organic acid) which is effective up to a temperature of 500°F.

HR-7 (Lignox or Kembreak) acts as a dispersant as well as a retarder and is the only retarder that should be used with high bentonite cement. HR-7 can be substituted for HR-4 where static formation temperature is less than 260°F. See Table 4-10.

TABLE 4-9
API Class H Densified Cement

Slurry properties

Water ratio, gal/sk	CFR-2 %	Slurry weight, lb/gal	Slurry volume, cu ft/sk	Thickening time [a] hours: minutes
4.3	1.0	16.5	1.05	3:03
3.8	1.0	17.0	0.99	2:51
3.4	1.0	17.5	0.93	2:12

12-hour Compressive strength—psi

Slurry weight, lb/gal	API curing conditions			
	110°F. 1,600 psi	140°F. 3,000 psi	170°F. 3,000 psi	200°F. 3,000 psi
16.5	2,075	4,000	7,800	9,035
17.0	2,850	6,535	8,375	10,025
17.5	3,975	6,585	8,550	10,675

[a] 8,000 ft squeeze-cementing schedule BHST–200°F.: BHCT–150°F.

Saturated saltwater (3.1 lb. sodium chloride/gal of water) when used as the mixing water will moderately retard setting time. Retardation should be sufficient for placement of API Class A cement to depths of 10,000 ft (230°F. static). Defoamer may be required to prevent foaming when saltwater and cement are mixed.

Evaporate salt is not pure NaCl but contains contaminates such as sulfates and magnesium chloride. Sufficient concentration of NaCl can probably not be obtained to have significant retarding effect.

Fluid Loss Control for Primary Cementing

In cementing long sections, at relatively deep depths, high pressures and high temperatures, loss of water from the slurry as it moves past permeable formations may be detrimental from the standpoint of bridging and alteration of slurry properties.

After the cement is in place, loss of filtrate from the "setting" slurry below a bridged zone may reduce hydrostatic head, and allow entry of formation fluids into the wellbore. This is particularly

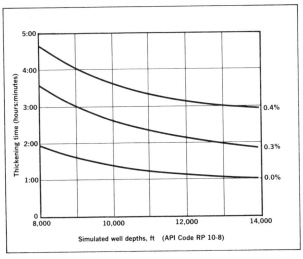

FIG. 4-3—Typical retardation with HR-4 retarder, Poz-mix cement, casing schedules.

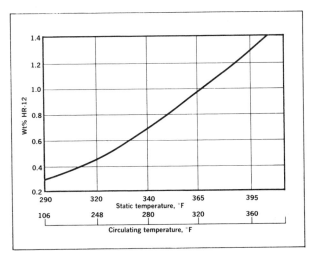

FIG. 4-4—HR-12 concentration to approximate a four-hour thickening time for casing conditions, API Class D or E cement.

TABLE 4-10
Percent HR-7 Retarder
To Approximate A 3-Hour Thickening Time

Well depth, ft	Bottom-hole static Temperature, °F.	12% Bentonite	25% Bentonite
Less than 4,000	Under 140	0.0	1.0
4,000–6,000	140–170	0.3–0.4	1.2
6,000–10,000	170–230	0.4–0.6	1.4
10,000–12,000	230–260	0.6–0.8	1.6

true where gas zones are involved.

Where these conditions are present, use of additives to reduce fluid loss of the primary cement slurry may be justified. Additives function by forming micelles, or films, and improving particle size distribution.

Several materials are used to reduce fluid loss. CFR-2, Diacel LWL, Halad 9, Halad 14. Most have other effects also, such as reduced viscosity and retardation.

CFR-2, although primarily a friction loss reducer, also provides moderate fluid loss control through a dispersing action. It has no retarding effect and can be used from 60°F. to 350°F. Its dispersing action permits use of lower water-cement ratios while maintaining reasonable slurry viscosities; thus, this densifying action provides reduced fluid loss.

Diacel LWL, CMHEC, is primarily a water loss control material. It is not effective in concentration below 0.3%. Above 0.7%, water ratio must be increased to offset slurry viscosity increase. At low temperature an accelerator (Diacel A) is needed to offset retardation. Maximum temperature limit is 170°F.

Halad 9, primarily designed for fluid loss control, has moderate retarding effect and can be used from 60° to 350°F. At low temperatures, 2% $CaCl_2$ may be needed to provide acceleration. It tends to increase slurry viscosity somewhat, thus slightly more mix water may be desirable (5.65 gal/sk rather than 5.20 gal/sk).

Halad 14 is designed for use at high temperatures, 170° to 400°F. It has the effect of reducing slurry viscosity and providing retardation as well as reducing fluid loss. Since it is normally used under critical conditions of depth and temperature, specific testing should be run to establish its properties with the particular cement composition to be used. Table 4-11 shows Halad 14 characteristics under conditions where it might be used. Note the effect of temperature on fluid loss.

Halad 22A is designed for use at very high temperatures, from 80°F. to in excess of 400°F. BHST. It can be used in salt cement slurries up to 18% NaCl or 5% KCl, where Halad 9 or 14 are less effective.

Slurry Viscosity Control

Use of cement friction-reducing additives aids removal of annular mud by the cement slurry on the basis of:

—Reducing displacement rate to provide turbulent flow in some portion of the annular cross section;

—At the same displacement rate, to increase the cross-sectional area of the annulus affected by turbulent flow;

—Friction reducers also lower placement pressures where restricted annular clearances are involved.

Friction reducers (Halliburton's CFR-2) are essentially dispersing agents which reduce the apparent viscosity of the slurry. Other materials such as salt (NaCl) or Halad 9, or HR-7 which act as

TABLE 4-11
Fluid Loss Class H Cement

Water—5.0 gal/sk

Slurry weight—15.8 lb/gal

Low temperature fluid loss

Halad-14 %	Slurry consistency, Bc 140°F.		Fluid loss cc/30 min 325-mesh screen 1,000 psi Room temperature
	Initial	20 minute	
0.75	1	2	64
1.00	1	2	38
1.25	1	2	30

High-temperature fluid loss

Halad-14 %	Slurry consistency, Bc 80°F. to 190°F.		Fluid loss cc/30 min 325-mesh screen 1,000 psi Temperature, 190°F.
	Initial	20 minute	
0.75	1	9	107
1.00	2	9	84
1.25	2	8	64

dispersants, also act to some extent as friction reducers.

Effect of Halliburton's CFR-2 is shown in Table 4-12. Flow rates shown are the minimum required to initiate turbulence.

Light-Weight Additives

Basic cements are ground to a fineness which, when mixed with the recommended amount of water, produce slurry weights in excess of 15 lb/gal. Light-weight additives are used both to permit the use of longer columns of cement without formation breakdown, and reduce the cost of the cementing material even where low weight is not a requirement.

Two methods of reducing weight are used:

1. Additives such as bentonite, attapulgite, pozzolans or sodium silicate which will permit use of increased mixing water and at the same time prevent water separation.

2. Additives such as Gilsonite having a low specific gravity (limited application).

Wyoming Bentonite (API Specification 10-A for use in Cement)—Bentonite has been used for many years to reduce slurry cost, slurry weight, and water separation. High percentages of bentonite reduce compressive strength and decrease thickening time. At temperatures above 230°F. bentonite promotes strength retrogression.

For use with cement, bentonite must meet API Spec 10-A which is designed to insure that the bentonite does not contain soda ash or acrylates to increase bentonite yield.

Water requirements are about 1.3 gal for each 2% bentonite up to 8–12%. At the higher bentonite percentages dispersants are usually used which give considerable latitude in water ratio. Average slurry properties of Wyoming bentonite in API Class H cement are shown in Table 4-13.

Prehydrating Bentonite and Attapulgite—The technique of prehydrating bentonite or attapulgite in fresh water or seawater reduces clay requirements and assures uniform blending where dry blending facilities are not available. One pound Wyoming bentonite prehydrated in fresh water provides about the same slurry properties as 3.6-lb dry blended.

Prehydrating in seawater, use of European bentonite, or dispersing with CFR-2 rather than HR-7 increases the bentonite needed to limit free water to 1%. Table 4-14 details water and clay requirements and shows slurry properties for various combinations.

Sodium Silicate—Small amounts (2–3%) of liquid sodium silicate added to Class G or H cement permit use of two to three times the normal mix water ratio, providing slurry densities less than 12.0 lb/gal and slurry yields greater than 2.5 cu ft/sk. For a given slurry density, sodium silicate slurries pro-

TABLE 4-13
API Class H Cement With Bentonite

SLURRY PROPERTIES

Bentonite Per Cent	API Water Requirements Gal./Sk.	Cu. Ft./Sk.	Slurry Weight Lbs./Gal.	Lbs./Cu. Ft.	Slurry Volume Cu. Ft./Sk.
0	4.30	0.58	16.4	123	1.06
2	5.49	0.73	15.5	115	1.22
4	6.69	0.89	14.7	110	1.38
6	7.88	1.05	14.1	105	1.55
8	9.07	1.21	13.6	101	1.73
10	10.27	1.37	13.2	99	1.90
12	11.46	1.53	12.9	96	2.07
14	12.66	1.69	12.6	94	2.24
16	13.86	1.85	12.4	93	2.41

COMPRESSIVE STRENGTH — PSI

Bentonite Per Cent	Curing Time Hours	95°F. 800 psi	110°F. 1,600 psi	140°F. 3,000 psi	170°F. 3,000 psi	200°F. 3,000 psi
0	8	500	1200	2500	4000	5450
	24	3000	4050	5500	6700	8400
2	8	250	720	1400	2000	2500
	24	1550	2350	3250	3630	3800
4	8	130	450	830	1200	1550
	24	980	1490	2000	2250	2400
6	8	90	380	560	800	1050
	24	650	1000	1400	1650	1800
8	8	75	200	380	560	750
	24	430	700	1025	1150	1250
10	8	74	150	260	380	500
	24	325	500	700	825	900
12	8	70	120	200	280	360
	24	225	355	500	600	675
14	8	60	95	150	200	250
	24	160	270	400	490	550
16	8	50	80	110	170	220
	24	130	245	350	400	475

TABLE 4-12
Effect of CFR-2 on Turbulent Flow Rate

COMPOSITION	Per Cent CFR-2	Hole Size (Inches) 6¾ — Casing Size OD (Inches) 2⅞	6¾ — 4½	8¾ — 5½	9⅞ — 7
		FLOW RATE — BPM			
Pozmix A Cement 14.1 lbs/gal	0.00	23.14	16.16	29.01	30.58
	0.50	8.51	6.86	11.18	12.17
	0.75	2.55	2.72	3.67	4.25
Pozmix S Cement 14.5 lbs/gal	0.00	22.58	15.91	28.38	29.98
	0.50	10.40	8.09	13.49	14.58
	0.75	3.61	3.42	4.99	5.68
API Class A 15.6 lbs/gal	0.00	18.18	13.58	23.29	24.93
	0.50	14.32	11.28	18.66	20.21
	0.75	6.57	5.86	8.91	9.93
API Class A 4% Gel 14.1 lbs/gal	0.00	25.17	17.58	31.54	33.26
	0.50	15.50	11.21	19.65	20.88
	0.75	6.58	5.30	8.65	9.41
API Class A 12% Gel 12.8 lbs/gal	0.00	23.55	16.45	29.51	31.12
	0.75	14.08	10.26	17.89	19.05
	1.00	2.93	2.88	4.10	4.67
API Class E 16.25 lbs/gal	0.00	10.95	9.38	14.66	16.19
	0.50	4.14	4.46	5.97	6.74
	0.75	1.10	1.55	1.73	2.14

TABLE 4-14
Prehydrated Bentonite and Attapulgite

Dry Blend % gel equiv.	Water gal/sk	Class G Cement lb/bbl water Bentonite or attapulgite			Slurry weight, lb/gal	Slurry volume, cu ft/sk
Fresh water		Wyo.	Euro.	Atta.		
4	8.0	6.0	8.0	—	14.0	1.55
6	8.9	7.3	9.7	—	13.6	1.67
8	10.0	8.5	11.3	—	13.2	1.82
10	11.3	9.8	13.0	—	12.8	2.00
12	12.8	11.0	14.6	—	12.4	2.20
16	17.1	13.5	18.0	—	11.6	2.79
Sea water		Wyo.		Atta.		
4	8.0	9.0 (12.0)		2.5 (5.0)	14.1	1.55
6	8.9	11.0 (14.0)		3.5 (5.5)	13.7	1.67
8	10.0	13.5 (16.5)		4.5 (6.0)	13.3	1.82
10	11.3	15.5 (18.5)		5.5 (6.5)	12.9	2.00
12	12.8	18.8 (21.8)		6.5 (7.0)	12.5	2.20
16	17.1	23.0 (26.0)		8.5 (8.5)	11.7	2.79

[1]Add ( ) lb bentonite or attapulgite if HR-7 is used.

[2]With bentonite add 3 more lb bentonite if CFR-2 is used to replace HR-7.

[3]CFR-2 and HR-7 should not be used together when prehydrating bentonite in seawater. European bentonite should not be prehydrated in seawater.

[4]Wyoming or European bentonite can be prehydrated in 1/2 the required volume of fresh water—then seawater added to make up the remaining water requirement.

vide higher compressive strengths than other high water ratio slurries.

Manufactured Light-Weight Cements—Cements such as Trinity Lite-Wate or Halliburton Light-Weight contain pozzolans or shale, sometimes with the addition of bentonite, to permit use of increased mix water without exceeding the desired limit of 1% free water after set.

Table 4-15 compares properties of various light density slurries. Also shown is the longer term compressive strength buildup of a typical sodium silicate slurry.

Heavy Weight Additives

A suitable weighting material for cement should have these characteristics:

—Low water requirement.

—No strength reduction of cement.

—No reduction of pumping time.

—Uniform particle size.

—Minimum slurry volume increase.

—Chemically inert.

The most suitable weighting materials are Hematite Ore, Barite, and Ottawa Sand.

Hematite ore (Sp. gr. 5.05) can produce slurry densities up to 22 lb/gal. (See table 4-16). It is chemically inert and requires little additional water. Ilmenite ore (Sp. gr. 4.6) emits some degree of radioactivity and is less desirable than hematite.

Barite has a specific gravity of 4.2. (See Table 4-17). Particle size is very fine; thus, additional water requirements (0.2 lb water per lb barite) tends to offset higher specific gravity. 18.0 lb/gal slurry is about maximum.

Sand has a specific gravity of 2.65, and a low water requirement. Maximum slurry weight is about 17.5 lb/gal. (See Table 4-18).

Salt saturation of cement slurry can add 0.5 to 1.0 lb/gal density.

Mix-water ratios can be reduced to increase density about 2.0 lb/gal. A friction reducer such as Halliburton CFR-2 will then be required to provide a pumpable viscosity.

Where high slurry densities and low mix-water ratios are desired, a recirculating-type mixer will provide better density control than the jet-type mixer.

TABLE 4-15
Properties of Low-Density Slurries

Water, gal/sk	Percent sodium silicate	Slurry Viscosity		Free water, percent	Slurry density, lb/gal	Yield, cu ft/sk
		Initial	20 min			
API Class H Cement						
4.3	0	6	8	2.50	16.4	1.06
5.2	0	3	4	8.00	15.6	1.18
9.0	2	10	8	0.36	13.4	1.68
11.8	2	8	8	0.96	12.5	2.06
14.7	2	7	7	1.36	11.8	2.45
17.5	2	6	6	1.76	11.4	2.82
17.5	3	3	3	0.68	11.4	2.82
API Class A Cement						
5.2	0	4	9	2.30	15.6	1.18
9.0	2	7	8	0.00	13.4	1.68
11.8	2	6	6	0.52	12.5	2.06
14.7	2	3	2	0.68	11.8	2.45
17.5	2	2	2	1.50	11.4	2.82
17.5	3	3	3	1.00	11.4	2.82
Halliburton Light Cement						
7.7	0	7	14	0.60	13.6	1.54
8.8	0	5	9	0.60	13.1	1.69
9.9	0	4	6	1.30	12.7	1.84
10.9	0	5	7	1.30	12.4	1.97
10.9	1	11	12	0.40	12.4	1.97
12.9	1	8	8	0.56	11.8	2.21
14.9	1	5	5	0.72	11.4	2.46
16.9	1	3	4	2.10	11.1	2.74
16.9	2	7	8	0.24	11.1	2.74

Compressive stength: Class H w/2% Sodium Silicate and 11.0 gal/sk water

Curing temperature, °F	Curing time (days)					
	1	3	7	14	28	60
60	Set	75	530	585	755	1070
80	185	330	640	750	970	1380
95	310	700	705	950	1065	1215
140	445	615	750	1105	900	765

TABLE 4-16
Slurry Properties API Class H with Hematite

Cement, pounds	Hematite, pounds	Water, gal/sk	Slurry volume, cu ft/sk	Slurry weight, lb/gal
94	—	4.5	1.08	16.2
94	12	4.5	1.12	17.0
94	28	4.5	1.17	18.0
94	46	4.5	1.23	19.0
94	63	4.5	1.30	20.0

TABLE 4-17
Slurry Properties API Class H with Barite

Cement, pounds	Barite, pounds	Water, gal/sk	Slurry weight, lb/gal	Slurry volume, cu ft/sk
94	—	4.5	16.2	1.08
94	22	5.1	17.0	1.24
94	55	5.8	18.0	1.46
94	108	7.1	19.0	1.83

TABLE 4-18
Slurry Properties API Class G with Sand

Cement, pounds	Sand, pounds	Water, gal/sk	Slurry weight, lb/gal	Slurry volume, cu ft/sk
94	—	5.2	15.60	1.18
94	5	5.2	15.76	1.21
94	10	5.2	15.91	1.24
94	15	5.2	16.07	1.27
94	20	5.2	16.20	1.30
94	32	5.2	16.50	1.37
94	56	5.2	17.00	1.52
94	85	5.2	17.50	1.70
94	123	5.2	18.00	1.93

Lost Circulation Additives

Lost circulation materials for cement may be classified as:

Granular—Inert materials designed to bridge at surface of borehole or within the formation. Effectiveness depends on proper particle size distribution to yield a low permeability bridged mass. Granulars work better in high solids fluid system, such as cement, than in lower solids drilling mud system. Typical granular materials are Gilsonite and Tuf-Plug (walnut hulls). (Fibrous wood materials retard setting time and are not used).

Lamellated—Inert flake-type materials designed to mat at the face of formation. Cellophane flakes are commonly used.

Semi-Solids and Flash Setting Cements—By either chemical or physical action these materials thicken rapidly to form plug. Examples are Cal-Seal, Bentonite-Diesel Oil Slurry, or Thixotropic Cement.

In laboratory tests granular materials work best in unconsolidated formations. Lamellated flake-type materials work best for slots. Semi-solids or flash-setting materials work best in large vugs or fractures.

Gilsonite is the most effective granular material. Optimum quantity is about 12 to 15 lb/sk cement.

Walnut Hulls are used in quantity on 2 to 8 lb/sk. Other nut hulls should be checked for effect on cement setting properties.

Cellophane Flakes (3/8-in. size) are used in quantity of 1/4 to 2 lb/sk.

Cal-Seal is used in quantities of 25 to 75 lb/sk of Portland cement (50 lb/sk common) to give setting time of 7 to 15 minutes. Cal-Seal will set up even while moving through porous zone.

Thixotropic cements develop gel strength quickly when not agitated. Thixotrophy is obtained by addition to Class G cement, of about 10% calcium sulfate and 4–16% bentonite, or about 1% of a sodium silicate-like polymer.

A recent approach to controlling lost circulation by reducing the hydrostatic fluid column pressure involves injecting a carefully calculated volume of Nitrogen into a portion of the mud column just before the cement is mixed and pumped into the casing.

CEMENT BONDING

Thinking on the subject of cement bonding has undergone changes in recent years. The paramount objective in primary cementing is to eliminate fluid communication between zones. Originally it was thought that ''bonding'' of the cement to the casing and to the formation was closely related to this objective. Bonding was studied in the lab on the basis of steel surface characteristics and pipe expansion effects; and downhole, through sound transmission indications.

Initially the primary concern was with the bond at the casing-cement interface, because this could be ''looked at'' with the sonic bond log. Later it was realized that the bond between the cement and formation (largely dependent on mud filter cake properties) was more difficult to achieve, and more likely to be the source of interzonal communication problems.

Currently it appears that effective displacement of annular mud by the cement to eliminate mud channels is the most important consideration in eliminating fluid communication between zones. As shown in Figure 4-5, with a thin tough mud filter cake fluid communication is unlikely. But as the mud filter cake grades into a softer thicker cake, and further to what might be termed a mud channel, communication becomes more likely.

Thus, the subject of cement bonding per se, while still important, is not the primary concern it was originally thought to be.

Bonding Measurements in the Laboratory

Bonding measurements in the laboratory are indicative of factors which affect the contact between the cement and casing, and cement and formation, but should be considered in light of the overpowering importance of eliminating mud channels.

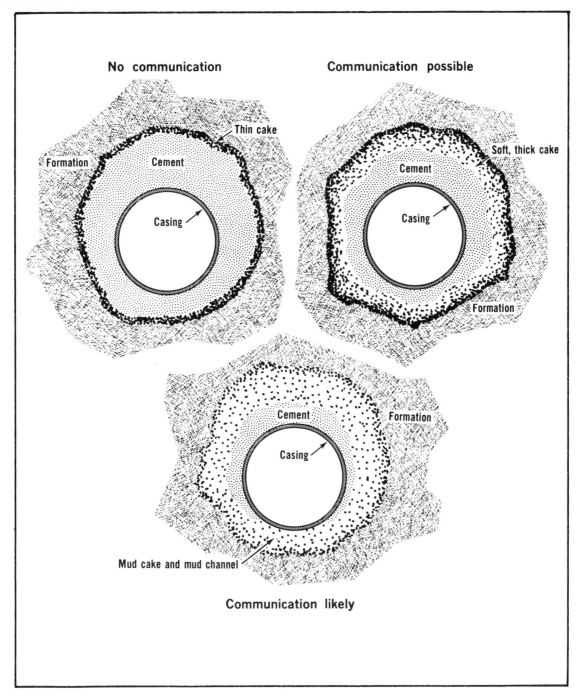

FIG. 4-5—How mud filter cake affects possibility of fluid communication between zones after cementing.

Shear bond mechanically supports pipe in hole. It is determined by measuring force required to initiate pipe movement, divided by cement casing contact surface area. (See Fig. 4-6.)

Hydraulic bond blocks migration of fluids. It is determined by applying liquid pressure at pipe-cement or formation-cement interface until leakage occurs.

Gas bond is measured in the same manner as hydraulic bond.

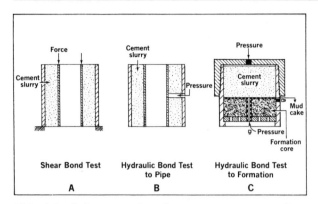

FIG. *4-6—Laboratory bonding measurements.*[5] *Permission to publish by the API Production Department.*

Pipe-Cement Bond

Higher cement compressive strength increases both shear and hydraulic bond.

Pipe finish typically affects bond strength as shown by lab tests in Table 4-19. With mill varnish on casing, bond strength is time-dependent (particularly below 140°F.) Lowest bond strength occurs about two days after cementing. This effect can be seen on sonic bond log. Oil wet pipe surfaces reduce hydraulic and shear bond strength.

The micro annulus situation as sometimes seen on the sonic bond log is primarily a function of pipe expansion or contraction due to application of heat or pressure. A true micro annulus condition does not necessarily mean fluid communication (see Production logging chapter). The following points are significant in regard to the tightness of contact between the cement and casing, however:

—Leaving pressure on casing during WOC time is harmful to "bond".

—Pumping plug down with light fluid is helpful to "bond".

—Heat of hydration of cement is harmful to "bond" due to casing expansion.

—Circulating while WOC is helpful to "bond" due to cooling.

—After cement is set, increased pressure inside casing increases "bond."

Hydraulic bond failure is time-dependent and is a function of the viscosity of the fluid causing failure. With water, field tests indicate that bond failure progresses at a rate of about 1.1 to 1.3 ft/min.

Formation-Cement Bond

Laboratory tests show that cement-formation hydraulic bond is largely influenced by the presence of mud filter cake. A tough mud cake slightly improves hydraulic bond strength compared with a soft mud cake. With clean dry formation hydraulic bond strength can exceed formation strength. With filter cake bond strength is greatly reduced.

Higher hydraulic bond is obtained on more permeable formations since cement slurry loses water to formation and develops a higher compressive strength.

Results of lab tests of formation cement bond on Indiana Limestone (1 md) and Berea Sandstone (100 md) are shown in Fig. 4-7.

TABLE 4-19
Bond Strength vs. Pipe Finish

Type of finish		Bond strength	
Steel	Shear psi	Hydraulic psig	Gas psig
New (Mill Varnish)	74	200–250	15
New (Varnish Chemically Removed)	104	300–400	70
New (Sandblasted)	123	500–700	150
Used (Rusty)	141	500–700	150
New (Sandblasted—Resin-Sand Coated)	2400	1100–1200	400+

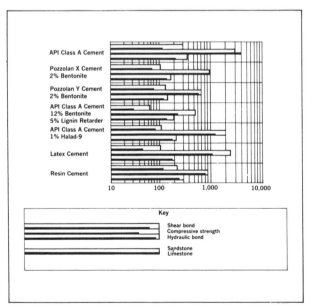

FIG. *4-7—Bonding properties of cement to formation, cement-squeezed, walls not cleaned.*[5]

FLOW PROPERTIES OF PRIMARY CEMENTS

Cement flow properties during primary cementing are important because of their effect on:

—Efficiency with which the cement displaces the annular mud column.

—Frictional pressure drop in the annulus, which adds to the hydrostatic pressure exerted on the formation.

—Hydraulic horsepower required to place the cement in a given time period.

Cement slurries are usually non-Newtonian fluids; thus, viscosity is a function of shear rate. In order to calculate pressure losses in a system, a mathematical description of "viscosity" or the shear stress vs. shear rate relation must be developed.

The Power Law concept more nearly describes the flow properties of cement, rather than the Bingham Plastic concept used to describe muds.

Power Law Equation: $S_s = K'(S_r)^{n'}$

S_s = shear stress (lbs/ft²)
S_r = shear rate (sec⁻¹)
n' = slope of log S_s vs. log S_r curve.
K' = intercept of log S_s vs. log S_r curve on shear stress axis.

In the Power Law concept, two slurry parameters must be determined in order to estimate frictional pressure loss and predict the flow velocity required to initiate turbulence. These are: flow behavior index (n'); and consistency index (K').

Fann dial readings taken at 600, 300, 200, and 100 rpm are converted to shear stress (pounds force/square foot), and plotted on the Y axis of log-log paper vs. Fann rpm's converted to shear rate (sec⁻¹) on the X axis. (n') is the slope of a straight line through the four points, and (K') is the intercept of the straight line at unity rate of shear (1.0 sec⁻¹).

$$\text{Shear Stress} \left(\frac{\text{lb force}}{\text{sq ft}} \right) = $$

$$\frac{\text{Fann Dial Reading} \times N \times 1.066}{100}$$

Shear Rate (sec⁻¹) = Fann rpm × 1.703

On the field model Fann Viscometer, with only 300 and 600 rpm, n' and K' can be calculated as follows:

$$n' = 3.32 \times \left(\log_{10} \frac{600 \text{ rpm Dial Reading}}{300 \text{ rpm Dial Reading}} \right)$$

$$K' = \frac{N \times (300 \text{ rpm Reading}) \times 1.066}{100 \times (511)^{n'}}$$

N = range extension factor of the Fann torque spring (usually 1.0)

Flow Behavior Index (n') and Consistency Index (K') values have been published for most cement slurries. Power Law curves for several slurry are shown in Fig. 4-8.

Where the Bingham Plastic parameters (Plastic Viscosity, PV, and Yield Point, YP) are known:

$$n' = 3.32 \times \log_{10} \left(\frac{2PV + YP}{PV + YP} \right)$$

$$K' = \frac{N \times (PV + YP) \times 1.066}{100 \times (511)^{n'}}$$

Newtonian cements do a better job of displacing mud. If $n' = 1.0$, the slurry is Newtonian, and viscosity is constant at all flow rates. For many cement slurries (particularly high gel type slurries) the n' value is considerably less than 1.0. Thus, apparent viscosity increases as flow velocity decreases. For most effective mud displacement by the cement slurry it is desirable that the cement slurry be as Newtonian as possible; i.e., n' should be close to 1.0.

Formulae For Making Flow Calculations

1. Apparent Viscosity

$$\mu_a = \frac{4.788 \times 10^4 \, K'}{(\text{Shear Rate})^{1-n'}}$$

μ_a = apparent viscosity, centipoise
K' = consistency index
n' = flow behavior index

$$\text{Shear Rate} = \frac{96 V}{D}, \text{ sec}^{-1}$$

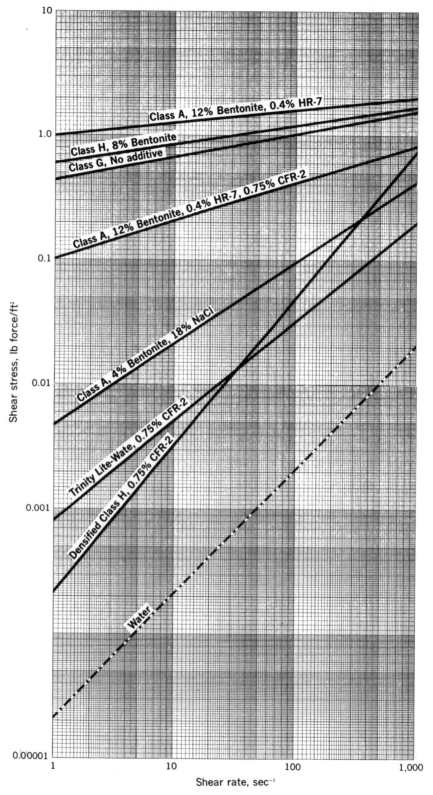

FIG. 4-8—Flow curves for typical cement slurries.

2. Displacement Velocity

$$V = \frac{17.15 \, Q_b}{D^2} = \frac{3.057 \, Q_{cf}}{D^2}$$

V = velocity, feet per second
Q_b = pumping rate, barrels per minute
Q_{cf} = pumping rate, cubic feet per minute
D = inside diameter of pipe, inches

For annulus,

$$D = D_o - D_i; \text{ or,}$$

$$D = \frac{4 \times \text{Area of Flow}}{\text{Wetted Perimeter}}$$

$$D^2 = D_o^2 - D_i^2$$

D_o = outer pipe id or hole size, inches
D_i = inner pipe od, inches

3. Reynolds Number

$$N_{Re} = \frac{1.86 \, V^{(2-n')} \rho}{K' (96/D)^{n'}}$$

N_{Re} = Reynolds number, dimensionless
ρ = slurry density, pounds per gallon

4. Frictional Pressure Drop

$$\Delta P_f = \frac{0.039 \, L \rho \, V^2 f}{D}$$

ΔP_f = frictional pressure drop; psi
L = length of pipe, ft
f = friction factor, dimensionless
(see Fig. 4-9)

For $N_{Re} < 2100$: $f = \dfrac{16}{N_{Re}}$

For $N_{Re} > 2100$ (Non-newtonian fluid):
$$f = 0.00454 + 0.645 \, (N_{Re})^{-0.7}$$

5. Velocity at Which Turbulence May Begin (N_{Re} = 2100)

$$V_c^{2-n'} = \frac{1129 K' (96/D)^{n'}}{\rho}$$

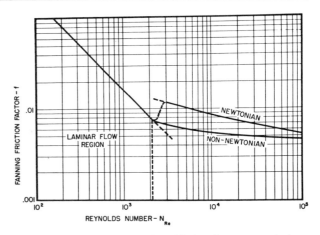

FIG. *4-9—Reynolds number-friction factor correlation.*

$$V_c = \left[\frac{1129 \, K' \, (96/D)^{n'}}{\rho} \right]^{1/(2-n')}$$

V_c = critical velocity, feet per second

6. Hydrostatic Pressure

$P_h = 0.052 \, \rho \, H$
P_h = hydrostatic pressure, psi
H = height of column, feet

DISPLACEMENT MECHANICS

The paramount objective in primary cementing is to place cement so as to eliminate mud channels. The most predominant cause of cementing failure appears to be channels of gelled mud remaining in the annulus after the cement is in place.

If mud channels are eliminated, almost any cement will provide an effective seal. Conversely, if mud channels remain after the cement is in place, no matter what the quality of the cement, there will not be an effective seal between formations. See Fig. 4-10.

Forces helping to displace mud are:

1. Drag stress of cement upon mud due to difference in flow rates and flow properties.
2. Drag stress of pipe upon mud and cement due to pipe motion—either rotation or reciprocation.
3. Buoyant forces due to density differences between mud and cement.

FIG. *4-10—Gelled mud remaining after cement was circulated in place in laboratory model study (Halliburton Services).*

Factors Affecting Annular Flow and Mud Displacement

In evaluating factors affecting displacement of mud, it is necessary to consider the flow pattern in an eccentric annulus condition with the pipe closer to one side of the hole than the other. Flow velocity in an eccentric annulus is not uniform. Highest flow rate occurs in the side of the hole with the largest clearance as shown in Fig. 4-11.

With a non-Newtonian fluid it is possible to have turbulent flow in the wide side of the annular cross section, and laminar or even a stagnant zone on the narrow side.

—As flow rate increases, the annular cross sectional area affected by turbulent flow increases.
—As the viscosity of the fluid decreases, the annular cross sectional area affected by turbulence increases.
—As casing standoff decreases the displacement rate needed to prevent a stagnant area on the narrow side of the annulus increases.
—As the viscosity and gel strength of the mud increases, the displacement rate needed to prevent a stagnant area increases.
—The more "Newtonian" the displacing fluid, the more effective the displacement.
—With the casing close to the wall of the hole it may not be possible to displace cement at a rate high enough to develop turbulent flow throughout the entire annular cross section.

Laboratory Studies Point the Way

A number of laboratory studies have been conducted in recent years to determine the relative importance of the forces helping to displace mud.

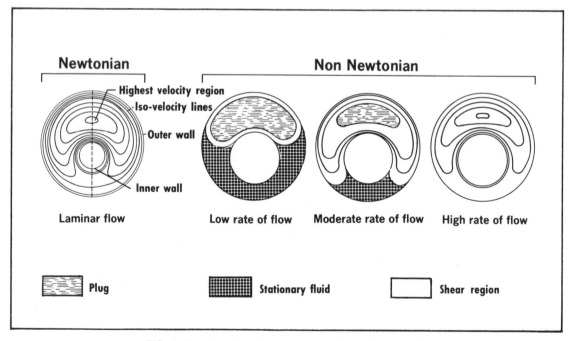

FIG. *4-11—Annulus flow patterns. After Piercy et al.*

The study reported in References 22 and 24 most closely simulates well conditions of temperature and wall cake effect.

Figure 4-12 shows the effect of pipe standoff (100% standoff = centered casing; 0% standoff = casing against the wall of the hole). Total annular mud is percent of the annular cross section filled with mud. It is channel mud plus filter cake mud. Channel mud is circulatable mud.

Figure 4-13 shows the effect of mud properties. "Thick mud" is a fresh water mud having a Plastic Viscosity of 46 cp and a Yield Point of 21 lb/100 sq ft. "Thin mud": PV = 20 cp; YP = 3 lb/100 sq ft. Laminar flow velocity = 90 ft/min; turbulent flow velocity = 255 ft/min; super turbulent flow velocity = 455 ft/min.

These data make the point that a "thin" mud can be displaced more readily, even at laminar flow rates, than a "thick" mud at super turbulent flow rates.

Figure 4-14 evaluates the effect of flow rate on displacement of a "thick mud" (PV = 43 cp; YP = 21 lb/100 sq ft) by a thin cement slurry. With other conditions equal high flow rates do a more effective job of mud removal.

Figures 4-15 and 4-16 evaluate effects of rotation and reciprocation with and without scratchers. Note that pipe movement is a very important factor. Further, scratchers, which basically act to break the gel of the mud, are a significant aid in mud removal, particularly in washed out sections.

Contact Time

Field studies[15] and laboratory studies[22,24] have shown the importance of "contact time," or the length of time cement moves past a point in the

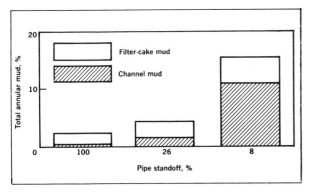

FIG. *4-12—Evaluation of pipe standoff.*[24] *Permission to publish by The Society of Petroleum Engineers.*

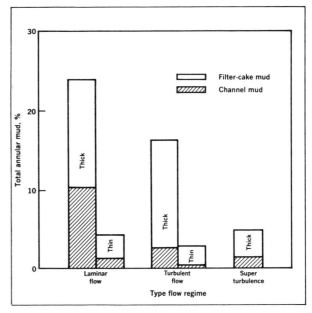

FIG. *4-13—Effect of mud properties.*[24] *Permission to publish by The Society of Petroleum Engineers.*

annulus in *turbulent flow.* Thus, if a mud channel is put in motion, even though its velocity is much lower than cement flowing on the wide side of the annulus, given enough time the "mud channel" will move above the critical productive zone.

Lab studies[24] show that contact time is less effective with the cement in laminar flow conditions

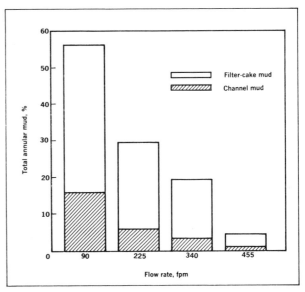

FIG. *4-14—Effect of flow rate.*[24] *Permission to publish by The Society of Petroleum Engineers.*

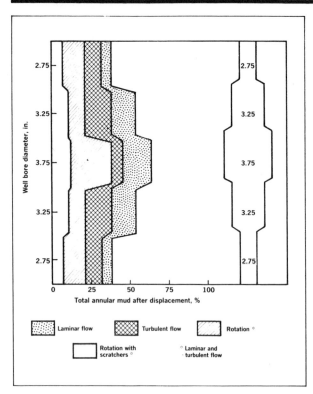

FIG. *4-15—Effect of rotation with and without scratchers.*[24] *Permission to publish by The Society of Petroleum Engineers.*

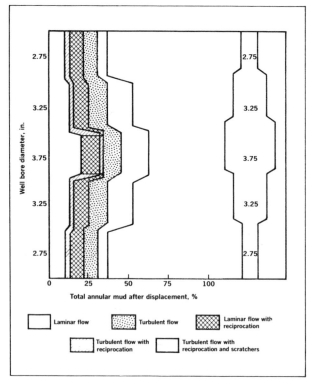

FIG. *4-16—Effect of reciprocation with and without scratchers.*[24] *Permission to publish by The Society of Petroleum Engineers.*

where apparently the cement does not exert sufficient drag stress on the mud to start the mud channel moving.

On a practical basis, contact time should exceed 10 minutes. At a given displacement rate contact time is directly proportional to volume of cement.

Computer Solutions

Computer solutions are helpful for planning of difficult primary cementing jobs—or for improving the economics where several similar jobs will be performed.

Slurry Flow Plan—with accurate well data on cement and mud properties, slurry flow plan provides following factors for a number of possible conditions:

1. Flow rate required for turbulence.
2. Hydraulic horsepower required.
3. Fill-up factor.
4. Contact time.
5. Location of DV equipment based on formation fracture gradient.

Rheologically Balanced Flow Plan—Again with accurate well data on cement and mud properties, hole diameter, standoff of eccentric casing, rheologically balanced flow plan can show:

1. Mud displacement efficiency for various mud and cement properties.
2. Minimum volume of cement required to provide desired fillup.
3. Effect of displacement rate on displacement efficiency.

Summary of Important Factors

All laboratory and field work performed in recent years point to the same factors as contributing to the success of primary cementing. All of the factors are aimed at removing mud from the annular cross section.

—Pipe centralization significantly aids mud displacement.

—Pipe movement, either rotation or reciprocation, is a major driving force for mud removal. Pipe motion with scratchers substantially improves mud displacement in areas of hole enlargement.

—A well conditioned mud (low PV and YP) greatly increases mud displacement efficiency.

—High displacement rates promote mud removal. At equal displacement rates a thin cement slurry in turbulent flow is more effective than a thick slurry in laminar flow.

—Contact time (cement volume) aids in mud removal if cement is in turbulent flow in some part of the annulus.

—Buoyant force due to density difference between cement and mud is a relatively minor factor in mud removal.

Practicalities

In a given situation it may not be possible, or even necessary, to maximize each of these factors. To some extent one factor may compensate for another. In the above list, however, centralization, pipe movement and mud conditioning are primary factors.

Careful evaluation of cementing success is the key.

If cementing failures are experienced, examination of cementing practices in light of these important factors should show where improvements in cementing practices should be initiated.

COST OF PRIMARY CEMENTING

Overall well cost should be optimized considering such factors as:

1. Cost of cement slurry—should be on a volume basis (cu ft of slurry behind pipe).
2. Cost of pumping equipment used in placing cement.
3. Cost of rig time in performing cementing operation and subsequent WOC time.
4. Cost of future remedial operations as a result of inadequate primary cementing.

Typical bulk cement and additive costs (Domestic U.S. April, 1976):

API Class A & H Cements*	$2.90/sk
Pozmix A	1.30/cu ft
0–4% Gel/sk (cu ft)	0.24
5–8% Gel/sk (cu ft)	0.48
9–12% Gel/sk (cu ft)	0.72

HR-4	0.52/lb
HR-7	0.52/lb
HR-12	1.45/lb
CFR-2	2.35/lb
Hi-Dense No. 3	11.21/100 lb
Gilsonite	0.16/lb
Calcium chloride	11.50/100 lb
Sodium chloride	3.00/100 lb
*Plus service charge (land)	0.55 cu ft
(offshore)	0.80

Comparative costs of typical Cement Slurries used in the Deep Anadarko Basin are shown in Table 4-20.

SPECIAL PROBLEM SITUATIONS—NEW DEVELOPMENTS
Cementing High Temperature Wells

Deep wells (below 15,000 feet) having static formation temperatures above 230°–250°F. are considered to be critical situations. Major problems involve displacement of mud by cement slurry, design of slurry to provide adequate rheological properties and pumpability time, attainment of desired slurry properties during the mixing process, and control of subsequent strength retrogression. The following factors should be considered:

Formation Temperature—Accurate knowledge of formation static or bottom-hole circulating temperature is the starting point for slurry design. A particular design problem exists with a long cement column due to temperature difference between the top and bottom of the slurry. Overstating temperature to provide "safety factor" is poor practice. Safety factor should be provided by adjusting pumpability time.

Slurry History—Anticipated temperature-pressure-time history of the cement slurry as it is mixed and pumped into place must be established as a basis for running Amoco Pressure Temperature Thickening Time Tests to fix additive requirements.

Laboratory Tests—Slurry design tests must be run with cementing materials, additives and mix water which will actually be used on the job. Use of Class H basic cement promotes uniformity, but laboratory slurry tests should be checked by field tests shortly before the job with materials from the location.

Fluid Loss—Slurry design should provide for controlled fluid loss (100 to 175 cc API 1,000 psi, 325-mesh screen at 190°F.) Viscosity reduction to

TABLE 4-20
Typical Cement Slurries—Deep Anadarko Basin

Slurry type	Properties		Cost	
	Wt-lb/gal	Yield cu ft/sk	sk	ft³
Class H 2% CaCl₂	15.6	1.18	$ 3.67	$3.11
*HLC 10% NaCl, 2% CaCl₂ 10 lb Gilsonite	12.7	1.93	4.71	2.44
Class H 10% NaCl, 2% CaCl₂	15.8	1.2	3.81	3.17
Class H (Densified) 10% NaCl, 1% CFR-2	16.4	1.10	5.73	5.20
50-50 Pozmix Cement 10% NaCl, 1.25% CFR-2	15.5	1.07	5.18	4.84
*HLC 5% NaCl	12.4	1.97	3.08	1.56
HLC 10% NaCl, .4% HR-4 10 lb Gilsonite, 0.75% CFR-2	12.5	2.14	6.43	3.00
HLC 4% gel, 0.5 to 1% CFR-2 6 lb Gilsonite, 2% HR-4 0.2% HR-4	11.5	2.81	5.29	1.88
Class H 30% silica flour 1% CFR-2; 1.2–1.5% HR-12 0.25 lb/sk NF-P	16.0–16.2	1.40	7.96	5.68
**80–20 Pozmix cement 2% gel, 18% NaCl 30% silica flour 2% Halad 14, HR-12 0.25 lb/sk NFP	14.5	1.84	10.52	5.17

*Halliburton Light Cement
**80% cement–20% Pozmix cement

permit turbulent flow at reasonable displacement rates should be considered.

Slurry Mixing—Batch mixing of cement slurry promotes uniformity of mixing, and permits actual tests of slurry properties and, if necessary, adjustments of properties before pumping slurry into well.

Strength Retrogression—To inhibit strength retrogression where formation temperatures are above 230°F., 30 to 40 percent by weight silica flour should be used. Water requirement for normal cement grade silica flour (less than 200-mesh particle size) is 4.8 gal/100 lb silica. For high weight cement slurries, coarse silica (60–140 mesh) can be substituted to reduce the additional water needed.

Mud Displacement—Adequate displacement of mud by the cement slurry is a major problem due to high temperature gelation of the mud, small annular clearances, and difficulty in moving a long heavy string of casing or liner. Aids in displacement of mud include:

1. Centralization of casing.
2. Movement of casing.
3. Reduction of mud plastic viscosity and yield point (PV < 12 cp; YP < 5 lb/100 ft²).
4. Laboratory model test simulating high temperature conditions show that invert-emulsion mud is displaced much more effectively than water base

mud where high temperature gelation is a problem.

5. Slurry displacement rates as high as possible into the turbulent flow range.

6. Use of sufficient volume of cement to insure that slower moving cement on the "narrow side" of the annulus rises to the desired height in the hole.

Steam injection wells normally have low temperature during placement of cement. Injection of steam (400° to 700°F.) later creates extreme stresses down hole on casing and cement. Best practices include:

1. Use of heavier casing—K-55, N-80, or P-105—plus special threads in some cases.

2. Injection of steam down tubing leaving empty annulus as an insulator.

3. Circulate cement to surface.

4. Cement bottom section of casing—pull tension on casing—cement remainder of casing through DV tool.

5. Use API Class A. G, or H cement (or Pozzolan cement) with 40% silica flour to prevent strength retrogression. See Table 4-21.

Geothermal steam wells present very high temperature conditions during placement of cement (300° to 700°F.). It is desirable that cement have good insulating properties to reduce heat loss to formation.

A satisfactory slurry can be designed using API Class G basic cement, with 40% silica flour to prevent strength retrogression and sufficient retarder to allow placement. Slurry should be pretested in laboratory to determine retarder requirements. See Table 4-22.

Expanded Perlite can be added to improve insulating properties (reduce "*K*" value) but it reduces compressive strength. See Table 4-23.

Fire flood wells are cemented under low temperature conditions, but can be subjected to very high temperature (750° to 2,000°F.).

TABLE 4-22
API Class G Cement
For Geothermal Steam Wells
(Cured at 440°F.: 1, 3, 7 Days)

% Silica flour	Perlite, cu ft/sk	Compressive strength—psi		
		1 day	3 days	7 days
0	0	545	545	425
40	0	7,330	11,025	10,010
40	1	3,690	3,580	3,975
40	3	1,690	1,734	1,825

TABLE 4-23
Thermal Insulating Values
API Class A Cement with Various Additives

Water, gal/sk	5.20	6.80	12.50	24.30	14.60
Bentonite %	—	—	2.0	2.0	—
Silica flour %	—	40.	30.	30.	30.
Perlite, cu ft/sk	—	—	1.0	3.0	—
Vermiculite %	—	—	—	—	30.
"*K*" Value—BTU/sq ft/hour/°F./ft					
24 hours	1.068	.955	.535	.653	.324
48 hours	1.005	.828	.420	.515	.268
120 hours	.803	.475	.400	.332	.324
144 hours	—	—	.319	.333	—
329 hours	—	—	—	—	.320

If temperatures higher than 750°F. are anticipated, and cracking of cement will be detrimental, a refractory-type cement (calcium-aluminate) containing 40 to 60% silica flour should be used through the hot zone. A cheaper cement can be used 150 ft above the hot zone. See Table 4-24.

Cementing in Cold Environments

Cementing in frozen formations such as the Permafrost in the northern areas of Canada, the

TABLE 4-21
API Class G Cement with Silica Flour
(Cured 3 days @ 80°F., thereafter @ 600°F.)

% Silica flour	Compressive strength—psi			
	0 days	1 day	14 days	28 days
40	2,610	3,380	3,375	3,165
50	2,385	3,212	3,015	2,925
60	2,160	2,950	2,780	2,865

TABLE 4-24
Calcium-Aluminate Cement
With Silica Flour for Fire Flood Application

% Silica flour	Compressive strength—psi 7 days curing time at oven temperature		
	700°F.	1,000°F.	1,500°F.
40	1,240	1,130	1,530
60	1,120	950	1,020

Arctic Islands and the Alaskan North Slope presents problems depending on the type of permafrost.

Consolidated Formation—A frozen consolidated formation that is unharmed by thawing can usually be cemented with a variety of slurries that will adequately set up at the curing temperature available.

Accelerated API Class A, C, or G Cement with a slurry temperature of 50°F. to 60°F., preferably densified, could be used. Subjected to 20°F. formation temperature, unheated Class G Cement with up to 4% $CaCl_2$ freezes before it sets up.

Subsequent thawing and hydration of the slurry results in great loss of strength and increase in permeability, thus slurry heating to maintain 40°F. is justified.

Figure 4-17 shows slurry temperature resulting from mixing cement of a particular temperature with water having a particular temperature. In situations where it is not possible to maintain slurry temperature above 40°F., gypsum-cement blend (Permafrost Cement) should be used.

High-aluminite cement (Ciment Fondu) has been used. Aluminite cement is quite expensive but will set up at low temperatures (20°F.–25°F.) to provide minimum compressive strengths of 300–500 psi in 24 hours. If allowed to freeze, however, permeability will be very high.

Unconsolidated Formations—In areas where the shallow formations consist of ice lenses, frozen muskeg, and unconsolidated sands and gravels, requirements for successful cementing are more difficult:

—Hole enlargement due to melting and erosion during drilling must be prevented.

—Cement must set at low temperatures without requiring excessive heating of mix water.

—Cement must set without excessive heat of hydration to prevent melting of permafrost.

—Cement sheath must be thick enough to provide insulation to prevent thawing of permafrost by circulation of mud during subsequent drilling.

—Displacement of mud by cement is very essential to eliminate freezable fluids in annulus.

Aluminite cement (Ciment Fondu) does not work well under these conditions because it liberates a large quantity of heat (90 BTU/lb of slurry) even when diluted with 50% fly ash.

The best answer to cementing unconsolidated permafrost appears to be a gypsum-cement blend (Halliburton tradename—Permafrost Cement). Heat of hydration is less than 20 btu/lb of slurry. Salt can be added to lower freezing temperature to about 20°F. Slurry properties can be adjusted to give two hours minimum pumpability at temperatures from 30°F. to 60°F. with a 16-hour compressibility strength of 500 psi at 20°F. See Figure 4-18.

Permeability of any cement is increased significantly if it is allowed to freeze before setting. Ciment Fondu allowed to freeze has a permeability to water of 50 to 60 md. Permafrost Cement under similar conditions has a permeability of about 25 md; thus, where permeability is important, the slurry must not be allowed to freeze.

Cementing Through Gas Zones

Gas communication through a cemented annulus was first recognized in the completion of gas storage wells. The problem became more evident in deep well completions across gas intervals, causing a pressure buildup in the annuli of the production and intermediate casings, and on occasion a blowout.

Laboratory model studies simulating downhole conditions have been helpful in understanding the

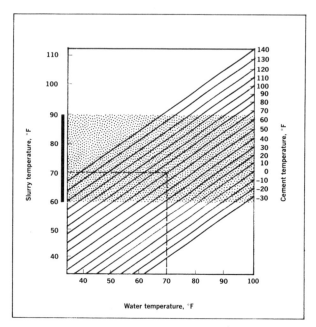

FIG. 4-17—*Slurry temperature for various temperatures of water and cement.*[19] *Permission to publish by The Society of Petroleum Engineers.*

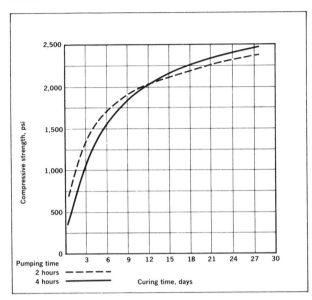

FIG. *4-18—Compressive strength of Alaskan permafrost cement, 20° to 80°F. Courtesy of Halliburton Services.*

occurrence of gas leakage and determining some of the factors responsible for gas migration in wells. These studies show that leakage is dependent upon hole conditions, mud-cement density, temperature, pressure differential and cementing composition.

Leakage appears to be related to the inability of the cement column to effectively transmit full hydrostatic pressure to the gas zone.

One explanation may be that (a) local dehydration of cement slurry opposite uphole permeable zones, or (b) higher temperatures off bottom due to the circulating process, may accelerate the initial setting process in a section of the cement column above the gas formations. As this upper portion of the cement column takes its initial set it begins to partially support the hydrostatic column above.

As cement filtrate is lost into lower zones the hydrostatic pressure on the gas zone is reduced allowing gas bubbles to enter the annulus and migrate upward disturbing the setting cement and developing communication "worm holes" for passage of additional gas.

To minimize gas leakage, consideration should be given to:

1. Sufficient mud circulation prior to cementing to insure that dispersed gas is not trapped in the system.

2. Mix cement slurry as heavy as hole conditions will tolerate.

3. Use filtration control slurry to keep cement from dehydrating.

4. Do not over-retard cement slurry. Have accurate well temperature information so that the cement can be designed to set as quickly as possible after placement.

A typical slurry composition designed for cementing deep abnormal pressure gas wells in South Texas consists of Class H Cement with these additives:

—25 lb/sk hematite ore—to increase density.

—30% silica 40–170 mesh—for strength retrogression and fluid loss control.

—1¼% CFR-2—for viscosity control.

—HR-12 retarder—as needed to permit placement under formation temperature conditions.

—0.3 NFP—antifoam powder to eliminate an entrainment.

Mixed with a low-water ratio this composition results in a slurry having a density of 19.0 lb/gal and a fluid loss of about 20 cc at 1,000 psi on 325-mesh screen for 30 minutes.

Salt Cement

Cementing Through Salt Sections—Salt-saturated cement (NaCl) has been used for many years to improve bonding through massive salt sections. Fresh water slurries will not bond to salt since water from the slurry dissolves away the salt at the interface.

Shale Sections—Currently there is a trend toward use of salt cement (10% to 18%) to improve bonding through shale sections where contact with fresh water from the slurry may cause disintegration of shales. Normally inhibition of clay disintegration can be obtained with less sodium chloride than that required for saturation.

Sixteen to 18% sodium chloride appears to be optimum since this concentration will inhibit clay hydration, but will not significantly affect thickening time or reduce the 24-hour compressive strength of the cement. See Figure 4-19.

Compatibility With Other Additives—Sodium chloride is generally compatible with all light weight and heavy weight additives, accelerators and retarders, and other special additives. Some fluid-loss

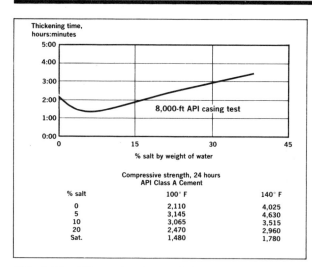

FIG. 4-19—Effect of salt on thickening time and 24-hour compressive strength. Courtesy of Halliburton Services.

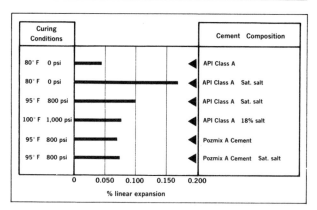

FIG. 4-20—Linear expansion of salt cement, 28-day curing. Courtesy of Halliburton Services.

additives are adversely affected by the chloride ion; thus, to obtain minimum fluid loss a slight increase in fluid-loss additive is required. See Table 4-25.

Expansion—Salt-saturated cement undergoes a slightly greater expansion upon setting than API Class A cement, as shown on Figure 4-20.

Rheology Improved—Sodium chloride acts as a dispersing agent to improve rheological properties and to reduce the displacement rate required for turbulent flow. Sodium chloride also increases slurry weight and volume slightly. See Figure 4-21.

Blending Problems—Bulk blending of dry granulated salt with cement greatly simplifies mixing problems. NFP is recommended to minimize foaming. Mixing cement with saltwater causes excessive foaming; if dry blending equipment is not available a defoamer such as tributylphosphate (Halliburton's

NF-1) should be used in a concentration of 1 pint/10 bbl of saltwater.

Fine granulated salt (20 to 100-mesh size) should be used in bulk blending. For a salt saturated slurry 3.1 lb of NaCl/gal of fresh mix-water is required. Figure 4-22 can be used to determine the amount of salt for other degrees of saturation. Normally speaking every 1/2 lb of salt added/gal of water is equivalent to 6 wt % salt in water.

Substitution of Potassium Chloride—Potassium chloride can be substituted for sodium chloride in cement slurries to provide similar effects at lower concentrations. At high concentrations (above 10%) excessive slurry viscosities may result, however.

TABLE 4-25
Effect of Salt on Fluid Loss—Class A Cement

	Filtration Rate (cc/30 minutes) 325-mesh screen at 1000 psi	
% Salt	1.0% Additive	1.2% Additive
0	80	35
5	84	44
10	80	40
15	78	48
Sat.	197	58

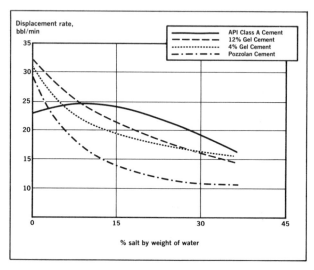

FIG. 4-21—Displacement rate for turbulent flow (8 3/4-in. hole, 5 1/2-in. casing. Courtesy of Halliburton Services.

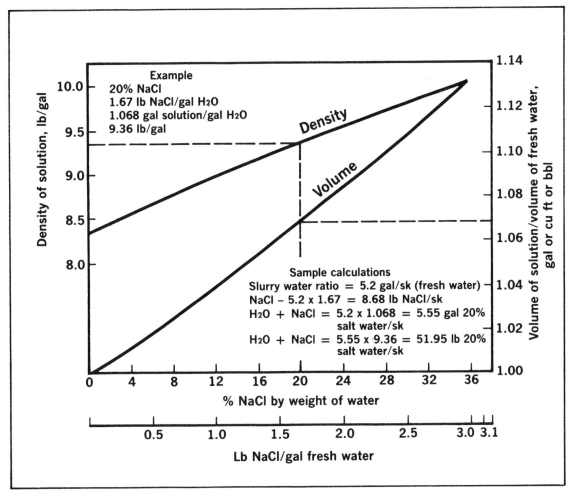

FIG. *4-22—Properties of sodium chloride solutions at 68°F, based on percent salt by weight of water for salt cement slurries. Courtesy of Halliburton Services.*

Delayed Setting Cement

With proper fluid loss control and retarder, cement slurry can be designed to remain fluid for periods up to 36 hours. Delayed set permits pipe to be run into unset cement. Principal applications are:

1. Multiple tubingless completions to improve primary cement job between strings.
2. Liner cementing where small clearance and small volume of cement makes satisfactory cementing difficult.

Expansive Cements

API Class A, G, or H cement cured under moist conditions and recommended water ratios expand slightly when set. Pressure decreases expansion, whereas temperature increases expansion, particularly from heat of reaction.

Chemical additives beneficial in increasing cement expansion include salt (NaCl), calcium or sodium sulfates, and pozzolan. Gas-forming agents (aluminum, iron, zinc, or magnesium powder) may be beneficial at low hydrostatic pressure.

For oil well use, salt cement (18% to saturation) appears to be a good way to realize benefits from expansion plus providing other benefits. (See Figure 4-20). Cement plus gypsum (Cal Seal) and Pozmix cement also have application in oil wells.

Commercial expanding cements (trade name Chem Comp) made specifically for concrete and construction purposes have 0.05 to 0.15% expansion according to their manufacturer. This has been used in oil wells, but is expensive and benefits are questionable.

Densified Cement

Densified cement—or cement mixed with less than normal water (about 3.5 gal/sk rather than 5.2 gal/sk) and with friction reducer (CFR-2) added to improve pumpability—has several desirable properties including:

—High compressive strength.
—Relatively low fluid loss.
—Resistance to mud contamination.

Typical properties are shown in Table 4-26.
Ability to tolerate mud contamination is shown by the data in Table 4-27.

Nylon-Fiber-Reinforced Cement ("Tuf Cement")

The addition of small quantities of selected fibers to most oil well cementing compositions minimize shattering due to shock waves created by the forces of the perforator. Nylon or polypropylene fibers ($\frac{1}{8}$ to $\frac{1}{4}$ lb/sk) will transmit these localized stresses more evenly throughout the cement and thus improves the impact and shatter resistance of the set cement.

Studies on fiber-reinforced cementing slurries illustrate these advantages:

—Cement will deform with a minimum of shattering.
—Fibers reinforce the cement by transferring the stresses evenly throughout the set cement.
—Less chance for damage to cement and casing during perforating or subsequent remedial work.
—Better zone isolation because of increased cement integrity in the wellbore.

TABLE 4-26
API Class A-B-G or H Cement

Weight, lb/gal	% CFR-2	Fluid loss, 1000 psi-screen	Compressive strength, 100°F.—3000 psi
15.6	0.0	1000 cc+	1465 psi
15.6	1.0	286 cc	1445 psi
16.5	1.0	192 cc	3625 psi

TABLE 4-27
Effect of Mud Contamination on Strength of Cement

% Mud contamination	12 hr-compressive @ 230°F.	
	15.6 lb/gal	17.6 lb/gal*
0	2910	7919
10	2530	5005
30	1400	2910
60	340	2315

*Densified cement w/CFR-2

Thixotropic Cement

Thixotropic cement consists of Class A, G, or H cement with the addition of bentonite, Cal-Seal and calcium chloride to provide rapid buildup of gel strength as soon as movement of the slurry stops.

Primary application is in shallow low temperature (50°–120°F.) zones where natural or induced fractures otherwise permit loss of whole cement to the formation. Typical properties are shown in Table 4-28.

PRIMARY CEMENTING PRACTICES
Practical Considerations Before Cementing

Mud Conditioning—Mud should be conditioned before running casing at pumping rate equal or greater than when drilling in order to clean hole. Drill pipe rotation aids hole cleaning. Plastic viscosity and yield strength of mud should be as low as possible.

After casing is on bottom, mud should be circulated and pipe rotated to break gel strength.

Centralization— Effective centralization is a critical factor in obtaining a good primary cement job.

In straight hole use one centralizer per joint, 200 ft above and below pay zone. Use one centralizer every third joint in remainder of cemented zone. See Table 4-29 for API specifications for centralizers.

Centralizers should be placed in gauge sections of the hole as determined by caliper or other logs.

In crooked hole centralizer placement depends on hole deviation. See Figure 4-23.

Starting force—maximum force should be less than weight of joint of 40-ft casing between centralizers.

Scratchers—Wall cleaners or "scratchers" prob-

TABLE 4-28
Properties of Thixotropic Cement—Class A
(Thickening Time 60 Min)

Water gal/sk	Bentonite %	Cal-Seal lb/sk	CaCl$_2$ %	Slurry wt lb/gal	Compressive strength psi @ 80°F.		
					6 hr	12 hr	24 hr
7.8	4	8	2.0	14.4	160	475	1120
13.0	12	10	1.0	12.7	45	145	460
15.6	16	10	1.0	12.3	30	110	205

TABLE 4-29
API Specifications for Centralizers

Casing size, in.	Wt. of casing, lb/ft	Minimum restoring force, lb
4½	11.6	464
5½	15.5	620
7	26.0	1,040
9⅝	40.0	1,600

Starting force—maximum force should be less than weight of joint of 40-ft casing between centralizers.

ably help in eroding wall cake and improving bond, although the need for wall cleaners in this regard is difficult to prove in the field. In deep holes complete wall cake removal may increase chance of sticking the pipe.

Recent lab tests closely simulating downhole conditions indicate that the important function of scratchers is not to "remove the wall cake," but to break the gel of the mud and to mix up the mud with the cement, thus reducing channeling.

On this basis scratchers should be given serious consideration. They should perhaps be renamed "stirrers" instead of scratchers to more properly reflect their actual downhole effect.

Floating Equipment—Float shoe (with back pressure valve) minimizes derrick strain, and prevents cement back-flow when pressure on casing is released after cement is in place.

Float collar should be located one or two joints above the float shoe to prevent mud-contaminated cement (which may collect below the top wiper plug) being placed outside the bottom casing joint.

Differential fillup equipment can often justify additional cost by reduced downtime to fill casing, and reduced pressure surge on formation.

Casing-Running Practices— Where lost circulation is not a problem casing can be run at a rate of 1,000 ft/hour or faster. Magnitude of pressure surges should determine running time where lost circulation is a possibility.

Volume of Cement—Circulate cement at least 500 ft above top producing formation. Corrosion protection or lost circulation zones may dictate more or less cement. Where possible a larger volume of cement, placed in turbulent flow, reduces possibility of mud channels remaining in the productive zone. Use of lower density slurries permit longer cement columns.

Hole washout must be considered to determine the cement volume to provide a desired length of fillup. With a caliper survey to determine annular volume between casing and hole, add 10% to 15% for cement volume. With no caliper survey, add 25% to 100% depending on experience.

Considerations During Cementing

Cement Mixing—Quality of the cement mixing operation correlates with job success. Weight of cement slurry should be monitored (and recorded) to insure that the correct quantity of mix water is used.

The last volume of cement mixed will be at the casing shoe; thus, particular care should be exercised to insure that it meets desired specifications.

Where densified or high weight slurries are used, particularly at displacement rates less than five barrels per minute, a recirculating-type jet mixer improves uniformity of slurry compared with the standard jet mixer.

For critical cementing jobs, batch mixing provides much greater slurry precision, and permits measuring and adjusting rheological or other properties

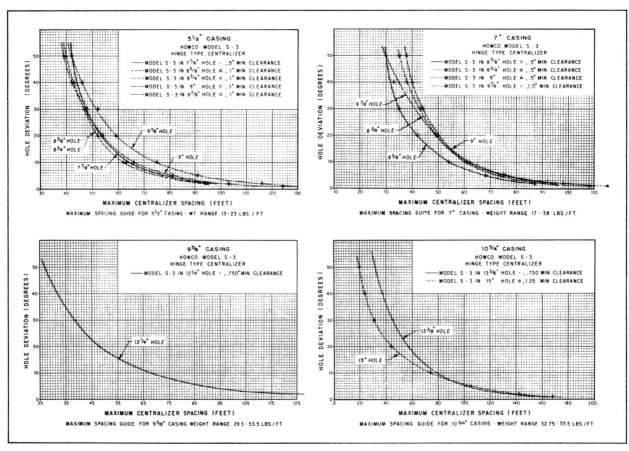

FIG. 4-23—Centralizer spacing in deviated hole. Courtesy of Halliburton Services.

of the slurry before it is pumped into the well. Several large batch tanks permit continuous operation.

Washes Ahead of Cement—Water if used in sufficient quantity is an excellent wash since it is cheap, easy to put into turbulence, and doesn't affect cement setting time. Fifty bbl of water (300–500 ft of annular fill) should be used unless hydrostatic head is reduced excessively.

Some mud thinners (quebracho, lignosulfonates) added to water may retard (or inhibit entirely) cement setting.

Dilute mix of Portland or Pozzolan cement for scavenging purposes is an excellent preflush since it is easy to put in turbulent flow and solid particles promote erosion of gelled mud and filter cake.

Acetic acid (10%) with corrosion inhibitor and surfactant may be beneficial. Acetic acid should not cause pitting corrosion.

Hydrochloric acid (5% to 10%) with corrosion inhibitor and surfactant is sometimes used; but pitting corrosion is possible. Acid should be separated from cement by water spacer.

Non-acid washes, such as mud thinner and surfactant (Halliburton Mud Flush), may be beneficial where large volumes of water would reduce hydrostatic head.

Thickening behavior can be caused when cement and some invert oil systems come in contact. Diesel oil with a water wetting surfactant should be used as spacer, followed by a flush in sufficient quantity to water-wet the pipe.

Cementing Wiper Plugs—Top plug separates mud and cement, and provides shut-off when cement is in place.

Bottom plug should be used to wipe mud off casing ahead of cement, as well as to separate mud and cement. Without bottom plug, mud film wiped by top plug accumulates ahead of top plug as shown in Table 4-30.

Two plug containers should be used to facilitate release of plugs without delay.

TABLE 4-30

Mud film thickness, in.	Fillup/1,000 ft in casing ahead of top plug 5½-in. casing	7-in. casing
1/16	50.6 ft	40.0 ft
1/32	25.5 ft	20.1 ft
1/64	12.6 ft	9.9 ft

Displacing Fluid Behind Top Plug—Mud is normally used on surface or intermediate casing although fresh water may be better, depending on mud program for drilling deeper.

For the production casing freshwater, saltwater, or seawater could be desirable, depending on completion program. Selection should minimize formation damage and completion time. Diesel oil might be used to reduce swabbing time.

Sugarwater or retarding additive is sometimes placed immediately above top plug in small diameter casing to inhibit setting of cement that may have bypassed top plug.

Casing Movement During Cementing—Casing movement rates high on the list of factors affecting successful mud displacement. Ideally casing should be reciprocated (or rotated) until the top plug reaches bottom. Pipe movement may be desirable after cement is in place.

Frictional drag, weight of the pipe, and differential sticking are factors acting to prevent casing movement. Differential sticking, often the important factor, is a function of:

—Contact area between pipe and permeable borehole wall.
—Pressure difference between mud column and formation.
—Available pulling force.
—Sticking coefficient.

Sticking coefficient is a function of the mud properties (primarily water loss) and the time that the casing remains stationary against a permeable formation. By reducing "stationary time" through fast cementing head hook-up procedures after casing is on bottom, differential sticking can often be eliminated.

Casing movement (reciprocation or rotation) during cement displacement should be slow as the cement first reaches bottom but should increase as the cement is displaced and the top plug reaches

bottom. Rate of reciprocation should be on a two-minute cycle over 15 to 20-ft intervals. Rotation can be done with hydraulic casing tongs to limit torque to that used to make up casing.

Considerations After Cementing

Casing Pressure—Casing pressure should be released after it is determined that back pressure valve in float collar and guide shoe are holding. This prevents casing expansion and improves bonding.

In small diameter casing (tubingless completions) it may be desirable to hold pressure on casing, as a means of applying additional tension (to prevent subsequent buckling) if casing is landed before cement takes initial set. Bonding may be adversely affected, but hoop expansion due to differential pressure is low in small diameter pipe.

Waiting-On-Cement Time—Reason for WOC time is to permit cement to attain sufficient strength to:

—Anchor the pipe and withstand shock of subsequent operation.
—Seal permeable zones, (and confine fracture pressures).

Tests show that tensile strength of 8 psi (compressive strength of 50 psi) is sufficient for anchoring in most situations. See Table 4-31.

Compressive strength of 130 psi with Portland cement provides sufficiently low permeability for sealing.

TABLE 4-31
Length of Casing Supported by a 10-ft Column of 8-psi Tensile Strength Cement

Casing OD, in.	Casing weight, lb/ft	Length supported by 10 ft of cement, ft
4½	9.5	367
	11.6	301
5½	15.5	275
	17.0	251
	20.0	213
7	20.0	271
	26.0	209
	32.0	170
10¾	40.5	206
	51.0	163

(After Bearden and Lane)

WOC time often based on time required for cement to attain 500 psi compressive strength. This allows safety factor of 2 to 5.

Lab and field correlations show that generally Portland cement achieves tensile strength of 8 psi in 1.5 times the period required to reach maximum temperature due to heat of hydration of the cement. As a practical method, temperature increase of the fluid inside the casing due to the setting cement can be observed on the rig floor by flow back of fluid from the casing.

Casing Landing Practices—Casing landing practices based on API Study Group Recommendations are outlined below:

In general, it is recommended that all casing be landed in the casing head at exactly the position in which it was hanging on the hook when the cement plug hit bottom. In other words, land the casing as cemented with the only movement of the pipe being that necessary to transfer the weight of the casing to the well head or casing hanger.

This recommendation applies to all wells where mud weights do not exceed 12.5 lb/gal and where:

—Standard design factors are used in tension and collapse.

—Wellhead equipment is available to permit hanging weight equal to the tensile strength of the casing on the hanger without damage to the casing.

—The joint strength, in compression, of the top section of the surface casing is sufficient to withstand the loads imposed by landing the casing as cemented plus the weight of the tubing, plus induced loads that may be brought about by future operations.

The ability of the surface formations surrounding the surface casing to help support the loads imposed on the top of the surface casing is an important factor to be considered in designing the top section of the surface casing to withstand these loads.

Special cases are those where casing is to be set:

—In extremely deep wells where the standard design factors are of necessity reduced.

—Where extreme top-hole operating pressures are anticipated.

—Where excessive mud weights are necessary.

—Where other unusual circumstances dictate special consideration.

In these cases it is the opinion of the committee that the landing practice should be based on calculations as outlined by Lubinski. Several operators have used Lubinski's work to arrive at rather simple formulas that are considered adequate for their use on an area or field-wide basis.

Testing of Primary Cement Job—Temperature survey, run two to four hours after cementing, is an excellent method of locating cement top.

Bond logs should normally be used only to make special evaluation studies.

Assume a cement job is satisfactory unless there is good evidence to the contrary.

REFERENCES

1. Morgan, B. E. and Dumbauld, G. K.: "A Modified Low Strength Cement," AIME Trans., (1950), Vol. 192.

2. Owsley, W. D.: "Twenty Years of Oil Well Cementing," J. Pet. Tech., September, 1953, p. 17.

3. Smith, D. K. and Carter, Greg: "Properties of Cementing Compositions at Elevated Temperatures and Pressures," AIME Trans., (1958), Vol. 213.

4. Ostroot, G. W. and Walker, W. A.: "Improved Compositions for Cementing Wells with Extreme Temperatures," J. Pet. Tech., March 1961, p. 277

5. Evans, G. W. and Carter, G. L.: "Bonding Studies of Cementing Compositions to Pipe and Formations," Drilling Production Practice (1962).

6. Walker, W. A.: "Cementing Compositions for Thermal Recovery Wells," J. Pet. Tech., February, 1962, p. 139.

7. Bleakley, W. B.: "A Really Engineered Cement Job," Oil and Gas Journal, (February 12, 1962).

8. Slagle, Knox A.: "Rheological Design of Cementing Operations," J. Pet. Tech., March, 1962, p. 323.

9. Pettiette, Roy and Goode, John: "Primary Cementing of Multiple Tubingless Completions," Southern District API, (March, 1962).

10. Slagle, K. A. and Smith, D. K.: "Salt Cement for Shale and Bentonitic Sands," J. Pet. Tech., February, 1963, p. 187.

11. Harris, Francis and Carter, Greg: "Effectiveness of Chemical Washes Ahead of Squeeze Cementing," Mid-Continent District API, (March, 1963).

12. Carter, L. G. and Evans, G. W.: "A Study of Cement Pipe Bonding," J. Pet. Tech., Feb. 1964, p. 157.

13. Buster, John L.: "Cementing Multiple Tubingless Completions," Southwestern District API, (March, 1964).

14. Carter, L. G. and Evans, G. W.: "New Technique for Improved Primary Cementing," Southwestern District API, (March, 1964).

15. Brice, J. W., Jr., and Holmes, R. C.: "Engineering Casing Cementing Programs Using Turbulent Flow Techniques," J. Pet. Tech., May 1964, p. 503.

16. Parker, P. M., Ladd, B. J., Ross, W. M., and Wahl,

W. W.: "An Evaluation of a Primary Cementing Technique Using Low Displacement Rates," SPE Denver (1965).

17. McLean, R. H., Manry, C. W., and Whitaker, W. W.: "Displacement Mechanics in Primary Cementing," J. Pet. Tech., Feb. 1967, p. 251.

18. Childers, Mark A.: "Primary Cementing of Multiple Casing," J. Pet. Tech., July 1968, p. 751.

19. Maier, L. F., Carter, M. A., Cunningham, W. C., and Bosley, T. G.: "Cementing Practices in Cold Environments," J. Pet. Tech., Oct. 1971, p. 1,215.

20. Garvin, Tom, and Slagel, Knox A.: "Scale Displacement Studies to Predict Flow Behavior During Cementing," J. Pet. Tech., Sept. 1971, p. 1,081.

21. Carter, Greg, and Slagle, Knox A.: "A Study of Completion Practices to Minimize Gas Communication," J. Pet. Tech., Sept. 1972, p. 1,170.

22. Clark, Charles R., and Carter, Greg L.: "Mud Displacement with Cement Slurries," J. Pet. Tech., July 1973, p. 775.

23. Graham, Harold L.: "Rheology-Balanced Cementing Improves Primary Success," Oil and Gas Journal, December 18, 1972.

24. Carter, Greg L., and Cook, Clyde: "Cementing Research in Directional Gas Well Completions," London SPE (April, 1973), SPE Paper 4313.

25. Clark, Charles R., and Jenkins, Robert C.: "Cementing Practices for Tubingless Completions," Las Vegas, SPE, October, 1973, SPE Paper 4609.

26. "Specification for Oil Well Cements and Cement Additives," API Standard 10A Eighteenth Edition, (January, 1974).

27. "Running and Cementing Liners in the Delaware Basin, Texas," API Bulletin D-17, December, 1974.

28. Holley, J. A.: "Field Proven Techniques Improve Cementing Success," World Oil, August 1, 1976.

29. Smith, Dwight K.: "Cementing," Monograph Volume 4, Henry L. Doherty Series, SPE of AIME 1976.

30. API Recommended Practice for Testing Oil Well Cements and Cement Additives, Twentieth Edition, April, 1977.

Appendix
Cement Additives Comparison Chart

PRODUCT CLASSIFICATION	HALLIBURTON	DOWELL	B.J.	WESTERN	CHEMICAL OR MATERIAL DESCRIPTION
Accelerators	CaCl$_2$	S-1	A-7	CaCl$_2$	Calcium Chloride
	HA-5	D-43	A-8	WA-4	Blend of Inorganic Accelerators
	D-12	A-2	Diacel A	Diacel A	Diacel A
	Salt	Salt	A-5	Salt	Sodium Chloride-Granulated
H-TLW Blends	1:1 (etc.)	1:1(etc.)		1:1 Talc	
Fluid Loss	Halad 9,11,14	D-60	Aquatrol 13,15	CF-1	Low Temp. Fluid Loss Control
		D-59		CF-2	Low Temp. Fluid Loss Control
	Diacel LWL	D-8	R-6	Diacel LWL	Carboxymethyl Hydroxyethyl
Liquid		D-73			Cellulose
Turbulence Inducers	CFR-2	D-65,45	Turbo-Mix D-16	TF-4	Polymer
	CFR-1		Turbo-Mix D-30	TF-5	
Weighting Material	Barite	D-31	W-1	Barite	Barite
	Hi Dense 3	D-76	W-2	WM-2	Hematite
	Hi Dense 2	D-18	W-3	Ilmenite	Ilmenite
Spacers & Washes	Mud-Flush	CW7	Mud-Sweep	WMW-1	Mud Thinner-Spacer
	Sam 4	Oil-Base Mud Spacer	J-22 & D-4	ASP-4	Oil Base Spacer
				ASP-4	Water Base Spacer
Latex	LA-2	D-15, D-78	D-5	CLX-1	Latex Cement
Extenders	Howco Gel	Bentonite	B.J. Gel	Bentoment	Bentonite
	Gilsonite	Kolite	D-7	Gilsonite	Gilsonite
	Econolite		Lo-Dense	Thrifty-Lite	Anhydrous Sodium Meta-Silicate
	Pozmix A	Litepoz 3	Diamix A, G, M	Pozment A	Artificial Pozzolans
		Litepoz 1	Diamix A, M, G	Pozment N	Natural Pozzolans
	Pozmix 140	Litepoz 180	Thermoset		Pozzolan - Lime Mixtures
	Howcolight HLC	D-79	Lo Dense	Thrifty-Ment	H-Poz Blends
Anti-Foam	NF-P	D-46	D-6	AF-4	Powdered Anti-Foam Agent
	NF-1	D-47	D-6	AF-L	Liquid Anti-Foam Agent
Mud Decontaminant	Mud Kil-1	K-21	Firm Set I	Shur Set I	Mud Kill Patented by Gulf Oil
	Mud Kil-2	K-21	Firm Set II	Shur Set II	Mud Kill Patented by Gulf Oil
Silica Sand	Silica Flour (Reg.)	J-84	D-8	SF-3	325 Mesh Silica Flour
	Silica Flour (Coarse)	D-30		SF-4	Okla. #1 Sand
Thixotropic Cement	Thixotropic Cement	Reg. Fill-Up Cmt.		Thixoment	Thixotropic Slurries
Lost Circulation	Gilsonite	Kolite D2Y	D-7 Gilsonite	Gilsonite	
	Cellophane Flakes	D-29 Jel Flakes	Cello-Flake	Cell-O-Seal Kwik-Seal	
Retarders	Kembreak	Kembreak	Kembreak	WR-1	Low Temp. Retarder
	HR-4	D-22	Retroset 2	WR-2	Low Temp. Retarder (Calcium Lignosulfonate)
	HR-7	D-13	Retroset 5	WR-4	Low Temp. Retarder
	HR-12	D-28	Retroset 8	WR-6	High Temp. Organic Retarder
	Diacel LWL	D-8	Retroset 6	Diacel LWL	Diacel LWL Carboxymethyl Hydroxethyl Cellulose
	HR-20	D-99	R10, R11		High Temp. Retarder
		D-93		WR-7	Borax

Chapter 5 Well Completion Design

Factors influencing design
Conventional tubing configurations
Unconventional tubing configurations
Sizing production tubulars
Completion interval selection
Tubingless completion techniques

The individual well is much more than "just an expensive faucet." It is our only communication with the reservoir. The effectiveness of that communication is a large factor in reservoir drainage as well as overall economics. Wells represent the major expenditure in reservoir development. Oil wells, gas wells, and injection wells present unique problems depending on the specific operating conditions. The individual well completion must be designed to yield maximum overall profitability on a field basis.

FACTORS INFLUENCING WELL COMPLETION DESIGN

The ideal completion is the lowest cost completion (considering initial and operating costs) that meets or nearly meets the demands placed upon it for most of its life. To intelligently design a well completion, a reasonable estimate of the producing characteristics during the life of the well must be made. Both reservoir and mechanical considerations must be evaluated.

Reservoir Considerations

Reservoir considerations involve the location of various fluids in the formations penetrated by the wellbore, the flow of these fluids through the reservoir rock, and the characteristics of the rock itself.

Producing rate to provide maximum economic recovery is often the starting point for well completion design. Among other factors producing rate should determine the size of the producing conduit.

Multiple reservoirs penetrated by a well pose the problem of multiple completions in one drilled hole. Possibilities include multiple completions inside casing separated by packers, or several strings of smaller casing cemented in one borehole to provide in effect separate wells. Other possibilities include commingling of hydrocarbons from separate reservoirs downhole, or drilling several boreholes from one surface location.

Reservoir drive mechanism may determine whether or not the completion interval will have to be adjusted as gas-oil or water-oil contacts move. A water drive situation may indicate water production problems. Dissolved gas drive may indicate artificial lift. Dissolved gas and gas drive reservoirs usually mean declining productivity index and increasing gas-oil ratio.

Secondary recovery needs may require a completion method conducive to selective injection or production. Water flooding may increase volumes of fluid to be handled. High temperature recovery processes may require special casing and casing cementing materials.

Stimulation may require special perforating patterns to permit zone isolation, perhaps adaptability to high injection rates, and a well hookup such that after the treatment the zone can be returned to production without contact with killing fluids. High temperature stimulation again may require special cementing procedures, casing and casing landing practices.

Sand control problems alone may dictate the type of completion method and maximum production rates. On the other hand, reservoir fluid control problems may dictate that a less than desirable type of sand control be used. Sand problem zones always dictate a payoff from careful well completion practices.

Workover frequency, probably high where several reservoirs must be drained through one wellbore, often dictate a completion conducive to wireline or through-tubing type recompletion systems.

Artificial lift may mean single completions even where multiple zones exist, as well as larger than normal tubulars.

Mechanical Considerations

The mechanical configuration or "well hookup" is often the key to being able to deplete the reservoir effectively, monitor downhole performance, and modify the well situation when necessary.

The mechanical configuration of the well is the key to being able to do what ought to be done in the well from the standpoint of controlling the flow of reservoir fluids, oil, gas, and water.

Formation damage is related to the well hookup, both minimizing damage initially and relieving the effects of damage later.

Mechanically, well completion design is a complex engineering problem. Basic philosophy is to design to specific well conditions, field conditions, and area conditions.

1. *Maximize profit* considering the time value of money. Economics are sometimes best served by delaying expenditures, particularly in wells where servicing is frequent. The isolated well is the one you can afford to provide with maximum flexibility for the future.

2. *Keep the installation simple,* both from equipment and procedural standpoints—consider level of operator skill available.

3. *Overall reliability* depends on reliability of individual components and the number of components. Design out maintenance, limit moving parts, avoid debris traps. As complexity increases, provide alternatives.

4. *Anticipate all operating conditions,* and associated pressure and temperature forces.

5. *Safety* must be designed into the well. In offshore, populated, or isolated areas, automatic shut-in systems and well pressure control methods must be considered.

Basic decisions to be reached in designing the well completion are: (a) the method of completion, (b) the number of completions within the wellbore,

(c) the casing-tubing configuration, (d) the diameter of the production conduit, and (e) the completion interval.

Method of Completion

Basically there are two methods of completing a well, Openhole where casing is set on top of the producing interval and Perforated Casing, where casing is cemented through the producing interval and communication is established by perforating. Each method has areas of predominate usage depending on formation characteristics. Generally openhole has greater application in carbonate zones. But each has inherent advantages and limitations.

Open-Hole Completion—The casing set on top of producing zone. See Figure 5-1.

Advantages:

1. Adaptable to special drilling techniques to minimize formation damage, or to prevent lost circulation into the producing zone.

2. With gravel pack, provides excellent sand control method where productivity is important.

3. No perforating expense.

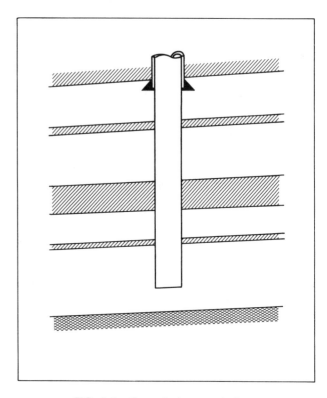

FIG. *5-1—Open-hole completion.*

4. Log interpretation is not critical since entire interval is open.

5. Full diameter opposite pay.

6. Can be easily deepened.

7. Easily converted to liner or perforated completion.

Limitations

1. Excessive gas or water production difficult to control.

2. Selective fracing or acidizing more difficult.

3. Casing set "in the dark" before the pay is drilled or logged.

4. Requires more rig time on completion.

5. May require frequent cleanout.

Perforated Completion — Casing cemented through producing zone and perforated. See Figure 5-2.

Advantages

1. Excessive gas or water production can be controlled more easily.

2. Can be selectively stimulated.

3. Logs and formations samples available to assist in decision to set casing or abandon.

4. Full diameter opposite pay.

5. Easily deepened.

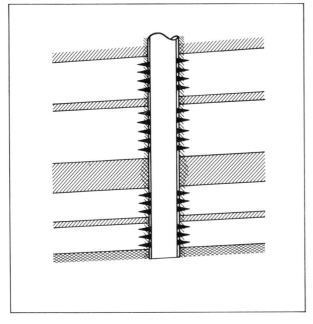

FIG. *5-2—Perforated completion.*

6. Will control most sands, and is adaptable to special sand control techniques.

7. Adaptable to multiple completion techniques.

8. Minimum rig time on completion.

Limitations

1. Cost of perforating long zones may be significant.

2. Not adaptable to special drilling techniques to minimize formation damage.

3. Log interpretation sometimes critical in order not to miss commercial sands, yet avoid perforating submarginal zones.

It should be recognized that a poor primary cement job in effect converts a "perforated casing" completion to an "openhole" completion. Continuous perforating with no "blank" zones between perforated intervals also converts a perforated casing completion to an openhole completion.

CONVENTIONAL TUBULAR CONFIGURATIONS
Single-zone Completion

Factors leading to selection of single-zone "conventional" completions, as opposed to miniaturized completions, multiple inside-casing completions, or multiple tubingless completions are: high producing rates—corrosive well fluids—high pressures—governmental policies—operator tradition.

Various hookups are possible depending on objectives. Basic questions concern use of tubing and packers. Many wells are produced without tubing. This possibility should always be considered.

Valid reasons for tubing may include:

1. Better flow efficiency.

2. Permit circulation of kill fluids, corrosion inhibitors or paraffin solvents.

3. Provide multiple flow paths for artificial lift system.

4. Protect casing from corrosion, abrasion, or pressure.

5. Provide means of monitoring bottom-hole flowing pressure.

Tubing should be run open-ended, and set above highest alternate completion interval to permit thru-tubing wireline surveys and remedial work.

A packer should be run only where it accomplishes a valid objective such as:

—Improve or stabilize flow.

—Protect casing from well fluids or pressure—
however, it should be recognized that use of a
packer may increase pressure on casing in the
event of a tubing leak.
—Contain pressure in conjunction with an artificial
lift system or safety shut-in system.
—Hold an annular well-killing fluid.

Where packers are used, landing nipples to permit
installation of bottom-hole chokes or safety valves
are sometimes desirable. Also a circulating device
is sometimes desirable to assist in bringing in or
killing the well.

In a high volume, annular flow well, where casing
can sustain shut-in well pressure, and a safety
shut-in capability is required, it may be desirable
to run a tubing string to bottom, but set a packer
and surface-controlled safety valve within several
hundred feet of the surface. Well fluids then flow
through both the annulus and tubing to a point
immediately below the packer. Here all flow is
brought into the tubing through the safety control
valve, and then back into both the annulus and
tubing to the surface. Thus, safety valve control
is maintained, but pressure restriction is minimized.

Effect of Tubing and Packer

Effect of tubing, with or without a packer, on
well pressure gradients under various conditions
should be recognized.

Tubing Without Packer (flowing well)—Figure 5-3
shows pressure gradient situation for an oil well
and a gas well with tubing "swung." The annulus
acts as separator, thus with a gas gradient in the
annulus, annular wellhead pressure is almost equal
to bottomhole pressure for the oil or the gas well.

For the gas well, wellhead annular pressure is
slightly greater with tubing than without tubing.
For the oil well, wellhead annular pressure is
considerably greater with tubing installed due to
the gas gradient in the annulus.

In the gas well the chances of a tubing leak,
with tubing swung, are nil; thus, there is no justi-
fication for a premium tubing joint. Chances of
a casing leak are essentially the same as if tubing
had not been installed.

With the oil well, chances of a tubing leak are
maximized. Pressure differential is from annulus
to tubing, and in the event of a shallow tubing

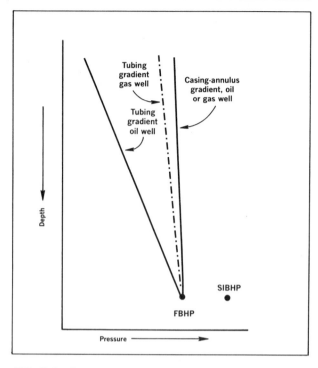

FIG. *5-3—Pressure gradient for wells with tubing
"swung."*

leak, the wellhead annular pressure will drop as
fluid level moves upward in the annulus.

Although a tubing leak is not disastrous, prolonged flow may erode casing and result in a casing
leak. Chances of a shallow casing leak are increased
with tubing swung due to the pressured annulus.

Tubing With Packer (flowing well)—Figure 5-4
shows tubing and annulus pressure gradients for
an oil well and a gas well with tubing set on a
packer. The annulus is filled with a liquid providing
a slight overburden above shut-in formation pressure.

For the oil well, differential pressure across the
tubing is now quite small, and the chance of a
tubing leak is nil. The same can be said for the
chance of a casing leak, assuming normal formation
pressures.

For the gas well, differential pressure across the
tubing increases to a maximum near the surface.
Chances for a near surface tubing leak are maximized due to an unfavorable situation as regards
to tubing load and temperature changes.

Effect of Tubing Leak (flowing well)—Figure 5-5
shows the effect of a tubing leak on the pressure
gradient in the annulus in an oil well and a gas

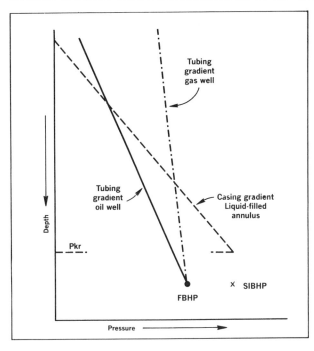

FIG. 5-4—*Pressure gradient for wells with tubing set on a packer.*

well with tubing set on a packer.

In the gas well, chances are that a collar leak will occur at a shallow depth since (1) pressure differential is greatest, (2) tubing tensile load is greatest, and (3) temperature fluctuations are greatest. Tubing pressure added on top of the pressure gradient of a high weight mud column can rupture casing downhole.

One solution to this gas well situation is to use a light liquid (water) in the annulus, then add pressure on top to more nearly match the tubing gradient. This reduces tubing leak probability, provides better retrievability of packers, and permits monitoring of casing pressure to better determine condition of tubing and casing.

In the oil well a near-surface tubing leak imposes additional bottom-hole pressure on the casing, but depending on flowing pressure in the tubing, is not usually a serious downhole threat to the casing.

In either case, oil or gas well, heavy mud is a poor packer fluid from a formation damage standpoint, and in most situations does not provide additional safety over a lighter liquid.

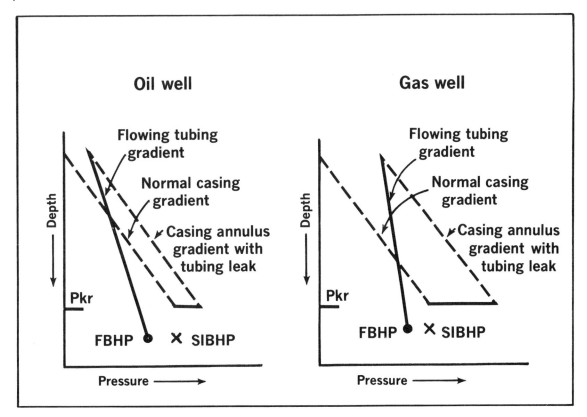

FIG. 5-5—*Effect of tubing leak on pressure gradient in well with tubing set on a packer.*

Multiple Zone Completion

Factors leading to selection of multiple completions are: higher producing rate, faster payout, and multi-reservoir control requirements. Numerous configurations are possible utilizing single or multiple strings of tubing.

Single String—Single Packer—There is both tubing and annulus flow (Figure 5-6). This is the lowest cost conventional dual.

Limitations

1. Upper zone cannot be produced through tubing, unless lower zone is blanked off.
2. Casing subject to pressure and corrosion.

3. Only lower zone can be artificially lifted.
4. Upper zone sand production may stick tubing.
5. Workover of upper zone requires killing lower zone.

Single String—Dual Packer—Again, there is both tubing and annulus flow (Figure 5-7). Advantage is that cross-over choke permits upper zone to be flowed through tubing.

Limitations

1. Casing subjected to pressure and corrosion.
2. Must kill both zones for workover of upper zone.

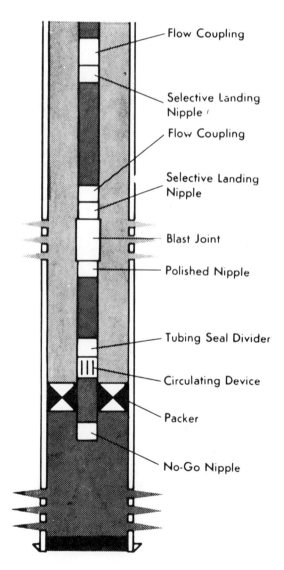

FIG. *5-6—Single string with single packer.*

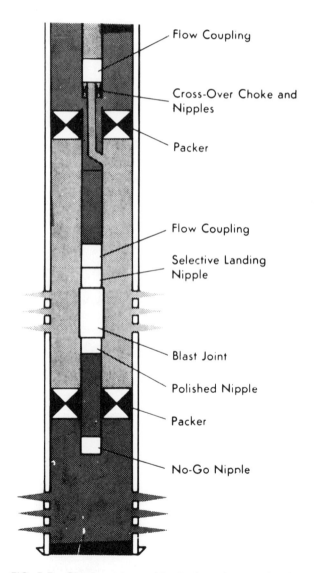

FIG. *5-7—Single string with dual packers, selective crossover.*

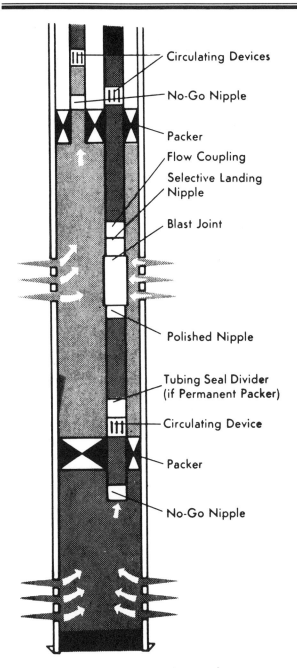

FIG. *5-8—Parallel string with multiple packers.*

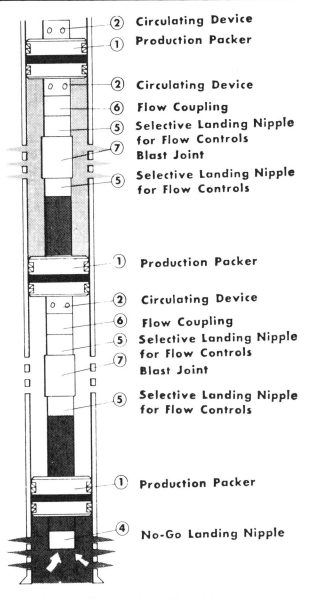

FIG. *5-9—Single string with multiple packers.*

Parallel String—Multiple Packer—This is shown in Figure 5-8.
Advantages

1. Can lift several zones simultaneously.
2. Concentric tubing and wireline workovers practical in all zones.

Limitations

1. High cost.
2. Susceptibility to tubing and packer leaks.
3. Hesitation to perform stimulation treatments or workovers of individual zones.

Single—Multiple Packer—Selective Zone—This is shown in Figure 5-9.
Advantage

1. Producing sections can be opened or closed by use of wireline.

Limitations

1. Difficulty of monitoring flow from individual zones.

2. Difficulty of treating or even reperforating individual zones unless well is killed and tubing is pulled.

UNCONVENTIONAL TUBULAR CONFIGURATIONS
Multiple "Tubingless" Completion

The multiple tubingless completion is an outgrowth of permanent well completions (PWC) and concentric tubing workover technology. It involves cementing several strings of pipe inside one wellbore, as shown in Figure 5-10. Originally this concept was applied to multiple strings of 2⅞-in. pipe; but, currently multiple strings of 3½-in. and even 4½-in. are used. The concept should not be thought of as being limited entirely to low volume producing or injection wells.

Advantages

1. Reduced cost—initial completion and future workover costs are reduced.

2. Each zone is independent and can be worked on without disturbing the other completions.

3. Communication between strings is easily located and eliminated.

4. Procedures are simplified.

Limitations

1. Restricted production rate.
2. Corrosion and paraffin control more critical.
3. Higher risk due to pressured well fluids.
4. High rate stimulation treatments are more difficult.
5. Long-zone sand control more difficult.

The application of multiple tubingless completions, along with procedures for running and cementing multiple strings of casing, are covered in detail in Appendix A of this chapter.

SIZING PRODUCTION TUBULARS

The size of the production string casing depends upon the diameter of flow conduit (single or multiple) needed to produce the desired flow stream, the method of artificial lift, if required, or specialized completion problems such as sand control.

Sizing of the production tubing depends primarily on the desired production rate. Maximum production rate in a given well depends upon:

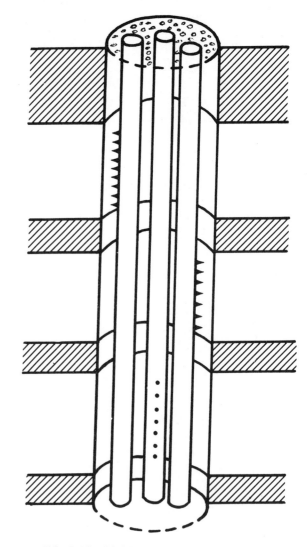

FIG. 5-10—Multiple tubingless completion.

—Static reservoir pressure.
—Inflow performance relation.
—Pressure drop in tubing.
—Pressure drop through the wellhead constrictions.
—Pressure drop through the flow line.
—Pressure level in the surface separating facilities.

Where maximum flow rate is an objective of well completion design all of these factors must be considered.

Inflow Performance Relation

The Inflow Performance Relation for a specific well represents the ability of that well to produce

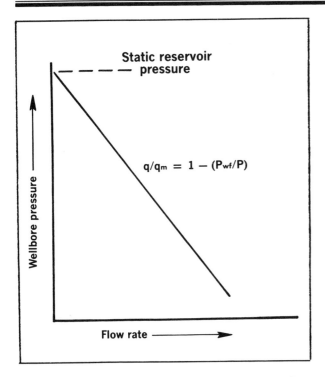

FIG. *5-11—Generalized inflow performance relation.*

fluids against varying bottom hole or "well intake" pressures. Sometimes this relation is assumed to be a straight line (Fig. 5-11). However, except for water drive reservoirs producing above bubble point pressure, flow rate usually drops off significantly from a straight line relation at higher wellbore pressure drawdown.

For a specific well the inflow performance relation often declines with cumulative production from the reservoir. For dissolved gas drive or gas drive reservoirs this decline may be rapid. The occurrence of formation damage or stimulation also affects the inflow performance relation.

Productivity index (bpd / psi pressure drawdown), a popular term used to describe well flow efficiency, represents only one point on the inflow performance curve. Thus, PI usually declines with:

1. Higher wellbore pressure drawdown.
2. Cumulative reservoir fluid withdrawals.
3. Degree of formation damage.

Assuming water-free oil production, P.I. can be estimated from reservoir parameters by:

$$PI = \frac{6 \times 10^{-4} \, k_o h}{u_o B_o}$$

k_o = oil relative permeability, md
h = zone height, feet
u_o = oil viscosity, centipoise
B_o = formation volume factor

The inflow performance relation can best be determined by Isochronal testing of the specific oil (or gas) well. However, simplified techniques are satisfactory for most purposes of well completion design (i.e.: sizing flow conduits). For more detailed study, Kermit Brown[10] is suggested.

Vogel developed the general inflow performance relation of Figure 5-12 through a computer study involving a wide range of parameters applicable to dissolved gas drive reservoirs. Having one well test measuring static reservoir pressure and a flow rate with corresponding bottom hole flowing pressure, Figure 5-12 predicts flow rate at any other BHFP. Future IPR's can be estimated by displacing the curve downward in proportion to the decline of static reservoir pressure.

Strictly the Vogel work applies to a dissolved gas drive reservoir; however, for practical purposes it can be used for any type of reservoir.

Referring to Figure 5-12,

$$\frac{q}{q_m} = 1 - 0.20 \frac{p_{wf}}{\bar{p}} - 0.80 \left(\frac{p_{wf}}{\bar{p}} \right)^2$$

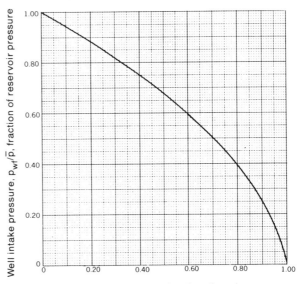

FIG. *5-12—Inflow performance relation (Vogel).*[11] *Permission to publish by The Society of Petroleum Engineers.*

q = producing rate at given p_{wf}
q_m = producing rate when $p_{wf} = 0$
p_{wf} = well intake pressure
$\bar{p}$ = shut-in reservoir pressure

Example Problem

Static res. pres: 2500 psi
Well test: BHFP = 2,000 psi
              Rate = 400 bpd
What flow rate @ BHFP of 1600 psi?

$$\text{Flow rate} = \frac{.55}{.33} \times 400 = 667 \text{ bpd}$$

In his computer study Vogel did not consider formation damage. Standing[12] extended the usefulness of Vogel's work presenting Figure 5-13 involving flow efficiency. Thus, if FE is known from pressure buildup tests (or can be estimated) the effect of formation damage, or elimination of formation damage can be estimated.

In Figure 5-13,

$$FE = \frac{\text{Ideal drawdown}}{\text{Actual drawdown}}$$

$$= \frac{\bar{p} - p_{wf} - \Delta p_s}{\bar{p} - p_{wf}}$$

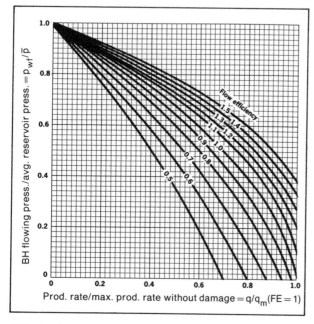

FIG. 5-13—Inflow performance relation, modified by Standing.[12]

Example Problem

$\bar{p}$, Static reservoir pressure = 2500 psi
p_{wf}, Well intake pressure = 2000 psi
q_o, Flow rate = 400 bpd
FE, flow efficiency = 0.7

—What would the well make if $p_{wf} = 1600$ psi?

$$\text{Flow rate} = \frac{.40}{.23} \times 400 = 690 \text{ bpd}$$

—What would the undamaged well make if $p_{wf} = 1600$ psi?

$$\text{Undamaged flow rate} = \frac{.54}{.23} \times 400 = 935 \text{ bpd}$$

Pressure Drop in Well Tubing

Pressure loss in two-phase or three-phase vertical flow is difficult to calculate since the average density and velocity of the fluid is usually unknown due to gas break-out and fluid slip.

Poettmann and Carpenter[8] developed empirical correlations which can be used to approximate multiphase vertical flow. Generally this correlation applies to 2⅜-in. to 3½-in. od tubing; flow rates greater than 400 bpd; GOR less than 1,500 cu ft/bbl; and viscosity less than 5 cp.

Since this original work, Dun and Ros, Hagedorn and Brown, and Beggs and Brill have developed additional vertical flow correlations aimed at improving the accuracy of pressure loss calculations. Most are applicable to all conditions including annular flow. Also vertical flow relations can be used in holes deviated up to 15° to 20° from vertical.

Figure 5-14 shows typical vertical pressure traverse curves from the Hagedorn and Brown correlation. Use of the curves to obtain flowing pressure drop where surface pressure is known is briefly:

1. Pick proper curve to fit situation, i.e., flow rate, WOR, pipe size, etc.
2. Drop vertical line from surface pressure intersecting gas-liquid ratio to determine "pseudo depth."
3. To this pseudo depth add well depth to determine "pressure depth."
4. Move horizontally from pressure depth to proper GLR and read bottom-hole pressure.

These are included here to provide a "feel" for their relative importance.

Effect of tubing size is shown in Figure 5-15 for a flow rate of 200 stb/d of 35° API oil. Although larger tubing sizes show an advantage in lowering bottom-hole pressure, the income from the increased production rate may not offset the increased cost of the larger tubing. Also as tubing becomes larger, flow velocity decreases and may let gas break through the liquid. Resulting liquid fallback and accumulation may kill the well.

Effect of tubing size on gradient reversal is shown in Figure 5-16. With small diameter tubing producing high gas-liquid ratios (gas lift wells) this gradient reversal effect is common. As the pressure on the mixture flowing up the tubing decreases near the

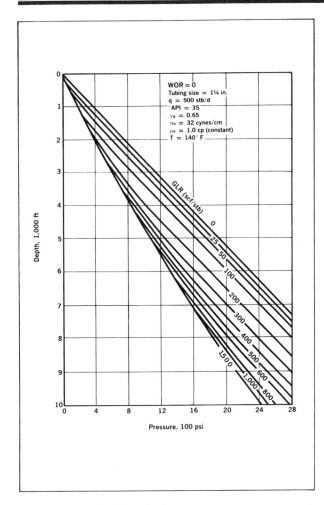

FIG. *5-14—Typical vertical pressure traverse curves.* [10] *Permission to publish by The Petroleum Publishing Co.*

5. Subtract surface pressure from BHP to determine pressure drop in tubing.

With gas in the flow stream the effect of increasing surface pressure is to increase pressure loss in the tubing. Thus, back pressure against the formation is increased due to (1) the higher pressure loss in the tubing and (2) the higher surface pressure.

Curves similar to Figure 5-14 are useful for most engineering work where approximate pressure drop calculations are required. Computer solutions of various correlations permit more detailed look at the effect of changing variables.

Brill, Doerr, Hagedorn and Brown studied the effect of certain variables on multiphase vertical flow, and presented the following relationships.

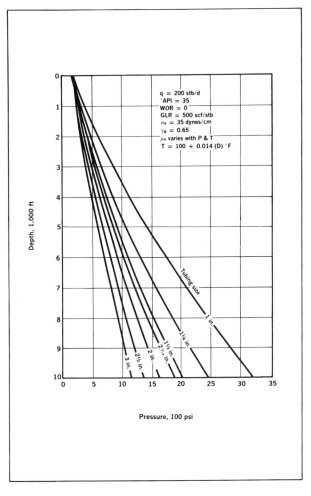

FIG. *5-15—Effect of tubing size.* [10] *Permission to publish by The Petroleum Publishing Co.*

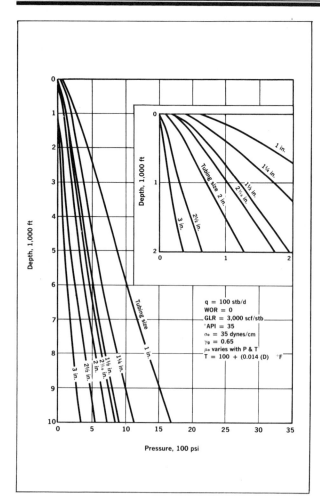

FIG. *5-16—Effect of tubing size on gradient reversal.*[10]
*Permission to publish by The Petroleum Publishing
Co.*

surface, the pressure gradient due to density decreases. However, with reduced pressure the flow
velocity increases, and the pressure gradient due
to friction loss increases at a greater rate than the
density gradient decreases. The result is an increased pressure loss near the surface.

Effect of surface flow rate is shown in Figure
5-17 for 2-in. tubing. At low rates (50 stb/d) these
correlation probably break down due to the so-called
"heading" effect. Attempts to describe this heading
effect mathematically or to predict the rate at which
it occurs have not yet been successful.

Effect of gas-liquid ratio is shown in Figure 5-18
for a flow rate of 200 stb/d of 35° API oil in 1¼-in.
tubing. As the gas-liquid ratio increases the flowing
bottomhole pressure required to produce the rate

decreases. However, a point is reached where
further increases in gas-liquid ratio actually increase
bottom-hole pressure.

Effect of liquid density is shown in Figure 5-19,
when oil viscosity is held constant at 1 cp. As API
gravity increases, flowing pressure decreases. It
should be noted that as API gravity increases the
amount of solution gas at a given pressure level
increases. This increases the liquid holdup factor,
however, which in turn increases density and tends
to offset the higher gas-liquid ratio.

Effect of liquid viscosity is shown in Figure 5-20,
for 1¼-in. tubing and a flow rate of 200 stb/d.
Free gas viscosity is assumed to be 0.02 cp while
liquid viscosity varies with temperature and solution
gas.

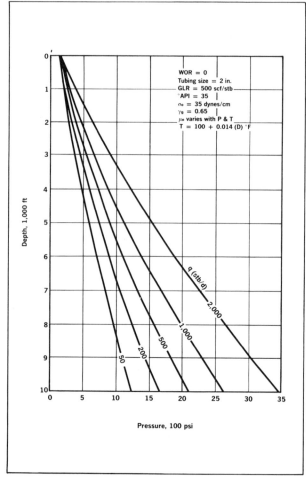

FIG. *5-17—Effect of flow rate.*[10] *Permission to publish
by The Petroleum Publishing Co.*

Effect of liquid surface tension is shown in Figure 5-21 for 1¼-in. tubing producing 200 stb/d 35° API oil. Larger surface tension results in greater liquid holdup, higher density, and, therefore, high flowing bottom-hole pressures.

Effect of kinetic energy is often neglected, but can become important with small diameter tubing with high gas-liquid ratios and low pressure levels, Figure 5-22.

Pressure Drop Through Tubing Restrictions and Wellhead

In high rate producing situations pressure drop through the wellhead and through tubing restrictions can be significant. Figure 5-23 shows one typical situation producing 3,600 bopd from 7,700 ft through

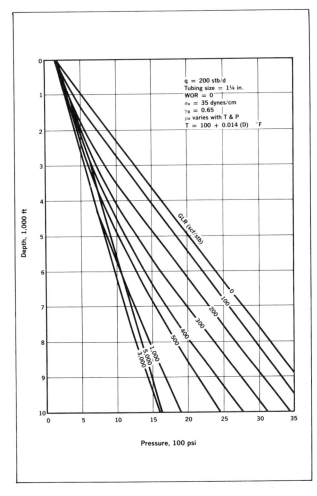

FIG. *5-18—Effect of gas-liquid ratio.*[10] *Permission to publish by The Petroleum Publishing Co.*

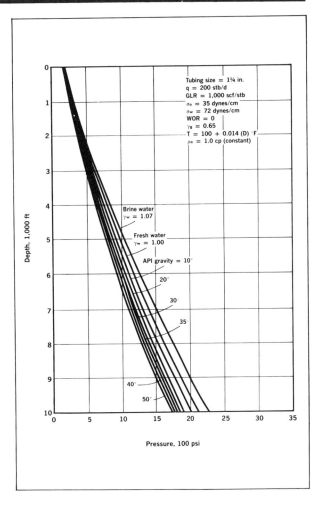

FIG. *5-19—Effect of API gravity.*[10] *Permission to publish by The Petroleum Publishing Co.*

3½-in. tubing. Tubing restrictions include a sliding side door assembly and a ball-type safety valve.

Pressure Drop in the Flow Line

The problem of horizontal two-phase flow is as complex as that of vertical two-phase flow. A number of correlations have been presented based on empirical data. Probably the best currently is one by Eaton and Brown described in Reference 9. Working curves for this correlation are presented in Reference 10.

A typical example of pressure drops with high flow rates is shown in Figure 5-23.

COMPLETION INTERVAL

Selection of the completion interval is dictated by a number of interrelated factors. In thin multi-reservoir fields the choice is usually obvious. In thicker multi-lense zones comprising single reservoirs considerable study of geologic and reservoir fluid flow mechanisms may be required for intelligent interval selection.

Reservoir Drive Mechanism

With reasonable vertical permeability, obviously a well in a water drive reservoir should be completed near the top of the zone, and a well in a gas drive reservoir should be completed near the bottom.

Reservoir Homogeneity

Reservoir homogeneity must be considered in selecting individual well completion intervals. Full

advantage should be taken of shale barriers to aid in control of unwanted fluids and to eliminate workover expense.

In sandstone reservoirs depositional environment controls the continuity of shale breaks. Shale or clay breaks in barrier bars can be correlated over a long interval. Clay breaks in fluviatile or tidal channel sediments are difficult to correlate and usually converge. Different types of environments—i.e. channel fills, barrier bars, point bars—have characteristic permeability distribution vertically and laterally.

Similar statements could be made regarding carbonate deposition—even reef type reservoirs (sometimes thought to act like a big tank) can have barriers which totally restrict vertical movement through a portion of the reservoir. By good choice of the completion interval above or beneath a barrier

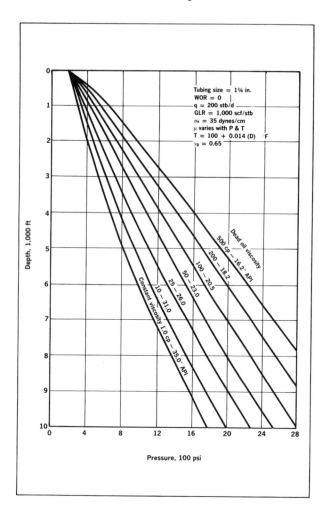

FIG. 5-20—Effect of viscosity.[10] Permission to publish by The Petroleum Publishing Co.

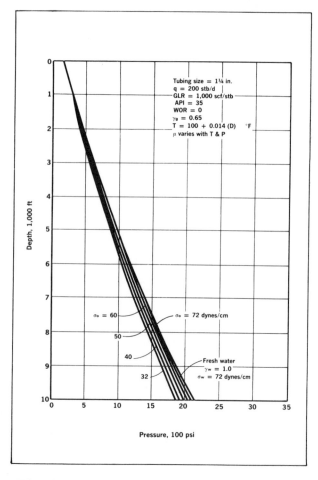

FIG. 5-21—Effect of surface tension.[10] Permission to publish by The Petroleum Publishing Co.

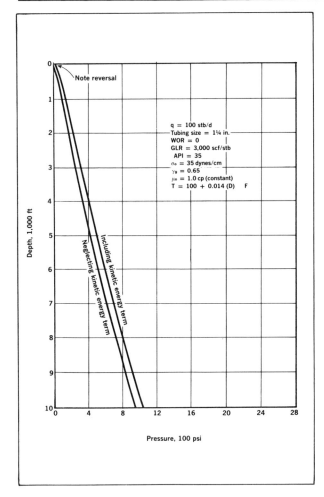

FIG. *5-22—Effect of kinetic energy.*[10] *Permission to publish by The Petroleum Publishing Co.*

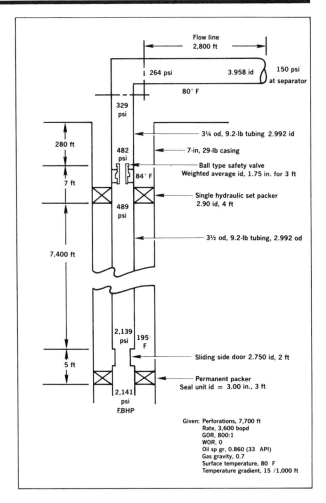

FIG. *5-23—Pressure at various points in high-flow-rate-well system.*

with favorable properties break-through of water or gas into the well may be retarded.

Producing Rate

In a massive reservoir with good permeability and vertical communication, opening a limited interval will provide high capacity and, at the same time, facilitate future workovers.

In a multi-zone field having marginal stringers, many possibilities are available. These include commingling separate reservoirs. Generally an optimum approach is to open only the number of sands that can be effectively stimulated.

Stimulation Needs

Stimulation is simplified by limiting the number of perforations and the intervals open. If several

stringers are to be produced, perforations should be "pinpointed" with as much blank section as possible between zones.

Future Workover Possibilities

It is often sound practice to open only a limited interval initially, and complete in other zones later, if needed, by means of low-cost workovers.

Cost

Minimum perforation density and interval enhances economics through reduced perforating cost, facilitating limited entry or ball sealer stimulation, and simplifying future squeeze cementing operations to control water or gas entry.

REFERENCES

1. Huber, T. A., Allen, T. O., and Abendroth, G. F.: "Well Completion Practices," API, Los Angeles, 1950.

2. Huber and Tausch: "Permanent-Type Completions," AIME Trans., Vol. 198, 1953.

3. Althouse, W. S., Jr. and Fisher, H. H.: "The Selection of a Multiple Completion Hookup," *Pet. Tech.*, December, 1958, p. 12.

4. Corley, C. B., Jr. and Rike, J. L.: "Tubingless Completions," Paper 926-4-6, Southwestern District API, March, 1959.

5. Willingham, J. E.: "Tubingless Completions in the West Texas Area," Paper 906-4-6, Southwestern District API, March, 1959.

6. Bleakley, W. B.: "What It Takes to Make a Good Well Completion," *Oil and Gas Journal*, June 11, 1962.

7. Murphy, L. A.: "Completion Developments and Trends Worth Watching," *Petroleum Engineer,* January, 1964.

8. Poettman, F. H. and Carpenter, P. G.: "The Multiphase Flow of Gas, Oil, and Water Through Vertical Flow Strings with Application to Design of Gas Lift Installations," Drilling and Production Practice, p. 257, 1952.

9. Eaton, Ben A. and Brown, Kermit E.: "The Prediction of Flow Patterns, Liquid Holdup and Pressure Losses Occurring During Continuous Two-Phase Flow in Horizontal Pipe Lines," The University of Texas, Petroleum Engineering Department," October, 1965.

10. Brown, Kermit E.: *Gas Lift Theory and Practice,* The Petroleum Publishing Co., 1972.

11. Vogel, J. V.: "Inflow Performance Relationship for Solution-Gas Drive Wells," J. Pet. Tech., January, 1968.

12. Standing, M. B.: "Inflow Performance Relationships for Damaged Wells Producing by Solution Gas Drive, *J. Pet. Tech.,*" Nov., 1970, p. 1,399.

Appendix

Tubingless Completion Techniques and Equipment

Permanent well completions (PWC), concentric tubing workovers, and tubingless completions should all be considered as a series of well completion developments. Large diameter single completions made without tubing in the Middle East and North Africa are outside the scope of this development.

PERMANENT WELL COMPLETION (PWC)

The overall concepts of PWC had the objective of eliminating the necessity of pulling tubing during the life of a well. Figure 5-24 illustrates the basic arrangement for permanent well completions with and without a packer. An essential feature of PWC is setting the bottom of the tubing open-ended and

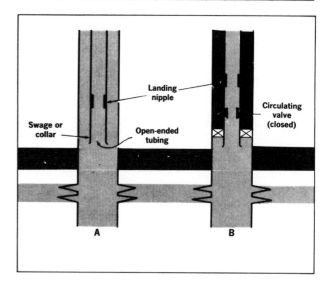

FIG. *5-24—Permanent-type well completion for single zone, without packer (A), or with packer (B).*

above the highest anticipated future completion zone.

Primary developments needed to make the system feasible form the basis of current well completion technology. These developments include:

1. The through tubing perforator, and along with it the concept of underbalanced perforating (differential pressure into the wellbore) to provide debris-free perforations.

2. A concentric tubing extension run and set on a wireline to permit circulating to the desired point in the well. Later the wireline tubing extension was replaced by the use of full string of small diameter tubing which could be run through the normal producing tubing using a small conventional workover rig. Use of this small tubing is termed Concentric Tubing Workover.

3. Low fluid loss, below frac pressure, squeeze cementing, to provide slurry properties so that cement can be placed at the desired point (in the perforations or in a channel behind the casing) and the excess cement subsequently reverse circulated out of the well.

4. Logging devices, gaslift valves, bridge plugs, and other necessary tools designed to be run through tubing on a wireline or electric line.

TUBINGLESS COMPLETIONS

This system usually involves the cementing of one or more strings of 2½-in. or 3½-in., as produc-

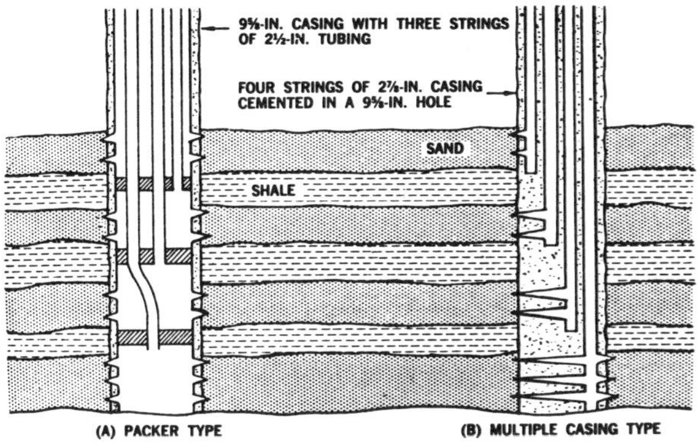
9⅝-IN. CASING WITH THREE STRINGS OF 2½-IN. TUBING
FOUR STRINGS OF 2⅞-IN. CASING CEMENTED IN A 9⅝-IN. HOLE
SAND
SHALE
(A) PACKER TYPE
(B) MULTIPLE CASING TYPE

FIG. *5-25—Standard and tubingless completion in multi-pay field.*[10] *Permission to publish by The Society of Petroleum Engineers.*

tion casing in a single borehole. Figure 5-25 shows a comparison between conventional and tubingless completion in a multi-pay field.

The original effort was aimed at reducing initial investment. However, the major economic benefits have been in reducing well servicing and workover costs, with particular application to triple completions in lenticular multi-reservoir fields, and to dual offshore wells.

This type of completion is not necessarily restricted to low-return, low-volume, short-lived wells. Single or multi-pay gas fields are excellent candidates for tubingless completions.

Hole and casing size should be designed to obtain optimum rate of return over the life of the well. Typical casing-hole size combinations for 2-in. or $2^1/_2$-in. tubingless completions in the U.S. are shown in Table 5-1.

TABLE 5-1
Popular Casing-Hole Size Combinations in the U.S.

Type completion	*Surface casing in.*	*Hole below surface casing in.*	*Production casing in.*
Single	7 to $8^5/_8$	$6^1/_4$ to $7^7/_8$	2 or $2^1/_2$
Dual	$8^5/_8$	$7^7/_8$	2 or $2^1/_2$
Dual Offshore*	—	$11^1/_2$	$3^1/_2$
Triple	$9^5/_8$	$8^5/_8$	2 or $2^1/_2$
Quadruple	$10^3/_4$	$9^5/_8$	2 or $2^1/_2$

*Deviated wells.

Casing and Cementing Practices

The following discussion presents details of casing and cementing practices generally used with tubingless or multiple tubingless completions. The mechanical problems of running multiple strings require special casing handling and blowout control equipment. Cementing of multiple tubingless wells is an important concern because the annular cross section with several strings encourages channeling of cement, and ineffective displacement of mud. Good cementing practices discussed under "Primary Cementing" must be adherred to. Pipe centralization, relative pipe movement, and low PV and YP drilling fluid properties are important.

Production Casing—Non-upset *J*-55 casing can be used to about 7,500 feet. Deeper wells should use *J*-55 with a full strength connection such as Buttress or Armco Seallock or a combination of upset *J*-55 and non-upset *J*-55 depending on strength requirements. Both ends of the collars must be beveled. Couplings should be floated on to provide uniform make-up torque on both ends of the collar.

Running and Cementing Practices (based on South Louisiana)—These are:

1. Thoroughly condition mud to optimum properties prior to pulling out to run casing.
2. In case of dual completion, make up and run both strings simultaneously, employing dual slips and elevators. On a triple completion run first two strings together, followed by third string.
3. Use reciprocating wireloop scratchers and bow string centralizers throughout the productive interval on all strings.
4. Use a torque-turn device to assure a tight uniform makeup of tubing joints and to eliminate pressure testing each joint.
5. Make up cementing heads on the handling joints while they are on the pipe rack to allow fast hook-up and initiation of reciprocation as soon as pipe is on bottom. Quick initiation of reciprocation reduces chances of sticking.
6. Maintain reciprocation for at least one full cycle while mud is circulated and conditioned. Employ strain gauges below the cementing heads to assure an even pull on each string and to assure that one string is not trying to stick.
7. Cement slurry should be as "Newtonian" as possible to improve mud displacement. Densified Class H or a mixture of lightweight and Class H cement have desirable flow properties. Batch mix cement for better uniformity.

8. Pump cement down two strings simultaneously (5–6 BPM) while maintaining reciprocation throughout the period cement is being pumped and up to 30 minutes after wiper plugs are bumped.

9. Use slip-type tubing hangers to assure that each tubing string is hung in tension.

10. Test all zones for communication by pressure testing each zone after perforating.

Completing Tubingless Wells

Normal practice is to release the drilling rig after casing is cemented. As a rule, perforating, stimulation, or other well completion operations are carried out without a rig. However, if a rig is required, a small completion or workover rig is moved on the well.

Because of size limitations, careful analysis should be made of available perforators to insure adequate hole size and penetration. Whenever feasible the well is perforated with a differential into the wellbore to avoid swabbing and to improve perforation productivity. Some of the most commonly used perforators are the scallop and strip guns. The scallop-type gun provides less casing splitting.

When perforating multiple tubingless completions, it is necessary to locate adjacent casing strings and rotate the gun by surface control so as to avoid

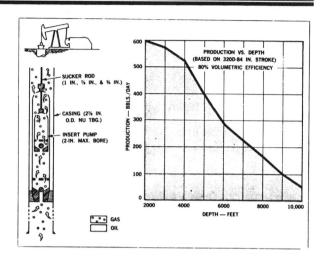

FIG. 5-27—Casing pumping installation, 2⅞-in. od casing.[10] Permission to publish by The Society of Petroleum Engineers.

perforating adjacent casing strings. See Figure 5-26.

For single completions, run radioactivity log and collar log for depth correlation. For multiples, run radioactive collars in each string near perforated intervals. Then run gamma ray—neutron log in one string to locate radioactive collars in all strings. Each casing string can then be perforated with depth correlation with previously located radioactive collars.

Logging tools are available for through casing logging down to the 2-in. size. Open hole logs are available for holes down to about 4½ in. in diameter.

Artificial Lift

The artificial lift system should produce fluid volumes required to optimize profits and cash flow rate of return from a tubingless completion program. Five methods are available to artificially lift wells having 2⅞-in. casing:

1. Casing Pumps—Figure 5-27 depicts a casing pump installation and theoretical volumes produced from various depths from 2⅞-in. od casing.

Factors to consider when planning a casing pump installation for a tubingless completion are: (1) Casing is subject to rod wear; (2) all gas must pass through pump; and (3) if appreciable sand, scale, or paraffin is anticipated, pump sticking may cause expensive workover jobs or possible loss of well.

2. Rod Pumping Inside Macaroni Strings—The most widely used method of pumping miniaturized completions is the insert pump illustrated in Figure

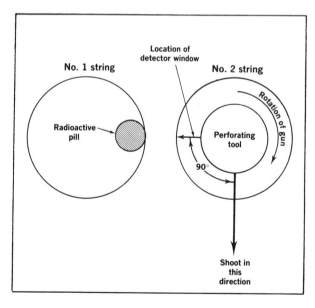

FIG. 5-26—Orienting perforating technique in dual tubingless completion.

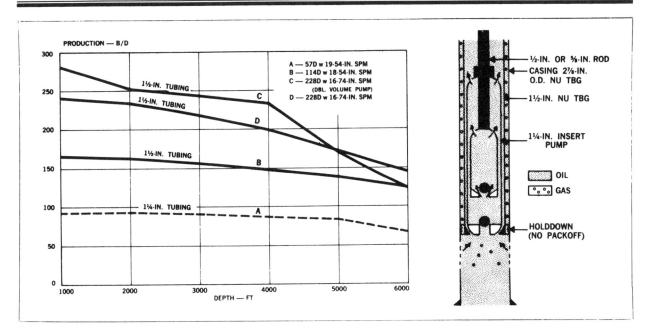

FIG. *5-28—Insert casing pump inside macaroni string.*[10] *Permission to publish by The Society of Petroleum Engineers.*

5-28. A larger volume can be obtained by running a larger pump than an insert-type on the macaroni

string. If gas is no particular problem, efficiency can be improved by running a bottom hold-down and a top pack-off seal.

Advantages of pumping inside macaroni string over casing pumping: (1) Reduces casing wear, (2) gas can be vented, thereby increasing pump efficiency, (3) corrosion and paraffin inhibitors can be circulated down casing-tubing annulus, and (4) sand can be confined to tubing, thus facilitating fishing if pump becomes stuck.

3. Hollow Sucker Rod Tubing Pumping—Figure 5-29 illustrates rod pumping of miniaturized completions with hollow rods or sucker rod tubing. This system eliminates the hollow polished rod and flexible hose used in some installations.

The hollow rod system has restricted volume compared with a casing pump. Pumping efficiency is aided by the ability to vent gas. Casing and tubing wear, similar to that with casing pumping, limits the use of hollow rods.

4. Subsurface Hydraulic Pumps—Figure 5-30 shows a conventional insert pump in which the macaroni string must be pulled to service the pump. The graph in Figure 5-30 shows pump volumes with a $2\frac{1}{2}$-in. standard pump. Standard $1\frac{1}{4}$-in. free-type pumps are also applicable in miniaturized completions. High pump rates, 300 to 400 bpd, are being pumped from 10,000 to 12,000 ft in low gas-fluid

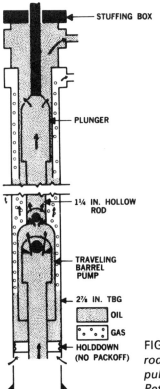

FIG. *5-29—Use of sucker-rod tubing.*[10] *Permission to publish by The Society of Petroleum Engineers.*

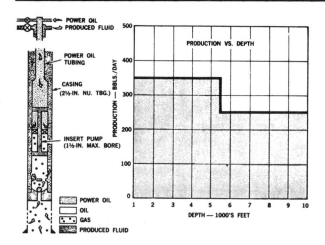

FIG. *5-30—Hydraulic pumping in miniaturized completion.*[10] *Permission to publish by The Society of Petroleum Engineers.*

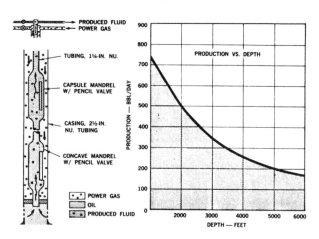

FIG. *5-31—Gas lift in miniaturized completion.*[10] *Permission to publish by the Society of Petroleum Engineers.*

ratio wells. Corrosion and paraffin can be controlled by additives to power oil.

5. Gas Lift—Gas lift is the most used artificial lift method in South Texas, where there are several thousand miniaturized completions and where gas is readily available.

Standard concentric and eccentric gas lift valves are available in tubing sizes down to 1¼-in. Figure 5-31 shows theoretical production through gas lift casing pencil valves in an eccentric installation.

Workover of Tubingless Wells

Although the original objective of tubingless or miniaturized completions was to reduce initial well costs, experience has proved the major long range saving to be in workover of multizone completions. All the necessary operations of well workover can be accomplished (i.e.: squeeze cementing perforating or sand consolidation) using wireline, electric line or concentric tubing techniques.

Multiple tubingless completions have several unique advantages in well servicing and well workover. Communication between zones as a result of tubing and packer leaks is eliminated; plug-back can be accomplished without a rig; by isolating a workover to a single completion, formation damage to other completions in a well and need for subsequent stimulation to overcome damage is eliminated; and during workover or servicing of an individual completion of a multiple well, there is no loss of production from remaining completions.

REFERENCES

1. Allen, T. O. and Atterbury, S. H., Jr.: "Effectiveness of Gun Perforating," Trans., AIME, 1954, 201, pp. 8–14.

2. Tausch, G. H. and McDonald, Price: "Permanent-Type Completions and Wireline Workovers," *Petroleum Engineer*, Vol. 28, No. 10, B-39, September, 1956.

3. Corley, C. B., Jr. and Rike, J. L.: "Tubingless Completions," API Paper 926-4-6, March, 1959.

4. Pistole, Harry and True, Martin E.: "A Challenge to the Operators and Designers of Oil Field Equipment," ASME 13th Annual Petroleum Mechanical Engineering Conference, Denver, Colorado, September 21–24, 1958.

5. Huber, T. A. and Corley, C. B., Jr.: "Permanent-Type Multiple and Tubingless Completions," *Petroleum Engineer*, February and March, 1961.

6. Lebourg, M. P. and Bell, W. T.: "Perforating of Multiple Tubingless Completions," *J. Pet. Tech.*, May 1960, p. 88.

7. Childers, Mark A.: "Primary Cementing of Multiple Casing," *J. Pet. Tech.*, July, 1968, pp. 751–762.

8. Frank, Wallace J., Jr.: "Improved Concentric Workover Techniques Offshore," *J. Pet. Tech.*, April, 1969, pp. 401–408.

9. Crosby, George E.: "Miniaturized Completions Can Be Artificially Lifted," *Petroleum Engineer*, pp. 54–59, February, 1969.

10. Rike, J. L. and McGlamery, R. G.: "Recent Innovations in Offshore Completion and Workover Systems," *J. Pet. Tech.*, January 1970, pp. 17–24.

11. Clark, Charles R. and Jenkins, Robert C.: "Cementing Practices for Tubingless Completions," SPE No. 4609, September 30, 1973.

Chapter 6 Tubing Strings, Packers, Subsurface Control Equipment

Tubing connections
Tubing string design
Inspection, handling, and running practices
Production packers
Considerations in packer selection
Effects of pressure and temperature changes
Application and operation of various packer types
Subsurface flow control systems

Tubing Strings

Proper selection, design, and installation of the tubing string is a critical part of a completion program. The tubing must be sized so that producing operations can be carried out efficiently; it must be designed against failure from tensile forces, internal and external pressures, and corrosive actions; and it must be installed in a pressure-tight and undamaged condition.

A number of grades of steel and types of tubing connections have been developed to meet demands of greater depths and new completion techniques. API Standard 5A, "Specification for Casing, Tubing, and Drill Pipe," and API Bulletin 5C2, "Bulletin on Performance Properties of Casing and Tubing," contain detailed specifications for oil well tubular goods. Bulletin 5C1, "Care and Use of Casing and Tubing," contains recommended makeup torque for API connections. See Tables 6-2 through 6-5.

Steel Grades

Standard API steel grades for tubing are J-55, C-75, C-95, N-80, and P-105. Grades C-75 and C-95 are intended for hydrogen sulfide service where higher strength than J-55 is required. The C grade steel is heat treated to remove martensitic crystal structure and produce a hardness not higher than 22 Rockwell C.

Numbers in the grade designations indicate mini-mum yield strength in 1000 psi. Grade of new pipe can be identified by color bands as follows: J-55 green; K-55 two green; C-75 blue; N-80 red; C-95 brown, P-105 white.

Tubing Connections

Standard API Coupling Connections—Two standard API coupling tubing connections are available:

The API Non-Upset, Tubing Connection (NU) is a 10-round thread form, wherein the joint has less strength than the pipe body.

The API External Upset Tubing Connection (EUE) is an 8-round thread form wherein the joint has greater strength than the pipe body. For very high pressure service the API EUE connection is available in 2⅜-, 2⅞- and 3½-in. sizes having a long thread form (EUE long T & C) wherein the effective thread is 50 percent longer than standard.

Extra Clearance Couplings—Where extra clearance is needed, API couplings can be turned down somewhat without loss of joint strength. Special clearance collars are usually marked with a black ring in the center of the color band indicating steel grade.

Extra clearance coupling-type thread forms have been developed for non-upset tubing which (unlike the API NU connection) have 100 percent joint strength. The National Tube Buttress connection and the Armco Seal Lock connection are examples.

Standard and turned-down diameters of several

coupling-type connections are shown in the following table:

Standard and Turned-down Coupling Sizes

	Coupling outside diameter—In.	
Thread Form	Standard	Special clearance
(2⅜ in.)		
API NU—10-round	2.875	2.642
API EUE—8-round	3.063	2.910
National Tube buttress	2.875	2.700
Armco Seal Lock	2.875	2.700
(2⅞ in.)		
API NU—10-round	3.500	3.167
API EUE—8-round	3.688	3.460
National Tube buttress	3.500	3.220
Armco Seal Lock	3.500	3.220

Integral Joint Connections—Several integral-joint thread forms are available from various manufacturers which provide extra clearance. Some can be turned down to provide even greater clearance. These joints usually carry a premium price and must, therefore, be justified by special conditions. An API integral-joint connection (10-round thread form) has recently been adopted for small diameter tubing, sizes (1 in., 1¼ in., 1½ in., and 2¹/₁₆ in.).

Connection Seals—In order to form a seal with any well-designed connection certain specific make-up requirements must be met. Most connections use a metal-to-metal seal which requires that the mating pin and box surfaces be forced together under sufficient stress to establish a bearing pressure exceeding the differential pressure across the connection. This principle is illustrated in Figure 6-1.

The API Round Thread connection forms several metal-to-metal seals between the tapered portions of pin and box surfaces (Figure 6-2). The small void (0.003-in. clearance) between the crest and root of mating threads must be filled with thread compound solids in order to transmit adequate bearing pressure from one threaded surface to the other.

Buttress and 8-Acme connections are similar to API Round Thread connections in that adequate bearing pressure between mating thread surfaces must be established, and voids must be filled with thread compound solids to transmit bearing loads across void spaces.

So-called "metal-to-metal" connections (Hydril and Extremeline) have large smooth metal-to-metal sealing surfaces. Threads have relatively large clearance, and do not act as seals. Armco Seal Lock has both a sealing thread and smooth metal sealing surface.

Resilient teflon rings are used in several connections to act as a supplementary seal and to provide corrosion protection.

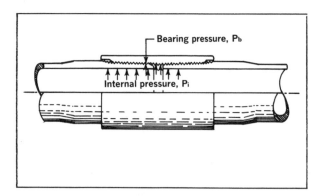

FIG. 6-1—Bearing pressure must exceed internal pressure to form pressure seal.[2] Permission to publish by The Society of Petroleum Engineers.

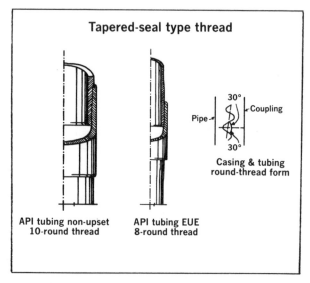

FIG. 6-2—API round-threaded connections.[2] Permission to publish by The Society of Petroleum Engineers.

Make-up of API Threaded Connection

The stresses induced in the connection during make-up and subsequent service determine the success of the connection as a sealing element.

Normal Conditions—Traditional field practice often dictates that API tapered thread connections are made up on one of the following bases:

1. Make-up position rules of thumb: "last tool mark on the pin must be 'buried' by the collar"; "two turns past hand tight (50 ft-lbs).

2. Make-up torque recommendations, "Minimum, optimum, or maximum," as shown in API Tubing Tables 6-4 and 6-5.

With good thread cleaning and doping practices, these techniques prove satisfactory for "normal" situations.

Critical Situations—To obtain maximum leak resistance with the API tapered thread connection the pin end of connection must be made up to the point of yielding. The problem of make-up is thus to screw the connection up sufficiently to provide the needed seal without permanently damaging the connection. Figure 6-3 shows the relation between "Make-up" (turns past hand tight), Hoop Stress in the coupling, and Radial Pressure between pin and coupling.

Where maximum seal is required, make-up torque is not a good indication of connection stress, be-cause it depends on friction between mating surfaces. Friction, for clean threads, is a function of thread dope lubricating characteristics. This is shown in Figure 6-4. Likewise turns or make-up position is not a reliable indication of pin stress due to size tolerance in machining the threads.

An extensive laboratory testing program by a major operator developed the torque-turn technique of tubing make-up which sets minimum criteria for both torque and make-up position which must be met to insure a seal under high-pressure conditions. For a particular connection and thread dope, the testing program established values for the following parameters:

Reference torque — Point of intimate contact— start of turns count.

Minimum torque — Lowest torque (ft. lbs.) needed for seal.

Low turns — Must get at least this number before reaching minimum torque—or connection is signaled "bad."

Minimum turns — Turns required with at least minimum torque for good seal.

Maximum turns — Must not exceed this value before reaching minimum torque or connection is signaled "bad."

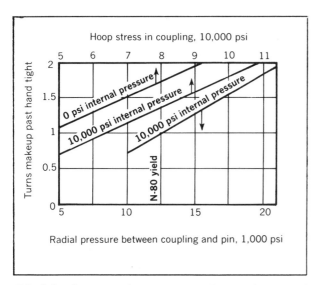

FIG. *6-3—Stresses due to connection makeup and internal pressure.*[2] *Permission to publish by The Society of Petroleum Engineers.*

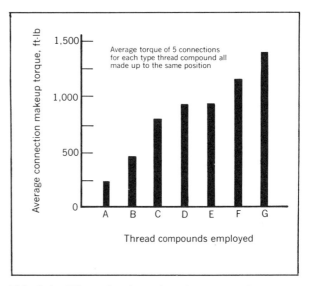

FIG. *6-4—Effect of various thread compounds on connection makeup torque.*[2] *Permission to publish by The Society of Petroleum Engineers.*

Special hydraulic tong equipment is available to measure torque and count turns, to record measurements, and to signal good or bad connections. For critical situations—high pressure oil or gas—or hostile environment conditions—field results of torque-turn method justifies consideration.

Design of Tubing Strings

Tubing string design is essentially the same as for casing. Tapered strings are becoming more common in deep wells, although uniform strings are desirable (but more expensive) due to difficulty of keeping tubing strings in proper order of weight and grade.

Uniform weight upset tubing strings reach their safe tensile limit in air as follows:

Tensile Setting Depths in Air for Upset Tubing (ft)[(a)(b)]

Grade	Safety Factor		
	1.50	1.60	1.75
J-55	10,200	9,600	8,000
C-75	13,900	13,000	11,900
N-80	14,800	13,900	12,700
P-105	19,500	18,300	16,700

[(a)] Based on minimum yield strength times area of section under root of last perfect thread, or body of pipe whichever is smaller.

[(b)] Values shown apply to normal or special clearance couplings.

A tension design factor of 1.60 is common for uniform tubing strings.

Collapse design factor should not be less than 1.00 based on the pressure differential that may actually be applied; for example, a full annular fluid column and an empty tubing string.

Tubing should not be subjected to burst pressures higher than its rated pressure divided by 1.1, unless it has been previously tested to a higher pressure.

High Strength Tubing

High strength tubing is usually considered to be those grades with a yield strength above 80,000 psi, i.e., C-75, N-80, C-95 and P-105. C-75 and N-80 are included because their yield strength as manufactured, often exceeds 80,000 psi. High strength tubing, particularly P-105, presents several problems due to decreased ductility and increased sensitivity to sharp notches or cracks.

Physical Properties of Steel

Yield and Tensile Strength—The tensile test provides basic design information on the strength of materials. This test subjects a standard specimen to a gradually increasing load. At relatively low loads (the elastic range) elongation is directly proportional to the load applied and permanent deformation does not occur. As load increases, a point is reached where elongation occurs with no increase in load.

This is the yield point. The load at this point divided by the specimen cross-sectional area is the yield strength. Further increases in load cause permanent deformation (plastic range) and finally the specimen breaks. The load at the breaking point determines the tensile strength or ultimate strength.

The numerical value of tensile strength must be used with care, since it is determined under very restrictive conditions of uniaxial loading, and may not relate closely to the complex conditions of stress and environment encountered in service.

Ductility—Ductility is the ability of material to plastically deform without fracturing. Material with high ductility will permit large deformation in the plastic range before breaking. Ductility is measured by percent elongation of a standard specimen and API standards specify elongation for each grade of tubing.

Toughness and Impact Strength—Material failure is usually classed as ductile fracture or brittle fracture. A ductile fracture occurs with plastic deformation prior to and during propagation of the crack. A brittle fracture occurs with little elongation and at a rapid rate. Toughness refers to the ability of a material to resist brittle failure. Toughness or impact strength is measured by the Charpy impact test, in which a swinging pendulum strikes and fractures a notched specimen.

API Physical Property Specifications—API physical property specifications for tubing cover only basic material properties and are shown in Table 6-1. The maximum yield strength limit and the minimum elongation specification are important factors in ensuring satisfactory mill control of tubing manufacture.

TABLE 6-1
API Physical Property Specifications for Tubing

| Grade | Yield Strength psi | | Minimum Tensile Strength psi | Min. Elongation % in 2 Inches | | Typical Hardness Rc |
	Minimum	Maximum		Strip [a]	Full Section [a]	
H-40	40,000	—	60,000	27	32	—
J-55	55,000	80,000	75,000	20	25	14
C-75	75,000	90,000	95,000	16	18	22
C-95	95,000	110,000	105,000	16	18	22
N-80	80,000	110,000	100,000	16	18	24
P-105	105,000	135,000	120,000	15	17	35

[a] Based on $e = 625,000 \dfrac{A^{0.2}}{\mu^{0.9}}$

e = Min. Elongation in 2 inches
A = Cross Sectional Area (in.^{2})
μ = Tensile Strength psi

Sensitivities of High-Strength Tubing

Notch Sensitivity—Failures of high-strength tubing are normally caused by: (1) manufacturing defects, (2) damage during transportation, handling and running, or (3) hydrogen embrittlement.

Any sharp-edged notch or crack in the surface of a material is a point of stress concentration which tends to extend the crack progressively deeper into the material much like driving a wedge. Low strength materials are soft and ductile and will yield plastically to relieve the stress concentration. High strength materials do not yield to relieve the stress concentration and, thus, fatigue or fail rapidly when subjected to cyclic stresses.

Hydrogen Embrittlement—In the presence of moisture only a trace of H_2S is needed to cause hydrogen embrittlement. The embrittlement process (sometimes called sulfide corrosion cracking), is not fully understood, but apparently results from the penetration of hydrogen atoms into the lattice structure of a high-strength steel. The hydrogen atom is smaller than the lattice structure and can migrate into the steel similarly to fine sand passing through a gravel pack.

When two hydrogen atoms meet within the steel lattice they combine to form a hydrogen molecule with a resulting increase in size. The stress created by this increase in size can part the grains of a low ductility steel structure. Low strength high ductility steels will yield to relieve the stress without failing; thus are more suitable for use in a hydrogen sulfide environment.

It is generally accepted that below a hardness of 22 Rockwell C embrittlement or sulfide cracking does not take place. Figure 6-5 shows correlation

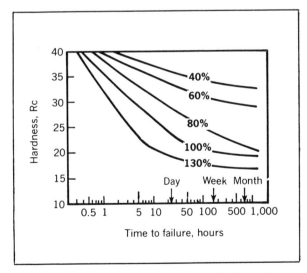

FIG. *6-5—Approximate correlation of failure time to hardness and applied stress for carbon steel in 5% NaCl solution containing 3,000 ppm H_2S (from CORROSION, 22(8), Aug. 1966).*

between hardness, applied stress (% of yield stress) and time to failure.

P-105 tubing with a hardness of about 35 Rc could be expected to fail due to embrittlement within a few days if loaded to even 50% of its yield stress in the presence of H_2S and water. Recent experience shows that the problem of embrittlement is reduced at temperatures above 150°F.

Embrittlement of high strength tubing has been reported as a result of acid used in mill cleaning operations, improperly inhibited formation acidizing operations, or sulfides formed by packer fluid degradation due to thermal, bacterial, or electro-chemical effects.

Tubing Inspection

Methods—Visual inspection of any tubing string regardless of grade should always be mandatory before tubing is run in a well. Defects recognizable by visual inspection include (1) mill defects such as seams, slugs, pits or cracks; (2) poorly machined threads; or (3) shipping or handling damage to the pipe body, coupling and threads. An alert crew can often prevent installation of pipe weakened by excessive slip or tong damage.

Hydrostatic pressure tests of tubing and connections as the string is being run in the hole are often considered good practice. The general procedure is to test only the connection unless there is definite indication of a leak within the joint body. Test pressures are usually 80 percent of internal yield pressure based on 100 percent wall thickness and minimum yield strength values. A successful pressure test is not positive proof of the lack of mill defects, since these may show up only after a number of cycles of pressure or temperature change. Where the Torque Turn Method of make up is used hydrostatic testing is probably not necessary.

Electromagnetic Search Coil inspection methods include two types which will identify defects as follows:

1. Corrosion pits and transverse defects—Tuboscope Company's Sonoscope; Plastic Applicator's Scanograph.
2. Corrosion pits, transverse defects and longitudinal defects—Tuboscope's Amalog; Plastic Applicator's Scanalog.

Since the conductor coils must be in contact with the pipe these electromagnetic inspections are ques-

tionable in the upset portion of tubing or in threaded areas.

Magnetic Particle inspection methods induce a transverse magnetic field in the tubing, and magnetic particles dusted on the tubing line up to point out longitudinal defects. Special end-area magnetic particle inspection methods have been developed to evaluate both longitudinal and transverse defects in coupling and upset areas that could not be detected by electromagnetic search coil inspections.

Magnetic particle inspections are considered to be less positive than electromagnetic search coil inspections because: (1) effectiveness is reliant on operator's attitude and efficiency, (2) environmental conditions (wind, rain, and light), (3) inspection is limited to outside of pipe, and (4) transverse defects are not detected (except with the end-area technique).

Inspection Criteria—In spite of mill inspections, new pipe containing serious defects is often received in the field. As a result, one major operator established these inspection criteria:

1. Inspections, other than visual inspection, of new J-55 tubing are generally uneconomical because failures caused by mill defects are uncommon in these grades.
2. New C-75 and higher grade tubing should have Tuboscope's Amalog (or equivalent) inspection. Defects from 5 to 12½ percent of wall thickness should be removed, and joints with defects greater than 12½ percent should be rejected.
3. New C-75 and higher grade couplings should be boxed separately and should have magnetic particle inspection.
4. Used C-75 or higher grade couplings should be visually inspected.
5. Used N-80 or higher grade tubing should have Tuboscope's Sonoscope (or equivalent) inspection when service conditions indicate that corrosion pits, transverse cracks or service-induced defects are present.

Tubing Handling Practices

High-strength tubing can be easily damaged by inadequate shipping and handling practices, or by careless use of slips and tongs. Improper use of tongs and slips is probably responsible for more critical damage than all other things combined. Proper handling practices are absolutely essential for high-strength materials.

Mill Coupling Buck-On Practices—Field experience shows that mills often do not obtain proper thread make-up. One major operator found that 85% of collar leaks were in the mill end of the collar. Some operators order C-75 and higher grade couplings boxed separately to insure proper doping and make-up. Where mills are using the Torque Turn make-up method the practice of separate boxing of collars is probably not necessary.

Loading, Transportation, and Unloading—P-105 and higher grade pipe should be unloaded with a gin-pole truck using a spreader, and nylon or webbing slings which will not scratch the pipe. It must not be unloaded by the common practice of rolling from trucks to pipe racks. Woodsills should be placed between the pipe rack and first row of pipe, and between each row of pipe to minimize pipe contact. Pipe should be racked in a stairstep fashion away from the catwalk so that pipe from the top row never falls more than one pipe diameter as it is rolled to the pipe rack.

Equipment for Use with High-Strength Tubing

Tongs—Use of power tongs is considered a necessity to obtain adequate make-up. Tong dies are available which (1) are contoured to fit the outside diameter of the pipe, (2) have larger surface areas to contact the pipe, and (3) have teeth separated by cross-hatches to minimize slippage and notch depth.

Elevators and Spiders—Collar-type tubing elevators with level bearing surfaces are satisfactory for running high-strength tubing with API non-beveled couplings fully made up. Picking up heavy tubing strings with collar-type elevators where collars are made up only "hand tight" may cause partially engaged threads to fail.

Integral joint or collars with beveled underside must, of course, be handled with slip-type elevators.

Satisfactory hand-set rotary slips are available for running high-strength tubing; however, a worn rotary table or bushings can provide uneven support for rotary slips causing non-uniform contact between tubing and rotary slips which can damage the tubing.

Tubing Running Practices

Adequate thread cleaning to assure removal of all sand, dirt, and dried thread dope is the key to proper connection make-up and pressure-tight strings. The following procedure describes a field-proven technique for proper thread cleaning and running practices:

1. Remove the thread protectors.
2. Clean the threads with kerosene and a wire brush; satisfactory cleaning requires complete removal of all dope, dirt, sand, and other foreign material to 100 percent bare, shiny steel. The use of kerosene in a compressed air spray gun operating off the rig air system or from a portable unit is also a satisfactory method and will accelerate job completion. If steam is available, steam cleaning is an excellent method.
3. Dry the threads with clean rags or compressed air.
4. Re-install clean thread protectors on the dry pin ends.
5. Apply thread compound to the male threads at the box end of the tube and to the female threads in one end of each clean coupling.
6. Install and make up tubing couplings manually with about 300 foot-pounds of torque using special friction-type tongs to eliminate notching. (Installing the couplings before picking up the tubing minimizes the possibility of dropping the string, in the event slippage occurs between the pipe and the elevators as the joint is lowered in the derrick.)
7. Wash all dirt off the ramp and catwalk.
8. Roll one joint of pipe at a time from the upper tiers onto wooden sills placed across the catwalk. Roll pipe slowly and maintain control at all times to prevent colliding of joints.
9. Steel thread protectors should remain on the pin ends of the pipe while picking up pipe from the catwalk.
10. While picking up each joint from the catwalk with a plaited pickup cable and air hoist or cathead, use a snub rope attached to the pin end to enable one man to restrain lateral motion of the joint and minimize contact between the pipe and ramp. Minimizing contact between the pipe and ramp by snubbing should permit the use of a clean plastic bucket over the box end as a dirt deflector. A rope hold-back should also be used at the "V" door to catch the lower end of the joint as it swings on to the derrick floor.
11. With the box end of the joint resting on the derrick floor, remove the plastic bucket and use dry compressed air or clean, dry rags to remove

sand or other foreign material from the dry box threads.

12. After the traveling block pulls the joint up to a vertical position using a plaited pickup cable, remove the pin-end thread protector. Some new pipe has loose mill scale inside which should be allowed to fall out prior to stabbing to minimize contamination of clean threads. Use dry compressed air or clean, dry rags to remove mill scale or other foreign material which may accumulate on the dry pin-end threads while picking up the joint. Apply a light coat of thread compound to clean pin-end threads and make up the joint.

Torque make-up should be measured by a direct reading Martin-Decker or equivalent power tong hydraulic torque gauge. A pressure gauge on air tongs is unsatisfactory, because the torque exerted for a given pressure is dependent on the condition of the air motor and the amount of lubricant used. The pointer of a hydraulic torque gauge should always return to zero when tong torque is zero to permit accurate measurements at low and high torques. Accurate torque measurement requires that the backup cable connect to the tong arm at a 90-degree angle and in the same horizontal plane; periodic leveling of the tong suspension system may be necessary to permit even die contact with the pipe.

Make-up torque values for API tubing connections are in the "Tubing Tables" at the end of this section. It should be remembered that in the case of API connections, torque values are a guide only and that proper make-up should move the coupling face to or past the "last scratch" (transverse plane passing through the last mark on the pipe OD made by thread-cutting tools). Manufacturers' current recommendations should be used for making up all other connections.

A satisfactory method of establishing the correct torque is to record the torque measurement necessary to place the coupling face as described above on the first 10 or 12 joints and then use this torque reading for the remainder of the string. Should a particular length require make-up torques much higher or lower than other lengths of the same size, weight, grade, and having the same type connection, back out the length and visually inspect for defects, thread damage, and foreign material on the threads and remove the joint from the string if necessary. Couplings should not be hammered to break out joints except as a final resort. Hammered couplings should be replaced before running the pipe.

REFERENCES:

1. Hilliard, Harold: "How Corrosive Environments Affect Drill Stem," *Petroleum Engineer*, p. 84 (Sept. 1968)

2. Weiner, P. D.; Sewell, F. D., Jr.: "New Technology for Improved Tubular Connection Performance," J. Pet Tech., March 1967, p. 337.

3. Weiner, P. D.—True, Martin—"A Method of Obtaining Leakproof API Threaded Connections in High Pressure Gas Service—API Paper 926-14-17 March 5, 1969.

API PUBLICATION LIST

Standards

Spec 5A: Specification for Casing, Tubing, and Drill Pipe.

Spec 5AC: Specification for Grade C-75 and C-95 Casing and Tubing.

Spec 5AX: Specification for High-Strength Casing, Tubing, and Drill Pipe.

Std 5B: Specification for Threading, Gaging, and Thread Inspection of Casing, Tubing, and Line Pipe Threads.

Recommended Practices

RP 5C1: Recommended Practice for Care and Use of Casing and Tubing.

Bulletins

Bul 5A2: Bulletin on Thread Compounds.

Bul 5C2: Bulletin on Performance Properties of Casing and Tubing.

TABLE 6-2
Tubing Minimum Performance Properties

Nom. (in.)	OD (in.)	T&C Non-Upset (lb/ft)	T&C Upset (lb/ft)	Int. Jt. (lb/ft)	Grade	Wall Thickness (in.)	Inside Dia. (in.)	Drift Dia. (in.)	Coup. Non-Upset (in.)	Coup. Upset Reg. (in.)	Coup. Upset Spec. (in.)	Int. Jt. Drift Dia. (in.)	Int. Jt. Box OD (in.)	Collapse Resistance (psi)	Internal Yield Pressure (psi)	JYS T&C Non-Upset (lb)	JYS T&C Upset (lb)	JYS Int. Jt. (lb)
3/4	1.050	1.14	1.20		H-40	.113	.824	.730	1.313	1.660				7,680	7,530	6,360	13,300	
	1.050	1.14	1.20		J-55	.113	.824	.730	1.313	1.660				10,560	10,360	8,740	18,290	
	1.050	1.14	1.20		C-75	.113	.824	.730	1.313	1.660				14,410	14,120	11,920	24,940	
	1.050	1.14	1.20		N-80	.113	.824	.730	1.313	1.660				15,370	15,070	12,710	26,610	
1	1.315	1.70	1.80	1.72	H-40	.133	1.049	.955	1.660	1.900		.955	1.550	7,270	7,080	10,960	19,760	15,970
	1.315	1.70	1.80	1.72	J-55	.133	1.049	.955	1.660	1.900		.955	1.550	10,000	9,730	15,060	27,160	21,960
	1.315	1.70	1.80	1.72	C-75	.133	1.049	.955	1.660	1.900		.955	1.550	13,640	13,270	20,540	37,040	29,940
	1.315	1.70	1.80	1.72	N-80	.133	1.049	.955	1.660	1.900		.955	1.550	14,550	14,160	21,910	39,510	31,940
1-1/4	1.660			2.10	H-40	.125	1.410					1.286	1.880	5,570	5,270			22,180
	1.660	2.30	2.40	2.33	H-40	.140	1.380	1.286	2.054	2.200		1.286	1.880	6,180	5,900	15,530	26,740	22,180
	1.660			2.10	J-55	.125	1.410					1.286	1.880	7,660	7,250			30,500
	1.660	2.30	2.40	2.33	J-55	.140	1.380	1.286	2.054	2.200		1.286	1.880	8,500	8,120	21,360	36,770	30,500
	1.660	2.30	2.40	2.33	C-75	.140	1.380	1.286	2.054	2.200		1.286	1.880	11,580	11,070	29,120	50,140	41,600
	1.660	2.30	2.40	2.33	N-80	.140	1.380	1.286	2.054	2.200		1.286	1.880	12,360	11,810	31,060	53,480	44,370
1-1/2	1.900			2.40	H-40	.125	1.650					1.516	2.110	4,920	4,610			26,890
	1.900	2.75	2.90	2.76	H-40	.145	1.610	1.516	2.200	2.500		1.516	2.110	5,640	5,340	19,090	31,980	26,890
	1.900			2.40	J-55	.125	1.650					1.516	2.110	6,640	6,330			36,970
	1.900	2.75	2.90	2.76	J-55	.145	1.610	1.516	2.200	2.500		1.516	2.110	7,750	7,350	26,250	43,970	36,970
	1.900	2.75	2.90	2.76	C-75	.145	1.610	1.516	2.200	2.500		1.516	2.110	10,570	10,020	35,800	59,960	50,420
	1.900	2.75	2.90	2.76	N-80	.145	1.610	1.516	2.200	2.500		1.516	2.110	11,280	10,680	38,180	63,960	53,780
2-1/16	2.063			3.25	H-40	.156	1.751					1.657	2.325	5,590	5,290			35,690
	2.063			3.25	J-55	.156	1.751					1.657	2.325	7,690	7,280			49,070
	2.063			3.25	C-75	.156	1.751					1.657	2.325	10,480	9,920			66,910
	2.063			3.25	N-80	.156	1.751					1.657	2.325	11,180	10,590			71,370
2-3/8	2.375	4.00			H-40	.167	2.041	1.947	2.875					5,230	4,920	30,130		
	2.375	4.60	4.70		H-40	.190	1.995	1.901	2.875	3.063	2.910			5,890	5,600	35,960	52,170	
	2.375	4.00			J-55	.167	2.041	1.947	2.875					7,190	6,770	41,430		
	2.375	4.60	4.70		J-55	.190	1.995	1.901	2.875	3.063	2.910			8,100	7,700	49,450	71,730	
	2.375	4.00			C-75	.167	2.041	1.947	2.875					9,520	9,230	56,500		
	2.375	4.60	4.70		C-75	.190	1.995	1.901	2.875	3.063	2.910			11,040	10,500	67,430	97,820	
	2.375	5.80	5.95		C-75	.254	1.867	1.773	2.875	3.063	2.910			14,330	14,040	96,560	126,940	
	2.375	4.00			N-80	.167	2.041	1.947	2.875					9,980	9,840	60,260		
	2.375	4.60	4.70		N-80	.190	1.995	1.901	2.875	3.063	2.910			11,780	11,200	71,930	104,340	
	2.375	5.80	5.95		N-80	.254	1.867	1.773	2.875	3.063	2.910			15,280	14,970	102,990	135,400	
	2.375	4.60	4.70		P-105	.190	1.995	1.901	2.875	3.063	2.910			15,460	14,700	94,410	136,940	
	2.375	5.80	5.95		P-105	.254	1.867	1.773	2.875	3.063	2.910			20,060	19,650	135,180	177,180	
2-7/8	2.875	6.40	6.50		H-40	.217	2.441	2.347	3.500	3.668	3.460			5,580	5,280	52,780	72,480	
	2.875	6.40	6.50		J-55	.217	2.441	2.347	3.500	3.668	3.460			7,680	7,260	72,580	99,660	
	2.875	6.40	6.50		C-75	.217	2.441	2.347	3.500	3.668	3.460			10,470	9,910	98,970	135,900	
	2.875	8.60	8.70		C-75	.308	2.259	2.165	3.500	3.668	3.460			14,350	14,060	149,360	186,290	
	2.875	6.40	6.50		N-80	.217	2.441	2.347	3.500	3.668	3.460			11,160	10,570	105,570	144,960	
	2.875	8.60	8.70		N-80	.308	2.259	2.165	3.500	3.668	3.460			15,300	15,000	159,310	198,710	
	2.875	6.40	6.50		P-105	.217	2.441	2.347	3.500	3.668	3.460			14,010	13,870	138,560	190,260	
	2.875	8.60	8.70		P-105	.308	2.259	2.165	3.500	3.668	3.460			20,090	19,690	209,100	260,810	
3-1/2	3.500	7.70			H-40	.216	3.068	2.943	4.250					4,630	4,320	65,070		
	3.500	9.20	9.30		H-40	.254	2.992	2.867	4.250	4.500	4.180			5,380	5,080	79,540	103,610	
	3.500	10.20			H-40	.289	2.922	2.797	4.250					6,060	5,780	92,550		
	3.500	7.70			J-55	.216	3.068	2.943	4.250					5,970	5,940	89,470		
	3.500	9.20	9.30		J-55	.254	2.992	2.867	4.250	4.500	4.180			7,400	6,980	109,370	142,460	
	3.500	10.20			J-55	.289	2.922	2.797	4.250					8,330	7,950	127,250		
	3.500	7.70			C-75	.216	3.068	2.943	4.250					7,540	8,100	122,010		
	3.500	9.20	9.30		C-75	.254	2.992	2.867	4.250	4.500	4.180			10,040	9,520	149,140	194,260	
	3.500	10.20			C-75	.289	2.922	2.797	4.250					11,360	10,840	173,530		
	3.500	12.70	12.95		C-75	.375	2.750	2.625	4.250	4.500	4.180			14,350	14,060	230,990	276,120	
	3.500	7.70			N-80	.216	3.068	2.943	4.250					7,870	8,640	130,140		
	3.500	9.20	9.30		N-80	.254	2.992	2.867	4.250	4.500	4.180			10,530	10,160	159,090	207,220	
	3.500	10.20			N-80	.289	2.922	2.797	4.250					12,120	11,560	185,100		
	3.500	12.70	12.95		N-80	.375	2.750	2.625	4.250	4.500	4.180			15,310	15,000	246,390	294,530	
	3.500	9.20	9.30		P-105	.254	2.992	2.867	4.250	4.500	4.180			13,050	13,330	208,800	271,970	
	3.500	12.70	12.95		P-105	.375	2.750	2.625	4.250	4.500	4.180			20,090	19,690	323,390	386,570	
4	4.000	9.50			H-40	.226	3.548	3.423	4.750					4,060	3,960	72,000		
	4.000		11.00		H-40	.262	3.476	3.351		5.000				4,900	4,580		123,070	
	4.000	9.50			J-55	.226	3.548	3.423	4.750					5,110	5,440	99,010		
	4.000		11.00		J-55	.262	3.476	3.351		5.000				6,590	6,300		169,220	
	4.000	9.50			C-75	.226	3.548	3.423	4.750					6,350	7,420	135,010		
	4.000		11.00		C-75	.262	3.476	3.351		5.000				8,410	8,600		230,750	
	4.000	9.50			N-80	.226	3.548	3.423	4.750					6,590	7,910	144,010		
	4.000		11.00		N-80	.262	3.376	3.351		5.000				8,800	9,170		246,140	
4-1/2	4.500	12.60	12.75		H-40	.271	3.958	3.833	5.200	5.563				4,500	4,220	104,360	144,020	
	4.500	12.60	12.75		J-55	.271	3.958	3.833	5.200	5.563				5,720	5,800	143,500	198,030	
	4.500	12.60	12.75		C-75	.271	3.958	3.833	5.200	5.563				7,200	7,900	195,680	270,040	
	4.500	12.60	12.75		N-80	.271	3.958	3.833	5.200	5.563				7,500	8,430	208,730	288,040	

TABLE 6-3
Casing Minimum Performance Properties

OD Size (in.)	Weight Per Foot Nom. (Lbs.)	Wall (in.)	ID (in.)	Drift (in.)	Coupling OD (in.)	Collapse H40	Collapse J55	Collapse N80	Collapse P110	Internal H40	Internal J55	Internal N80	Internal P110	Short H40	Short J55	Short N80	Short P110	Long J55	Long N80	Long P110
4-1/2	9.50	.205	4.090	3.965	5.000	2770	3310	....	....	3190	4380	....	....	77	101	....	....	....	....	....
	11.60	.250	4.000	3.875	5.000	....	4960	6350	7560	....	5350	7780	10690	...	154	....	....	162	223	279
	13.50	.290	3.920	3.795	5.000	....	....	8540	10670	....	....	9020	12410	...	....	....	....	....	270	338
	15.10	.337	3.826	3.701	5.000	....	....	....	14320	....	....	....	14420	...	....	....	....	....	....	406
5	11.50	.220	4.560	4.435	5.563	....	3060	....	....	....	4240	....	....	...	133	....	....	....	....	....
	13.00	.253	4.494	4.369	5.563	....	4140	....	....	....	4870	....	....	...	169	....	....	182	....	....
	15.00	.296	4.408	4.283	5.563	....	5550	7250	8830	....	5700	8290	11400	...	207	....	....	223	311	388
	18.00	.362	4.276	4.151	5.563	....	....	10490	13450	....	....	10140	13940	...	....	....	....	....	396	495
5-1/2	14.00	.244	5.012	4.887	6.050	2630	3120	....	....	3110	4270	....	....	130	172	....	....	....	....	....
	15.50	.275	4.950	4.825	6.050	....	4040	....	....	....	4810	....	....	...	202	....	....	217	....	....
	17.00	.304	4.892	4.767	6.050	....	4910	6280	7460	....	5320	7740	10640	...	229	....	....	247	348	445
	20.00	.361	4.778	4.653	6.050	....	....	8830	11080	....	....	9190	12640	...	....	....	....	....	428	548
	23.00	.415	4.670	4.545	6.050	....	....	11160	14520	....	....	9880	13580	...	....	....	....	....	502	643
6-5/8	20.00	.288	6.049	5.924	7.390	2520	2970	....	....	3040	4180	....	....	184	245	....	....	266	....	....
	24.00	.352	5.921	5.796	7.390	....	4560	5760	6710	....	5110	7440	10230	...	314	....	....	340	481	641
	28.00	.417	5.791	5.666	7.390	....	....	8170	10140	....	....	8810	12120	...	....	....	....	....	586	781
	32.00	.475	5.675	5.550	7.390	....	....	10320	13200	....	....	10040	13800	...	....	....	....	....	677	904
7	17.00	.231	6.538	6.413	7.656	1450	....	....	....	2310	....	....	....	122	....	....	....	....	....	....
	20.00	.272	6.456	6.331	7.656	1980	2270	....	....	2720	3740	....	....	176	234	....	....	....	....	....
	23.00	.317	6.366	6.241	7.656	....	3270	3830	....	....	4360	6340	....	...	284	....	....	313	442	....
	26.00	.362	6.276	6.151	7.656	....	4320	5410	6210	....	4980	7240	9960	...	334	....	....	367	519	693
	29.00	.408	6.184	6.059	7.656	....	....	7020	8510	....	....	8160	11220	...	....	....	....	....	597	797
	32.00	.453	6.094	5.969	7.656	....	....	8600	10760	....	....	9060	12460	...	....	....	....	....	672	897
	35.00	.498	6.004	5.879	7.656	....	....	10180	13010	....	....	9240	12700	...	....	....	....	....	746	996
	38.00	.540	5.920	5.795	7.656	....	....	11390	15110	....	....	9240	12700	...	....	....	....	....	814	1087
7-5/8	24.00	.300	7.025	6.900	8.500	2040	....	....	....	2750	....	....	....	212	....	....	....	....	....	....
	26.40	.328	6.969	6.844	8.500	....	2890	3400	....	....	4140	6020	....	...	315	....	....	346	490	....
	29.70	.375	6.875	6.750	8.500	....	....	4790	5340	....	....	6890	9470	...	....	....	....	....	575	769
	33.70	.430	6.765	6.640	8.500	....	....	6560	7850	....	....	7900	10860	...	....	....	....	....	674	901
	39.00	.500	6.625	6.500	8.500	....	....	8810	11060	....	....	9180	12620	...	....	....	....	....	798	1066
8-5/8	24.00	.264	8.097	7.972	9.625	....	1370	....	....	....	2950	....	....	...	244	....	....	....	....	....
	28.00	.304	8.017	7.892	9.625	1640	....	....	....	2470	....	....	....	233	....	....	....	....	....	....
	32.00	.352	7.921	7.796	9.625	2210	2530	....	....	2860	3930	....	....	279	372	....	....	417	....	....
	36.00	.400	7.825	7.700	9.625	....	3450	4100	....	....	4460	6490	....	...	434	....	....	486	688	....
	40.00	.450	7.725	7.600	9.625	....	....	5520	6380	....	....	7300	10040	...	....	....	....	....	788	1055
	44.00	.500	7.625	7.500	9.625	....	....	6950	8400	....	....	8120	11160	...	....	....	....	....	887	1186
	49.00	.557	7.511	7.386	9.625	....	....	8570	10720	....	....	9040	12430	...	....	....	....	....	997	1335
9-5/8	32.30	.312	9.001	8.845	10.625	1400	....	....	....	2270	....	....	....	254	....	....	....	....	....	....
	36.00	.352	8.921	8.765	10.625	1740	2020	....	....	2560	3520	....	....	294	394	....	....	453	....	....
	40.00	.395	8.835	8.679	10.625	....	2570	3090	....	....	3950	5750	....	...	452	....	....	520	737	....
	43.50	.435	8.755	8.599	10.625	....	....	3810	4430	....	....	6330	8700	...	....	....	....	....	825	1106
	47.00	.472	8.681	8.525	10.625	....	....	4750	5310	....	....	6870	9440	...	....	....	....	....	905	1213
	53.50	.545	8.535	8.379	10.625	....	....	6620	7930	....	....	7930	10900	...	....	....	....	....	1062	1422
10-3/4	32.75	.279	10.192	10.036	11.750	870	....	....	....	1820	....	....	....	205	....	....	....	....	....	....
	40.50	.350	10.050	9.894	11.750	1420	1580	....	....	2280	3130	....	....	314	420	....	....	....	....	....
	45.50	.400	9.950	9.794	11.750	....	2090	....	....	....	3580	....	....	...	493	....	....	....	....	....
	51.00	.450	9.850	9.694	11.750	....	2700	3220	3670	....	4030	5860	8060	...	565	804	1080	....	....	....
	55.50	.495	9.760	9.604	11.750	....	....	4020	4630	....	....	6450	8860	...	....	895	1203	....	....	....
	60.70	.545	9.660	9.504	11.750	....	....	5860	....	....	....	....	9760	...	....	....	1338	....	....	....
	65.70	.595	9.560	9.404	11.750	....	....	7490	....	....	....	....	10650	...	....	....	1472	....	....	....
11-3/4	42.00	.333	11.084	10.928	12.750	1070	....	....	....	1980	....	....	....	307	....	....	....	....	....	....
	47.00	.375	11.000	10.844	12.750	....	1510	....	....	....	3070	....	....	...	477	....	....	....	....	....
	54.00	.435	10.880	10.724	12.750	....	2070	....	....	....	3560	....	....	...	568	....	....	....	....	....
	60.00	.489	10.772	10.616	12.750	....	2660	3180	....	....	4010	5830	....	...	649	924	....	....	....	....
13-3/8	48.00	.330	12.715	12.559	14.375	740	....	....	....	1730	....	....	....	322	....	....	....	....	....	....
	54.50	.380	12.615	12.459	14.375	....	1130	....	....	....	2730	....	....	...	514	....	....	....	....	....
	61.00	.430	12.515	12.359	14.375	....	1540	....	....	....	3090	....	....	...	595	....	....	....	....	....
	68.00	.480	12.415	12.259	14.375	....	1950	....	....	....	3450	....	....	...	675	....	....	....	....	....
	72.00	.514	12.347	12.191	14.375	....	....	2670	....	....	....	5380	....	...	....	1040	....	....	....	....
16	65.00	.375	15.250	15.062	17.000	630	....	....	....	1640	....	....	....	439	....	....	....	....	....	....
	75.00	.438	15.124	14.936	17.000	....	1020	....	....	....	2630	....	....	...	710	....	....	....	....	....
	84.00	.495	15.010	14.822	17.000	....	1410	....	....	....	2980	....	....	...	817	....	....	....	....	....
18-5/8	87.50	.435	17.755	17.567	19.625	630	630	....	....	1630	2250	....	....	559	754	....	....	....	....	....
20	94.00	.438	19.124	18.936	21.000	520	520	....	....	1530	2110	....	....	581	784	....	....	907	....	....
	106.50	.500	19.000	18.812	21.000	....	770	....	....	....	2410	....	....	...	913	....	....	1057	....	....
	133.00	.635	18.730	18.542	21.000	....	1500	....	....	....	3060	....	....	...	1192	....	....	1380	....	....

TABLE 6-4
Recommended Tubing Makeup Torque, API Connections[1]

Size: Outside Diameter in.	Nominal Weight lb. per ft.			Grade	Torque, ft-lb								
	Threads and Coupling		Integral Joint		Non-Upset			Upset			Integral Joint		
	Non-Upset	Upset			Opt.	Min.	Max.	Opt.	Min.	Max.	Opt.	Min.	Max.
1.050	1.14	1.20	—	H-40	140	110	180	460	350	580	—	—	—
	1.14	1.20	—	J-55	180	140	230	600	450	750	—	—	—
	1.14	1.20	—	C-75	230	170	290	780	590	980	—	—	—
	1.14	1.20	—	N-80	250	190	310	830	620	1040	—	—	—
1.315	1.70	1.80	1.72	H-40	210	160	260	440	330	550	310	230	390
	1.70	1.80	1.72	J-55	270	200	340	570	430	710	400	300	500
	1.70	1.80	1.72	C-75	360	270	450	740	560	930	520	390	650
	1.70	1.80	1.72	N-80	380	290	480	790	590	990	550	410	690
1.660	—	—	2.10	H-40	—	—	—	—	—	—	380	280	480
	2.30	2.40	2.33	H-40	270	200	340	530	400	660	380	280	480
	—	—	2.10	J-55	—	—	—	—	—	—	500	380	630
	2.30	2.40	2.33	J-55	350	260	440	690	520	860	500	380	630
	2.30	2.40	2.33	C-75	460	350	580	910	680	1140	650	490	810
	2.30	2.40	2.33	N-80	490	370	610	960	720	1200	690	520	860
1.900	—	—	2.40	H-40	—	—	—	—	—	—	450	340	560
	2.75	2.90	2.76	H-40	320	240	400	670	500	840	450	340	560
	—	—	2.40	J-55	—	—	—	—	—	—	580	440	730
	2.75	2.90	2.76	J-55	410	310	500	880	660	1100	580	440	730
	2.75	2.90	2.76	C-75	540	410	680	1150	860	1440	760	570	950
	2.75	2.90	2.76	N-80	570	430	700	1220	920	1530	810	610	1010
2.063	—	—	3.25	H-40	—	—	—	—	—	—	570	430	710
	—	—	3.25	J-55	—	—	—	—	—	—	740	560	920
	—	—	3.25	C-75	—	—	—	—	—	—	970	730	1210
	—	—	3.25	N-80	—	—	—	—	—	—	1030	770	1290

[1]For non-API connections, see manufacturers' literature
Reprinted from API Bulletin 5C1

TABLE 6-5
Recommended Tubing Makeup Torque, API Connections

Size: Outside Diameter in.	Nominal Weight lb. per ft. Threads and Coupling Non-Upset	Upset	Integral Joint	Grade	Torque, ft-lb Non-Upset Opt.	Min.	Max.	Upset Opt.	Min.	Max.	Integral Joint Opt.	Min.	Max.
2⅜	4.00	—	—	H-40	470	350	590	—	—	—	—	—	—
	4.60	4.70	—	H-40	560	420	700	990	740	1240	—	—	—
	4.00	—	—	J-55	610	460	760	—	—	—	—	—	—
	4.60	4.70	—	J-55	730	550	910	1290	970	1610	—	—	—
	4.00	—	—	C-75	800	600	1000	—	—	—	—	—	—
	4.60	4.70	—	C-75	960	720	1200	1700	1280	2130	—	—	—
	5.80	5.95	—	C-75	1380	1040	1730	2120	1590	2650	—	—	—
	4.00	—	—	N-80	850	640	1060	—	—	—	—	—	—
	4.60	4.70	—	N-80	1020	770	1280	1800	1350	2250	—	—	—
	5.80	5.95	—	N-80	1460	1100	1830	2240	1680	2800	—	—	—
	4.60	4.70	—	P-105	1280	960	1600	2270	1700	2840	—	—	—
	5.80	5.95	—	P-105	1840	1380	2300	2830	2120	3540	—	—	—
2⅞	6.40	6.50	—	H-40	800	600	1000	1250	940	1560	—	—	—
	6.40	6.50	—	J-55	1050	790	1310	1650	1240	2060	—	—	—
	6.40	6.50	—	C-75	1380	1040	1730	2170	1630	2710	—	—	—
	8.60	8.70	—	C-75	2090	1570	2600	2850	2140	3560	—	—	—
	6.40	6.50	—	N-80	1470	1100	1840	2300	1730	2880	—	—	—
	8.60	8.70	—	N-80	2210	1660	2760	3020	2270	3780	—	—	—
	6.40	6.50	—	P-105	1850	1390	2310	2910	2180	3640	—	—	—
	8.60	8.70	—	P-105	2790	2090	3490	3810	2860	4760	—	—	—
3½	7.70	—	—	H-40	920	690	1150	—	—	—	—	—	—
	9.20	9.30	—	H-40	1120	840	1400	1730	1300	2160	—	—	—
	10.20	—	—	H-40	1310	980	1640	—	—	—	—	—	—
	7.70	—	—	J-55	1210	910	1510	—	—	—	—	—	—
	9.20	9.30	—	J-55	1480	1110	1850	2280	1710	2850	—	—	—
	10.20	—	—	J-55	1720	1290	2150	—	—	—	—	—	—
	7.70	—	—	C-75	1600	1200	2000	—	—	—	—	—	—
	9.20	9.30	—	C-75	1950	1460	2440	3010	2260	3760	—	—	—
	10.20	—	—	C-75	2270	1700	2840	—	—	—	—	—	—
	12.70	12.95	—	C-75	3030	2270	3790	4040	3030	5050	—	—	—
	7.70	—	—	N-80	1700	1280	2130	—	—	—	—	—	—
	9.20	9.30	—	N-80	2070	1550	2590	3200	2400	4000	—	—	—
	10.20	—	—	N-80	2410	1810	3010	—	—	—	—	—	—
	12.70	12.95	—	N-80	3210	2410	4010	4290	3220	5360	—	—	—
	9.20	9.30	—	P-105	2620	1970	3280	4050	3040	5060	—	—	—
	12.70	12.95	—	P-105	4060	5050	5080	5430	4070	6790	—	—	—
4	9.50	—	—	H-40	940	710	1180	—	—	—	—	—	—
	—	11.00	—	H-40	—	—	—	1940	1460	2430	—	—	—
	9.50	—	—	J-55	1240	930	1550	—	—	—	—	—	—
	—	11.00	—	J-55	—	—	—	2560	1920	3200	—	—	—
	9.50	—	—	C-75	1640	1230	2050	—	—	—	—	—	—
	—	11.00	—	C-75	—	—	—	3390	2540	4240	—	—	—
	9.50	—	—	N-80	1740	1310	2180	—	—	—	—	—	—
	—	11.00	—	N-80	—	—	—	4560	3420	5700	—	—	—
4½	12.60	12.75	—	H-40	1320	990	1650	2160	1620	2700	—	—	—
	12.60	12.75	—	J-55	1740	1310	2180	2860	2150	3180	—	—	—
	12.60	12.75	—	C-75	2300	1730	2880	3780	2840	4730	—	—	—
	12.60	12.75	—	N-80	2440	1830	3050	4020	3020	5030	—	—	—

Reprinted from API Bulletin 5C1

Packers and Subsurface Control Equipment

PRODUCTION PACKERS

Production packers are generally classified as either retrievable-type or permanent-type. Recent packer innovations include the retrievable seal nipple packers or semi-permanent type, packer bore receptacle and "cement packer."

Packers are sometimes run when they serve no useful purpose, resulting in unnecessary initial investment and the possibility of high future removal cost. Routine use of packers should be limited to these situations:

1. Protection of casing from pressure (including both well and stimulation pressures) and corrosive fluids,
2. Isolation of casing leaks, squeezed perforations, or multiple producing intervals.
3. Elimination of inefficient "heading" or "surging".
4. Some artificial lift installations.
5. In conjunction with subsurface safety valves.
6. To hold kill fluids or treating fluids in casing annulus.

General Considerations in Packer Selection

Packer selection involves an analysis of packer objectives in anticipated well operations, such as initial completions, production stimulation, and workover procedures. The packer with the minimum overall cost that will accomplish the objective, considering both current and future well conditions, should be selected. Initial investment and installation costs should not be the only criterion. Overall packer cost is directly related to retrievability and failure rate and to such diverse factors as formation damage during subsequent well operations.

Retrievability will be greatly enhanced by utilizing oil or saltwater rather than mud for the packer fluid. Frequency of packer failures may be minimized by utilizing the proper packer for the well condition and by anticipating future conditions when setting the packer. The permanent packer is by far the most reliable and, properly equipped and set, is excellent for the high pressure differentials imposed during stimulation, or when reservoir pressures vary significantly between zones in multi-completions.

Weight-set and tension types of retrievable packers will perform satisfactorily when the force on the packer is in one direction only and is not excessive.

Purchase Price—Table 6-6 presents a range of packer costs. The most economical types are weight-set and tension packers. However, inclusion of a hydraulic hold-down with a weight-set packer will increase the initial cost 20 to 100%. Multistring hydraulic-set packers are usually the most expensive and also require many accessories.

Packer Mechanics—The end result of most packer setting mechanisms is to (1) drive a cone behind a tapered slip to force the slip into the casing wall and prevent packer movement, and (2) compress a packing element to effect a seal. Although the end result is relatively simple, the means of accomplishing it and subsequent packer retrieval varies markedly between the several types of packers.

Some packers involve two or more round trips, some wireline time and some eliminate trips by hydraulic setting. The time cost should be examined carefully, especially on deep wells using high cost rigs. In some cases higher first cost of the packer may be more than offset by saving in rig time, especially on high cost offshore rigs.

Sealing Element—The ability of a seal to hold differential pressure is a function of the rubber pressure or stress developed in the seal, i.e., the stress must exceed the differential pressure. In a packer sealing element, the stress developed depends on the packer setting force, and the back-up provided to limit extrusion.

The sealing element may consist of one piece or may be composed of multiple elements of different hardness. In a three-element packer, the

TABLE 6-6
Cost Comparison of Production Packer Types

Packer type	Tubing-casing size, in.	Typical cost
Weight set	2 × 5½	$ 650
Tension set	2 × 5½	600
Mechanical set	2 × 5½	1,000
Hydraulic set	2 × 5½	1,500
	2 × 2 × 7	3,800
Permanent[a]	2 × 5½	1,200
Semi permanent[a]	2 × 5½	1,500

[a]Electric line setting charge not included.

upper and lowermost elements are usually harder than the center element.

The lower durometer center seals off against imperfections in the casing, while the harder outside elements restrict extrusion and seal with high temperature and pressure differentials. Many packers also include metallic back-up rings to impede extrusion.

Where H_2S or CO_2 are present, seal materials and conditions must be carefully considered. Temperature is a primary factor. Below 250°F, Nitrile rubber can be used with metallic backup for static seals. Viton becomes marginal at 300°F. Recent tests show that a tubing-to-packer seal consisting of vee-type rings of Kalrez, Teflon, and Rylon in sequence with metallic backup is satisfactory under limited movement to 300°F and 10,000 psi differential pressure. Seal sticking is minimal.

Teflon resists H_2S or chemical attack up to 450°F, but extrusion can be a problem. With controlled clearance and suitable metallic backup to prevent extrusion, glass-filled Teflon has performed satisfactorily to 450°F and 15,000 psi differential pressure. Due to seal rigidity it may not perform well below 300°F, however.

Corrosive Well Fluids—Materials used in packer construction must be considered where well fluids contain CO_2 or H_2S in the presence of water.

Sweet corrosion: CO_2 and water cause iron carbonate corrosion, resulting in deep pitting. For ferrous materials low strength steels or cast iron are desirable to resist stress concentrations from pitting. Depending on economics, corrosion inhibitors may be required to protect exposed surfaces. Critical parts of production equipment can be made of stainless steel with 12% or higher chromium.

Sour corrosion: Even small amounts of H_2S with water produce iron sulfide corrosion and hydrogen embrittlement. NACE specifies materials for H_2S conditions be heat-treated to a maximum hardness of 22 Rockwell C to prevent embrittlement. AISI 4140 steel heat-treated to 25 Rockwell C has more usable strength than 22 Rockwell C and compares favorably with C-75 steel for H_2S service. Hardness has no effect on iron sulfide corrosion however. For critical parts where high strength is required K-Monel is resistant to both embrittlement and iron sulfide corrosion.

Bimetallic or galvanic corrosion resulting from contact of dissimilar metals should be considered. Usually this is not a problem, however, since steel is the anode or sacrificial member, and resulting damage is negligible due to the massive area of the steel compared to the more noble stainless or K-Monel.

Retrievability—Retrievability is a combination of several factors, related to packer design and packer use. Retrievable packers are released by either straight pull or by rotation. In deviated hole torque usually develops more downhole releasing force than pull, although sometimes it is necessary to manipulate the tubing to transmit the torque to bottom.

The packer sealing element should prevent solids from settling around the slips. Usually the bypass opens before the seal is released, to permit circulation to remove sand or foreign material.

High setting force is needed to provide a reliable seal under high differential pressures, but it should be recognized that seal extrusion can contribute to the retrieval problem. A jar stroke between release and pickup positions is an aid in packer removal.

The method of retaining slip segments is a factor in retrievability. Bypass area is also important. Where external clearance is minimized to promote sealing, internal bypass area must be sufficient to prevent swabbing off the sealing element in pulling out of the hole.

Fishing Characteristics—A permanent packer must be drilled out to effect removal. This usually presents little problem because all material is drillable. Recent expensive variations of permanent packers provide for retrieval but retain the removable seal tube feature. Removal of "stuck" retrievable packers usually results in an expensive fishing operation because components are "non-drillable." In comparing packers consider the volume of metal that must be removed and the presence of rings or hold-down buttons that may act as ball bearings to milling tools.

Through-Tubing Operations—Packers with internal diameters equal to that of the tubing should be utilized to facilitate through-tubing operations. Also tubing should be set so as to minimize buckling where through-tubing operations are anticipated.

Surface Equipment—Downhole Correlation—Setting a packer always requires surface action and in most cases either vertical or rotational movement of the tubing. Thus selection of the packer must be related to wellhead equipment. The well completion must be considered as a coordinated operation and the surface and downhole equipment selected to work together to insure a safe completion, especially in high pressure wells.

For example, a wrap-around tubing hanger with a smooth joint designed to allow vertical and rotational tubing movement under pressure may be used with J-tool or rotational set compression packers. However, if a well is to be washed-in and cleaned before the packer is set, a hydraulic-set packer may be needed for proper well control.

Effect of Pressure, Temperature Changes

Changes in temperature, and in pressure inside and outside of tubing sealed in a packer, depending on the type of packer and how it is set will:

1. Increase or decrease the length of the tubing with a packer permitting free motion of the tubing.

2. Induce tensile or compressive forces in the tubing and packer if free motion is not permitted within a packer which is rigidly attached to the casing.

3. Unseat a packer not rigidly attached to the casing.

4. Unseal a permanent-type packer where the seal nipple section is not long enough and the tubing is not latched into the packer.

Several effects must be evaluated to accurately determine the tubing movement or stress situation as shown in Figure 6-6. Lubinski[1] and Hammerlindl[7] present an excellent discussion of these effects,

show calculation procedures, and give examples based on field situations. The following simplified discussion concerns vertical uniform completions. For a combination completion that contains more than one size of tubing or casing, or more than one fluid in the tubing or casing refer to the Hammerlindl paper.[7]

Hooke's Law Effect—The pressures inside the tubing and in the annulus above the packer act on the differential area presented by the tubing and the packer mandrel to change the tubing length according to Hooke's Law. If tubing motion is limited by the packer, the force of the packer on the tubing also affects tubing length according to Hooke's Law:

$$\Delta L = \frac{L \Delta F}{E A_s} \qquad (1)$$

With a free motion seal where only pressure induced forces are acting:

$$\Delta F = (A_p - A_i)\Delta P_i - (A_p - A_o)\Delta P_o$$

Where:

ΔL = change in length due to Hooke's Law effect, in.
L = length of tubing—in.
F = force acting on bottom of tubing, lb.
E = modulus of elasticity (30×10^6 psi for steel)
A_s = cross-sectional area of tubing, in[2]
A_i = area based on inside diameter of tubing, in[2]
A_o = area based on outside diameter of tubing, in[2]
A_p = area based on diameter of packer seal, in[2]
P_o = pressure at packer seal in annulus, psi
P_i = pressure at packer seal in tubing, psi

See Fig. 6-7.
Notes:

1. ΔL, ΔF, ΔP_i or ΔP_o indicate change from initial packer setting conditions. It is assumed $P_i = P_o$ when packer is initially set.

2. For values of dimensional tubing and packer constants, see Table 6-8.

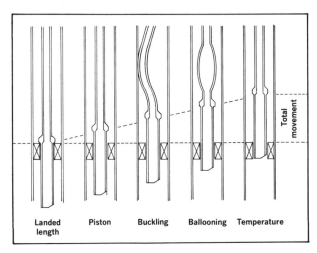

FIG. *6-6—Effect of various forces on tubing movement.*

Landed length   Piston   Buckling   Ballooning   Temperature

Total movement

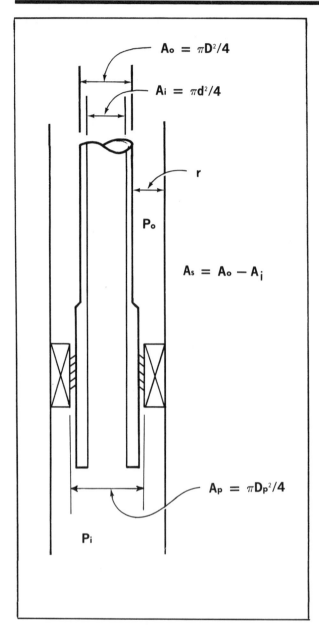

FIG. 6-7—Definition of terms.

Helical Buckling—Buckling or "corkscrewing" of the tubing above the packer may shorten the tubing. With a packer permitting free motion of the tubing, buckling is a result of the differential between the pressure inside and that outside the tubing acting on the full cross-sectional area of packer bore at the tubing seal. Where the packer limits tubing motion, the tubing weight set on the packer must also be considered.

Shortening due to helical buckling with a free motion seal and where the buckling affects only a portion of the length of the tubing, may be calculated by the following equation:

$$\Delta L_2 = \frac{r^2 A_p^2 (\Delta P_i - \Delta P_o)^2}{8\,E\,I\,(w_s + w_i - w_o)}$$

(2)

$$I = \frac{\pi}{64}\,(D^4 - d^4)$$

Force causing buckling: $F_f = A_p\,(\Delta P_i - \Delta P_o)$
(If F_f is negative, there is no buckling)
Length of tubing buckled: $n = \dfrac{F_f}{w}$

Where:

r = radial clearance between tubing and casing—in.
$w = w_s + w_i - w_o$ (See Table 6-8 for values)
w_s = weight of tubing—lb/in.
w_i = weight of fluid contained inside tubing—lb/in. (based on id of tubing).
w_o = weight of annulus fluid displaced by bulk volume of tubing—lb/in. (based on od of tubing).
D = tubing outside diameter—in.
d = tubing inside diameter—in.

Buckling is increased by higher pressure inside the tubing, by a larger ratio of casing to tubing diameter, by lower density fluid in the tubing, by a larger packer bore, and by an upward acting packer force.

With a free motion seal, buckling cannot occur if the pressure at the seal in the annulus (P_o) is greater than the pressure at the seal inside the tubing (P_i). Again it is assumed that $P_i = P_o$ when the packer was set initially.

Ballooning Effect—Radial pressure inside the tubing tends to increase tubing diameter and thereby shorten the tubing. Greater pressure outside the tubing lengthens the string due to "reverse ballooning."

Change in length due to ballooning or reverse ballooning can be calculated according to the following equation:

$$\Delta L_3 = \frac{\mu L^2}{E}\left(\frac{\Delta\rho_i - R^2\Delta\rho_o - \frac{1+2\mu}{2\mu}\delta}{R^2-1}\right)$$

(Density effect)

$$+ \frac{2\mu L}{E}\left(\frac{\Delta p_i - R^2\Delta p_o}{R^2-1}\right) \qquad (3)$$

(Surface pressure effect)

μ = Poisson's ratio (0.3 for steel)
R = tubing od/tubing id
$\Delta\rho_i$ = change in density of fluid inside tubing: lb/cu in.
$\Delta\rho_o$ = change in density of fluid outside tubing: lb/cu in.
Δp_i = change in *surface* pressure inside tubing
Δp_o = change in *surface* pressure outside tubing
δ = pressure drop in tubing due to flow psi/in. (usually considered as $\delta = o$)

Temperature Effect—A change in temperature due to producing hot fluids or injecting cold fluid changes the tubing length as follows:

$$\Delta L = LC\Delta T \qquad (4)$$

ΔL = change in length, (ft)
L = length of tubing string, (ft)
$C = 6.9 \times 10^{-6}$, coefficient of expansion of steel per °F
ΔT = temperature change, °F

Packer Setting Force—The initial compressive force or tensile force used in setting the packer has a direct bearing on the subsequent situation. Proper choice of the initial setting force must anticipate future conditions resulting from production, stimulation or remedial operations.

To convert tubing length change due to slack-off to compressive force, the following equation may be used:

$$\Delta L = \frac{LF}{EA_s} + \frac{r^2F^2}{8EI(w_s + w_i - w_o)} \qquad (5)$$

Permanent Buckling or "Corkscrewing"

1. *Due to Initial Slack Off:*—Excessive initial set-down weight or "slack off" on a packer may cause permanent buckling of the tubing even before pressures and temperatures are changed by completion or production operations. To investigate this possibility, the following equation may be used. If S_o exceeds the yield strength of the tubing grade, then permanent damage will result.

$$S_o = \frac{F}{A_s} + \frac{DrF}{4I} \qquad (6)$$

F = set-down force

2. *Due to Subsequent Operations:*—If stress induced in the tubing by pressure or temperature changes or by movements exceed the yield stress of the tubing material, then permanent deformation will result. Stress level should be checked at both the inner and outer wall of the pipe using the following equations. These stresses, S_i and S_o, should not be allowed to exceed the yield stress of the tubing.

$$S_i = \sqrt{3\left[\frac{R^2(P_i - P_o)}{R^2-1}\right]^2 + \left[\frac{P_i - R^2P_o}{R^2-1}+\sigma_a\pm\frac{\sigma_b}{R}\right]^2} \qquad (7)$$

$$S_o = \sqrt{3\left[\frac{P_i - P_o}{R^2-1}\right]^2 + \left[\frac{P_i - R^2P_o}{R^2-1}+\sigma_a\pm\sigma_b\right]^2} \qquad (8)$$

$$\sigma_a = \frac{F_a}{A_s}$$

For Free Motion Pkr.:

$$F_a = (A_p - A_i)P_i - (A_p - A_o)P_o$$

$$\sigma_b = \frac{Dr}{4I}F_f$$

For Free Motion Pkr.:

$$F_f = A_p(\Delta P_i - \Delta P_o)$$

Where packer exerts force on tubing, this force must be added to F_a or F_f.

Sign of σ_b in Equations 7 and 8 should be chosen to maximize S_i or S_o.

Table 6-7 shows examples of tubing movement in a permanent-type packer (3¼-in. bore) set at 10,000 ft in 7-in. casing. 2⅞-in. tubing with no locator sub was landed without latching into the packer; thus there is free motion of the tubing within the packer:

TABLE 6-7
Tubing Movement in Permanent Packer

Conditions:	Type of Operation			
	Swab	Production	Frac	Squeeze Cement
Initial Fluid	10 lb/gal Mud	10 lb/gal Mud	9.2 lb/gal Saltwater	30° Oil
Final Fluid:				
Tubing	45° Oil to 5000 ft	45° Oil	9.0 lb/gal frac fluid	15 lb/gal cmt
Annulus	10 lb/gal Mud	10 lb/gal Mud	9.2 lb/gal Saltwater	30°Oil
Final Pressure:				
Tubing	0	1000 psi	3000 psi	5000 psi
Annulus	0 psi	0 psi	1000 psi	1000 psi
Temp. Change	+10°F	+20°F	−50°F	−20°F
Tbg. Movement due to:				
Hooke's Law	+27.6 in.	+5.7 in.	−19.2 in.	−67.9 in.
Buckling	0	0	−3.6	−46.1
Ballooning	+13.4	−0.9	−9.7	−34.6
Temperature	+8.3	+16.6	−41.5	−16.6
Total Movement	+49.3 in.	+21.4 in.	−74.0 in.	−165.2 in.

Note: (−) indicates shortening, (+) indicates lengthening.

With the described conditions the permanent packer setup would not be suitable for fracing or high pressure squeeze cementing unless an unusually long seal nipple section had been provided.

Considering the high pressure squeeze cementing operation, if the tubing were latched into the permanent packer (with zero tubing pickup or slack-off) the forces on the tubing would result in an upward pull on the packer latch-in section of 70,000 lb. At the surface tensile force in the tubing would be 87,000 lb as compared to its weight in air of 64,000 lb.

Under these conditions the tubing would be buckled slightly; an initial pickup of 18,000 lb would be required to eliminate buckling. Depending on the tubing grade, the string might be permanently deformed by the latch-in procedure.

As a better alternative, if a locator sub were run, sufficient tubing weight could be slacked off on the packer to restrict tubing movement. Tubing slack-off of 72,000 lb would prevent tubing movement under the conditions of the high pressure squeeze job. However, excessive tubing slack-off is undesirable due to buckling and probably hinderance to use of through-tubing tools.

In deep wells, pressure fluctuations in normal producing operations can result in continual small movements of the seals and eventual failure. Best practice in this situation is to slack off sufficient tubing weight to prevent these small seal movements, but install sufficient sliding seal section to provide for upward movement of tubing during well treatments.

For a comprehensive understanding of pressure-temperature effects on packer hookups the Lubinski et al paper and the Hammerlindl paper, References 1 and 7, should be studied.

Unseating of Weight Set Packer

To determine whether or not a weight-set packer will be unseated by injecting in the tubing a simple force balance calculation, as shown in Figure 6-8, is sufficient. In this example, the well is equipped with 5½-in. casing, the packer is set with 7,000 lb tubing weight at 6,000 ft on 2⅜-in. od tubing. The annulus contains salt water. Acid is displaced down the tubing with crude oil. The resulting surface pressure of 1,000 psi imposes a 3,800-lb upward force on the bottom of the packer which will unseat it.

Unseating may be avoided by (1) applying pres-

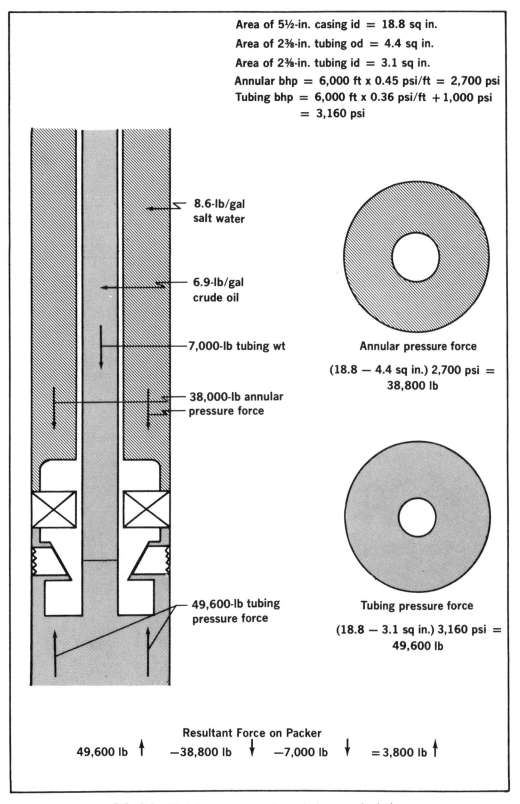

Area of 5½-in. casing id = 18.8 sq in.

Area of 2⅜-in. tubing od = 4.4 sq in.

Area of 2⅜-in. tubing id = 3.1 sq in.

Annular bhp = 6,000 ft x 0.45 psi/ft = 2,700 psi

Tubing bhp = 6,000 ft x 0.36 psi/ft + 1,000 psi
= 3,160 psi

8.6-lb/gal
salt water

6.9-lb/gal
crude oil

7,000-lb tubing wt

38,000-lb annular
pressure force

49,600-lb tubing
pressure force

Annular pressure force

(18.8 — 4.4 sq in.) 2,700 psi =
38,800 lb

Tubing pressure force

(18.8 — 3.1 sq in.) 3,160 psi =
49,600 lb

Resultant Force on Packer

49,600 lb ↑   —38,800 lb ↓   —7,000 lb ↓   = 3,800 lb ↑

FIG. *6-8—Weight-set packer force-balance calculation.*

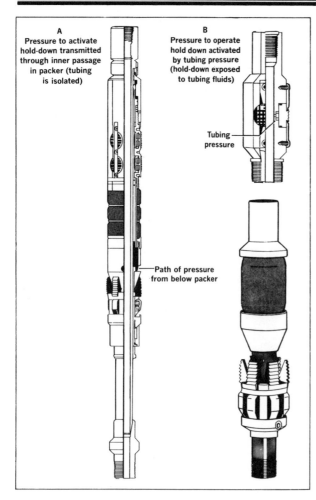

A
Pressure to activate
hold-down transmitted
through inner passage
in packer (tubing
is isolated)

B
Pressure to operate
hold down activated
by tubing pressure
(hold-down exposed
to tubing fluids)

Tubing
pressure

Path of pressure
from below packer

FIG. *6-9—Types of hydraulic hold-downs.*

sure to the casing-tubing annulus or (2) utilizing a hydraulic hold-down. Figure 6-9 pictures two types of hydraulic hold-downs.

Retrievable Packers

The following discussion describes briefly the operation and application of various "families" of Retrievable packers. The details of specific packers are best studied by reference to manufacturers literature and setting instructions.

Weight-set Packers—The weight-set packer is economical and ideally suited to low pressure situations where the annulus pressure above the packer always exceeds the tubing pressure below the packer. A hydraulic holddown is often included where pressure differentials from both directions are anticipated.

Weight-set packers (Figure 6-10) employ a slip and cone arrangement with the slips attached to a friction device such as drag springs or drag blocks. The friction device engages the casing and holds the slips stationary with respect to the remainder of the packer.

A "J" slot device permits vertical movement of the tubing and causes the cone to move behind the slips and anchor the packer in the casing. Tubing weight is then applied to expand the packing element. Frictional drag limits tubing weight and thus the setting force that can be applied. Release is effected by picking up tubing weight to pull the cone from behind the slips.

Tension-set Packers—Tension packers are essentially weight-set packers run upside down and set by pulling tension on the tubing. After a tension packer is set, a differential pressure from below

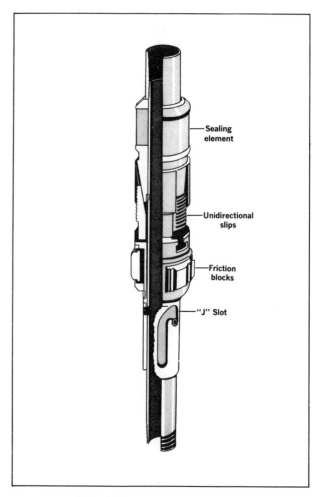

Sealing
element

Unidirectional
slips

Friction
blocks

"J" Slot

FIG. *6-10—Weight-set packer.*

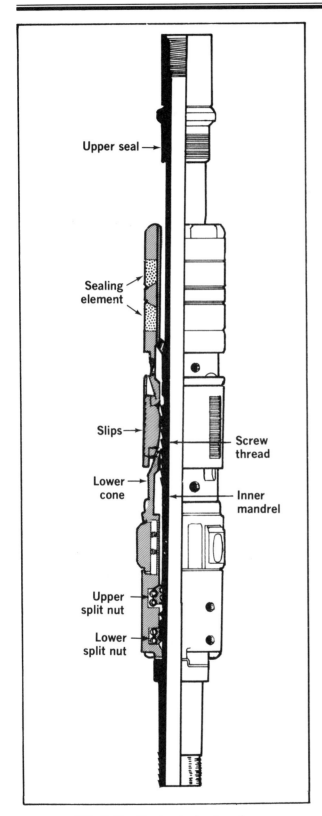

FIG. *6-11—Mechanical-set packer.*

increases the setting force on the packer and holds it in place automatically. This feature renders the tension packer particularly suitable for water injection wells or stimulation work.

Tension packers are frequently used in shallow wells where insufficient tubing weight is available to seal a weight-set packer. Being short and compact, they require a minimum of space for setting and little maintenance. Temperature should be taken into account in determining the initial setting force on the packer; injection or production of hot fluids can cause the tubing to elongate and release the packer.

Mechanical-set Packers—Tubing rotation plays a major role in setting and retrieving mechanical-set packers. Tubing rotation may either (1) simultaneously set the seals and slips in one continuous motion with a screw-thread and cone arrangement which forces the cones behind the slips and compresses the seals or (2) release the inner mandrel and allow tubing weight to drive the cones behind the slips and compress the sealing element. See Figure 6-11.

These packers generally incorporate a non-directional slip system which prevents movement from either direction thus eliminating the necessity for a hydraulic holddown. After the packer is set, tension may be applied to the tubing to reduce buckling and thus to facilitate passage of through-tubing tools.

Release is effected by righthand pipe rotation. The necessity to release the packer by rotating the tubing is the principal disadvantage of mechanical-set packers; the screwthread may be inoperative after extended periods of time, or solids settling on top of the packer may make it impossible to rotate the tubing.

Hydraulic-set Packers—Hydraulic-set packers utilize fluid pressure acting on a piston-cylinder arrangement to drive the cone behind the slips. The packer remains set by a pressure actuated mechanical lock. Release is accomplished by picking up the weight of the tubing string. A hydraulic holddown is required because the slips are unidirectional. Figure 6-12 pictures a single hydraulic-set packer and a schematic of the setting and releasing mechanism. Multi-string hydraulic packers are set and retrieved by essentially the same process.

A primary use of hydraulic-set packers is in multi-string conventional wells. Principal advantages are: (1) the tubing can be landed, christmas

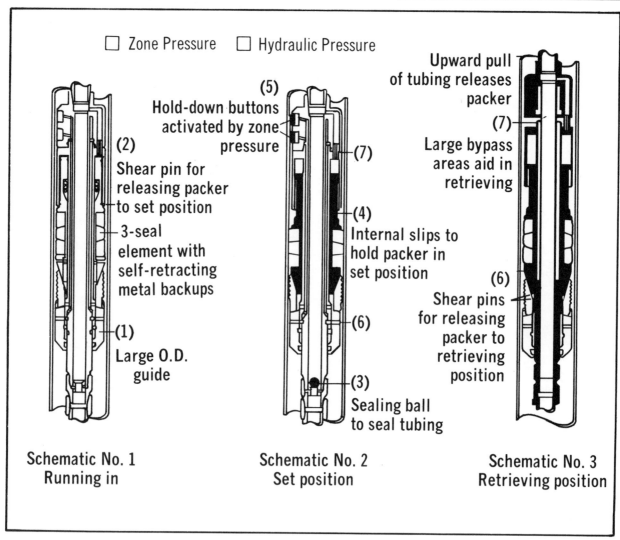

☐ Zone Pressure   ☐ Hydraulic Pressure

(5)
Hold-down buttons
activated by zone
pressure

(2)
Shear pin for
releasing packer
to set position

3-seal
element with
self-retracting
metal backups

(1)
Large O.D.
guide

(7)

(4)
Internal slips to
hold packer in
set position

(6)

(3)
Sealing ball
to seal tubing

Upward pull
of tubing releases
packer
(7)
Large bypass
areas aid in
retrieving

(6)
Shear pins
for releasing
packer to
retrieving
position

Schematic No. 1
Running in

Schematic No. 2
Set position

Schematic No. 3
Retrieving position

FIG. 6-12—*Single hydraulic-set packer, schematic of setting and releasing mechanism.*

tree installed, and well circulated with a light fluid or gas before setting the packer to initiate production without swabbing, (2) all strings can be landed in tension to enhance the passage of wireline tools and concentric tubing, (3) since tubing motion is not required, all strings in a multiple completion may be run and landed simultaneously with multiple slips and elevators before setting the packers, and (4) high setting force can be applied to hold large differential pressures.

Despite their advantages, the expense of hydraulic-set packers limits their use to situations where other packers are not applicable, such as in dual, triple, and quadruple-string completions, deviated

holes, or ocean floor completions.

Hydraulic Expansion Seal Packers—An interesting version of the hydraulic-set principle is utilized to effect a seal in open hole. An inflatable rubber element packer has been successfully employed to shut off bottom water and isolate high gas-oil ratio zones. Figure 6-13 shows the operating mechanics of the packer when utilized conventionally. Bull plugging the bottom of the mandrel and including an entry sub above the packer permits fluid production from above the packer only.

The packer is available with several different length seals. Selecting a long sealing element and setting the packer in a relatively gauge portion of

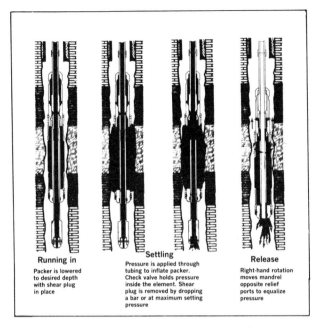

FIG. *6-13—Open-hole inflatable packer.*

Running in
Packer is lowered to desired depth with shear plug in place

Settling
Pressure is applied through tubing to inflate packer. Check valve holds pressure inside the element. Shear plug is removed by dropping a bar or at maximum setting pressure

Release
Right-hand rotation moves mandrel opposite relief ports to equalize pressure

the hole will increase the chances of obtaining an effective barrier. Due to the large diametral expansion the inflatable packer also has application inside casing where it may be necessary to set a packer below an obstruction.

Permanent Packers

Permanent packers utilize opposed slips with a compressible sealing element between the slips. They may be run on tubing or electric conductor cable. Wireline setting is a valuable asset where precise packer location is necessary. Since the tubing may be run separately from the packer, trip time is faster and replacement of the packing in the seal nipples is the only dressing required. The metallic back-up ring for the sealing element and the opposed-slip principle of this type packer is outstanding in applications involving high pressure differentials.

Figure 6-14A pictures a permanent packer. The tubing is sealed off in the bore of the packer by V-type chevron packing, attached to the seal nipples. Downward movement of the tubing is impeded by either a locator or anchor sub (Figure 6-14B). The anchor type permits the tubing to be latched into the packer, thus avoiding contraction. Right-

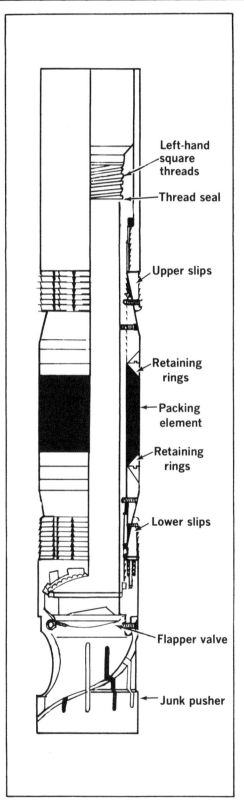

Left-hand square threads

Thread seal

Upper slips

Retaining rings

Packing element

Retaining rings

Lower slips

Flapper valve

Junk pusher

FIG. *6-14A—Permanent packer.*

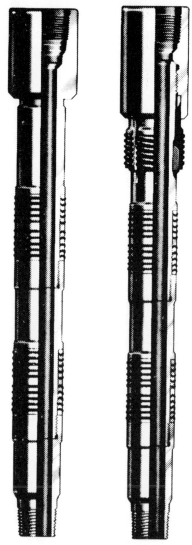

FIG. *6-14B—Seal assemblies, locator-type (left), and anchor-type (right).*

hand rotation releases the anchor assembly from the packer. As an alternative, any number of seals may be run to compensate for tubing movement with the locator sub.

Accessories are available to convert the packer to a temporary bridge plug to test, squeeze cement, or frac above the packer.

To set the packer by wireline methods, a setting tool and collar locator are attached to the packer and the entire assembly is run to the desired setting depth. An electrically detonated powder charge builds up gas pressure that is transmitted hydraulically through a piston arrangement to mechanical forces which set the packer. A release stud then shears and the setting tool is retrieved from the well. Seal nipples to pack off in the bore of the packer are run on the tubing.

A combination of right-hand pipe rotation, tension, and weight is utilized to set the packer on tubing. The seal nipple assembly is normally run with the packer, in this instance, to avoid an extra round trip.

The primary objection to permanent packers, their permanence, has been partially overcome by the development of through-tubing workover methods. Low-pressure squeeze cementing, dump bailers, through-tubing perforators, and concentric tubing techniques usually eliminate the necessity for packer removal.

It is usually possible to set a permanent packer above several alternate completion intervals and successively plugback without pulling the tubing string. If a desired completion interval is above the packer, the permanent packer serves as an excellent cement retainer to squeeze cement the zone below the packer.

Omission of the locator or latch-in sub permits running seals to pack off in any number of permanent packers set in the same wellbore. In this manner, permanent packers are adaptable as isolation packers.

Permanent packers are readily removed in two to three hours drilling time using a flat-bottom packer mill or about six hours with a rock bit. By comparison, removal of a stuck retrievable packer may require two or three days and considerable tool expense. Packer milling and retrieving tools are also available to recover the permanent packer by cutting the upper slips and catching the remainder of the packer.

Initial cost of permanent packers is somewhat higher than weight-set or tension packers, but less than hydraulic-set models. If accuracy within a few feet is not necessary, permanent packers are set most economically on tubing, thus avoiding per-foot depth and service charges. However, where a wireline truck is necessary for other services, such as perforating, the service charge is absorbed in the perforating operation.

Semi-permanent Packers—(Figure 6-15) Semipermanent packers are essentially the same as the permanent packers except that the slip and seal mechanisms are not drillable but are designed to be removed by straight upward pull from a retrieving

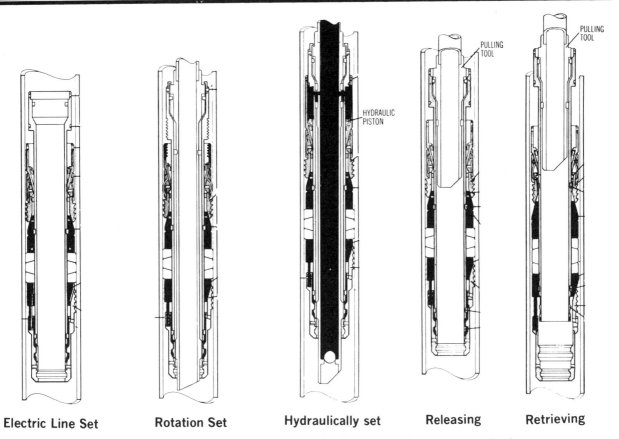

Electric Line Set   Rotation Set   Hydraulically set   Releasing   Retrieving

FIG. *6-15—Semipermanent packer, schematic of setting and releasing mechanism.*

tool run on tubing or a wireline. The setting procedure is the same as for permanent packers, i.e.: by electric wireline, by rotation or by hydraulics.

As shown in Figure 6-15 the retrieving tool carries cams that engage collets in the inner packer mandrel sleeve. Upward pull by retrieving tool shears pins or shear rings that allow the upper slips to move off the expander cone and the lower cone to move out of the slips.

Inner seal assemblies for these packers are the same as shown in Figure 6-14B for permanent packers.

Packer Bore Receptacle

In deep gas wells or other situations where casing or liner diameter is limited and maximum packer bore is desired, the Packer Bore Receptacle, Figure 6-16, may have application.

The seal arrangement of Figure 6-16 allows sufficient free upward tubing movement during stimulation treatments but permits tubing weight slack-off to eliminate seal movements during the producing life of the well. Produced fluids are not in contact with the intermediate casing or the polished sealing surface and the liner top is not exposed to pressure.

Cement Packers

The term "cement packer" has become accepted for tubing cemented inside conventional casing, which essentially creates a tubingless well. Cement is circulated into the tubing-casing annulus just as in cementing tubing in the open hole for tubingless wells. The cement packer replaces the conventional packer by sealing against vertical flow in the tubing-casing annulus.

The technique has these advantages:

1. Isolating leaking squeezed perforations and casing failures without remedial squeeze cementing.

2. Avoiding setting liners during deepening operations.

3. Minimizing the need for wireline completion equipment in multiple wells.

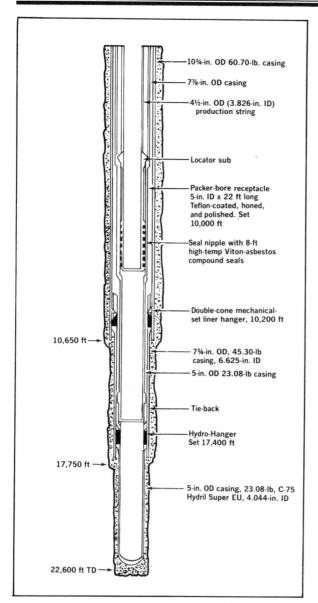

Locator sub

Packer-bore receptacle
5-in. ID x 22 ft long
Teflon-coated, honed,
and polished. Set
10,000 ft

Seal nipple with 8-ft
high-temp Viton-asbestos
compound seals

10¾-in. OD 60.70-lb. casing

7⅞-in. OD casing

4½-in. OD (3.826-in. ID)
production string

Double-cone mechanical-
set liner hanger, 10,200 ft

7¾-in. OD, 45.30-lb
casing, 6.625-in. ID

5-in. OD 23.08-lb casing

10,650 ft

Tie-back

Hydro-Hanger
Set 17,400 ft

17,750 ft

5-in. OD casing, 23.08-lb, C-75
Hydril Super EU, 4.044-in. ID

22,600 ft TD

FIG. 6-16—Deep gas well completion with maximum
bore from surface to TD.[6] Permission to publish by
The Society of Petroleum Engineers.

4. Eliminating communication repair jobs due to
tubing and packer leaks.

5. Permitting application of tubingless completion techniques during future servicing and workovers.

Figures 6-17, 6-18, and 6-19 illustrate typical applications. The workover objective in the first example was to deepen the well several hundred feet and complete for water injection in both the existing and deepened interval. Conventional methods would dictate setting and cementing a liner, re-running tubing on a packer and perforating. Utilizing the cement packer concept, tubing was merely cemented in the wellbore and the formation perforated through both strings of casing.

In the second example, a casing leak prohibited routine single-string dual completion of a conventionally cased well. Running and cementing two strings of tubing accomplished the desired results at minimum cost and eliminated the possibility of future problems with tubing or packer leaks should it be desired to stimulate.

Figure 6-19 illustrates the use of the cement packer idea to repair a leaking packer, or to recomplete above the upper packer in a dual packer well. Briefly the procedure is to wireline plug the short tubing string and perforate the short string with one or two holes immediately above the upper packer.

A concentric workstring is then used to circulate low fluid loss cement into the annulus above the upper packer. Excess cement can be reversed back, and the annulus cement held in place by pressure or by balancing fluids.

SUBSURFACE CONTROL EQUIPMENT

Subsurface control equipment includes: (1) safety valves which plug the tubing at some point below the wellhead should the surface controls become damaged or completely removed; (2) bottomhole chokes and regulators which reduce the wellhead flowing pressure and prevent the freezing of surface controls and lines by taking a pressure drop, downhole; and (3) check valves that prevent backflow of injection wells. The essential working elements of each of these devices can be installed or removed with a wireline.

Since all these tools are susceptible to erosion damage, the well should be brought in and thoroughly cleaned prior to installing a subsurface control device.

Safety Systems

Surface Safety Systems—The surface safety system is the first line of protection against minor mishaps in surface treating facilities. The surface system generally consists of normally closed valves held open by low pressure gas acting on a piston. If gas pressure is bled off, internal spring action

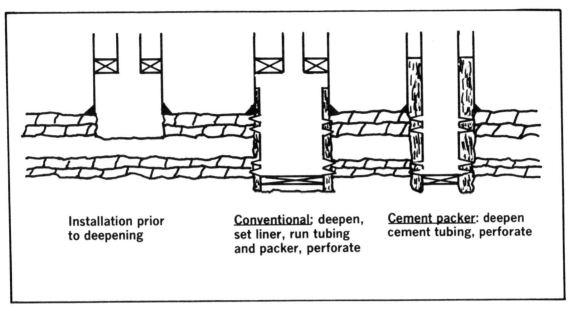

FIG. *6-17—Application of cement packer for deepening.*

closes valve against line pressure. Valves and controls are readily accessible for maintenance.

Low pressure gas can be tied into a network of sensors to detect abnormal conditions.

Catastrophe Systems—The catastrophe system is the downhole shut-in system activated (except for testing) only in the event of imminent disaster. In its most rigorous form this system consists of a near-surface packer or hanger and master valve, Figure 6-20, supplementing the surface Christmas tree and master valve. The downhole valve may be direct controlled (self controlled) or remote controlled from the surface. Remote control ap-

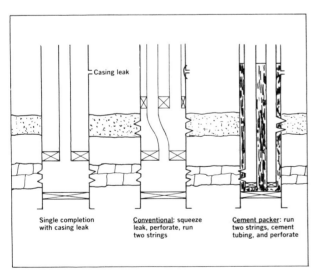

FIG. *6-18—Application of cement packer for dual completion.*

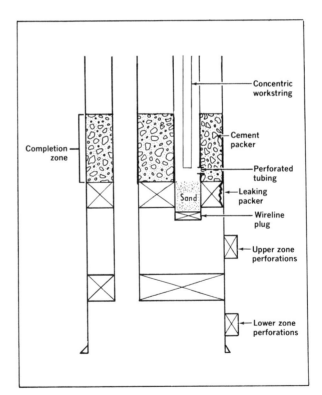

FIG. *6-19—Application of cement packer for top packer repair or recompletion above top packer.*

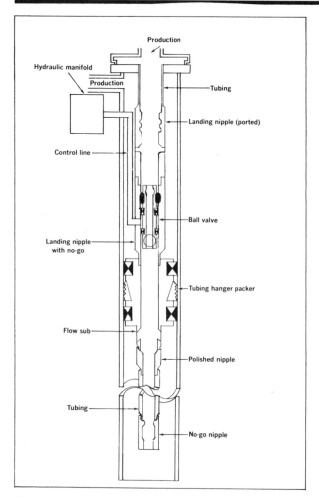

FIG. *6-20—Ideal downhole shut-in system, annular production.*

plication is becoming almost universal on new installations, because of more effective control of all wells on the platform.

Direct Controlled or self-controlled valves (storm chokes) are preset to close when conditions at the valve reach a particular criteria. Two basic types are available—one operated by differential pressure across the valve—Figure 6-21, the other operated by ambient pressure in a precharged bellows. Both are normally wireline installed in a tubing string landing nipple.

Differential Pressure Valves—Flow of well fluids through an orifice creates a pressure drop related to flow rate. The valve is held open by a preset spring tension, but as flow-rate increases pressure drop across the orifice eventually causes the valve to close. Various types of closure are available as shown in Figure 6-22.

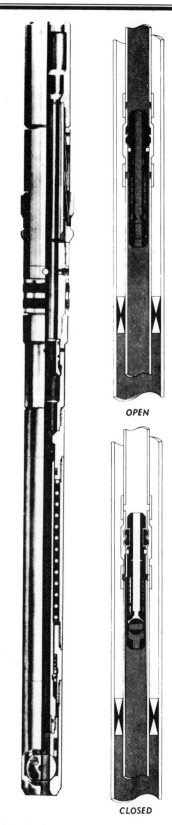

OPEN

CLOSED

FIG. *6-21—Subsurface controlled safety valves.*

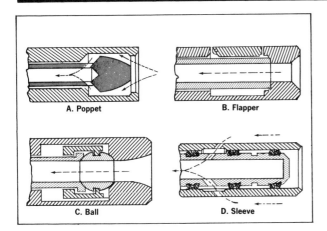

FIG. *6-22—Types of closures for differential-pressure valves.*[5] *Permission to publish by The Society of Petroleum Engineers.*

If a valve closes, it can be reopened by applying pressure to the tubing string or actuating the equalizing sub by running an equalizing prong on a wireline. When the pressure is equalized across the safety valve, the compressed spring causes the valve to open automatically.

Ambient Pressure Valve—Valve bellows is precharged to a pressure less than the pressure in the tubing under normal flowing conditions at the valve.

Under abnormally high flow rates the tubing pressure at the valve drops to a point permitting bellows charge pressure to close the valve. Shown in Figure 6-23, the ambient valve does not require a choke bean; thus, pressure drop across the valve is minimized.

Closing Calculations—Provided sufficiently accurate information is available on well flowing characteristics closing pressures can be calculated for direct controlled valves with reasonable accura-

cy. However, actual tests must be run under actual stabilized flowing conditions to insure proper operation.

Several conditions might occur that could prevent a direct controlled valve from closing when desired even though it was correctly set when initially installed in the well:

1. Maximum flow rate capability of well may decline to less than original value.

2. Paraffin or sand accumulation may restrict flow rate.

3. A catastrophe on the platform may result in a small wellhead leak—disastrous perhaps, but not causing sufficient flow rate to close valve.

For these reasons, many operators and regulatory bodies are reexamining the application of direct controlled valves.

Surface-Controlled Safety Valves

Surface-controlled safety valve systems are normally positioned slightly below the ocean floor or at 200 to 300 ft on land locations. Major items of equipment include: (1) a special landing nipple with an external 1/8-in. id (0.409-in. od) control line which is made up and run into the well with the tubing string, (2) the safety valve, (3) an exit assembly for the external line at the christmas tree, and (4) a surface control unit and related lines and pressure pilots.

Principal advantages of surface-controlled safety valves over subsurface models are (1) larger internal diameters for tubing-retrievable types which permit higher flow rates and the ability to lower wireline tools through the valve, (2) insensitivity to pressure and fluid surges, (3) more positive control because

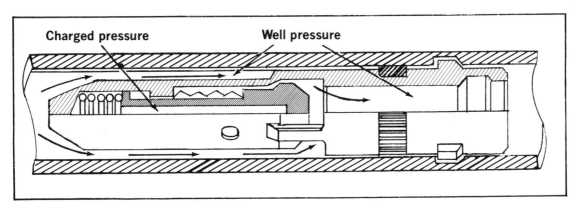

FIG. *6-23—Ambient-pressure valve.*[5] *Permission to publish by The Society of Petroleum Engineers.*

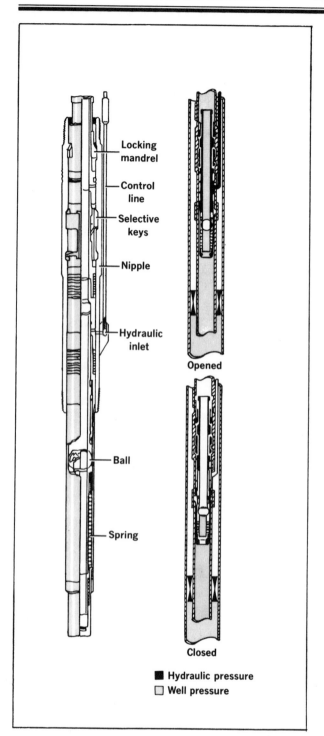

FIG. *6-24—Wireline-removable surface-controlled safety valve.*

operation does not depend on an orifice which may be damaged by sand, and (4) simplified testing. Disadvantages include more complicated design and higher cost.

Figure 6-24 pictures a wireline retrievable valve which is typical of most designs. It operates on a piston principle and is controlled from the surface by the aforementioned control manifold. Hydraulic pressure is exerted down the 1/4-in. line from the surface, enters the landing nipple and safety valve, and acts against a piston within the safety valve. This holds a spring in compression and maintains the flow tube in the down position, holding the closing mechanism open.

If surface controls fail, pressure exhausts from the hydraulic line at the surface, causing well pressure acting with the coil spring to move the flow tube upward and allow the valve to close. The valve is reopened by again applying pressure through the external line. It is also possible to pump through the valve when it is in the closed position to kill the well if necessary.

Figure 6-25A pictures a typical wellhead adapter necessary to accommodate the external line. A new style wellhead adapter permits retrieving the external line without pulling the tubing. Figure 6-25B

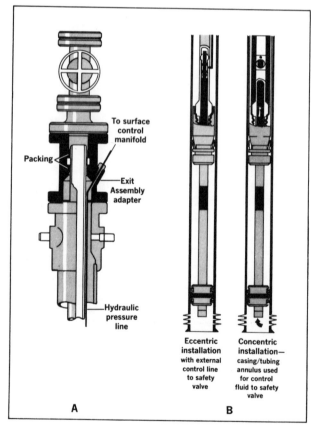

FIG. *6-25A—Exit assembly for control line.*
FIG. *6-25B—Pressure-control systems.*

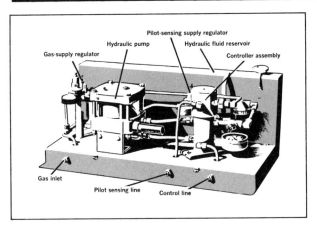

FIG. *6-26—Surface control unit.*

compares a conventional control line system with an annular pressure control system eliminating the control line and utilizing a tapered tubing string to provide a large diameter through the safety device. The safety device can be either wireline or tubing retrievable.

The control unit is normally placed at or near the wellhead. The main component of the unit is a pump which maintains a constant, pre-set pressure on the control line to hold the safety valve open. Figure 6-26 pictures one type of control unit and shows the pump, fluid reservoir, and miscellaneous equipment.

Operation Considerations

Excessive Setting Depth—To maintain surface control, hydrostatic head of fluid in the control line must present less valve opening force than the spring closing force available at the valve. Otherwise if tubing pressure at the valve dropped to zero hydrostatic control line pressure would keep the valve open even with no positive surface control line pressure.

If it is necessary to set the valve at a depth where this effect becomes a problem, response to surface control must be assured by using additional spring force, adjustment of piston areas or by providing a second small diameter "balancing" line from the surface.

Control Fluids—The following control fluids can be used:—treated water—water containing a water-soluble oil to provide lubrication and corrosion protection; diesel oil—add light oil to insure lubricity; and light-weight oil—good lubricity and flow

characteristics unless temperature drops below 50°F.

Gas should probably not be used since a control line failure near the valve would expose the valve to the hydrostatic head of the annular fluid rather than gas.

It is impossible to completely negate the possibility of solids settling out in parts of the valve. Periodic testing not only determines if the valve is working properly but also helps in reducing seal "striction" and keeping operating surfaces free of deposits.

Surface Sensing Systems—The primary problem in setting up a Sensing System is to design the system so that it actually "sees" a proper malfunction. The pressure pilot, the most common sensor, is intended to activate safety valves when pressure at some point in the system exceeds or falls below pre-set limits.

It must be remembered that a small leak can be "disastrous" but not actually lower flow line pressure from the wells. Also the high-low pressure malfunction can often be handled by surface safety shut-in equipment rather than downhole shut-in. Routine problems should not activate the primary downhole system.

Fire detectors include fusible plugs, or ultraviolet-light detectors. Proper placement is critical with either. Obviously fire requires downhole shut-in. Collision or storm damage is usually detected with fragile or brittle control lines. Downhole shut-in is warranted.

Bottom-Hole Chokes and Regulators

Bottom-hole chokes and regulators are utilized to eliminate the freezing of surface controls and lines due to the formation of hydrates by moving the point of pressure reduction and attendant temperature decrease to the lower portion of the wellbore. This allows the higher temperature at depth to reheat the flow stream before it reaches the surface. The lower surface flowing pressure is also advantageous in reducing the continuous surface pressure on the tubing and christmas tree.

Bottom-hole choke beans (Figure 6-27) are attached to the lower end of mandrels designed to be set in a landing nipple or anchored to the tubing wall. Type A bean is a spring-loaded ground seat bean that is recommended for high pressure or heavy fluid wells. The bean is mounted inside a cage and seats against the lower end of the mandrel. The bean is held on seat by the spring inside the bean

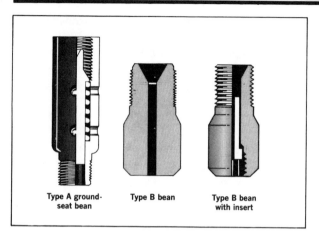

Type A ground-seat bean   Type B bean   Type B bean with insert

FIG. *6-27—Bottom-hole choke beans.*

cage. Type B bean incorporates a positive orifice and will accommodate any size choke up to the internal diameter of the mandrel.

A bottom-hole choke is susceptible to erosion and is also rate sensitive; however, if a well produces without sand and at a fairly constant rate, this positive choke assembly is superior to the bottom-hole regulator due to economy, simplicity, and minimum maintenance.

Bottom-hole Regulator (Figure 6-28)—This consists of a valve and a spring-loaded valve seat. A predetermined spring tension determines the pressure differential across the regulator. When this predetermined differential is reached, the valve seat moves off the valve to allow the well to flow.

The regulator differs from a bottom-hole choke in that it maintains a constant pressure differential across the valve, regardless of the flow rate. Well production is adjusted at the surface by a conventional choke. Theoretically, a regulator will also reduce the surface shut-in pressure. However, the sealing mechanism is not sufficiently reliable or durable for this purpose, particularly after the regulator has been installed for extended periods of time.

Regulators can be installed in all sizes of tubing, either in a landing nipple, slip-type mandrel, or collar lock mandrel. Pressure differentials up to 1,500 psi can be taken across a single regulator. If a larger reduction in surface pressure is desired, two regulators can be connected in tandem, or several separate regulators can be set at various depths in the tubing string.

Data required to properly calculate bottom-hole choke and regulator sizes include: (1) tubing size,

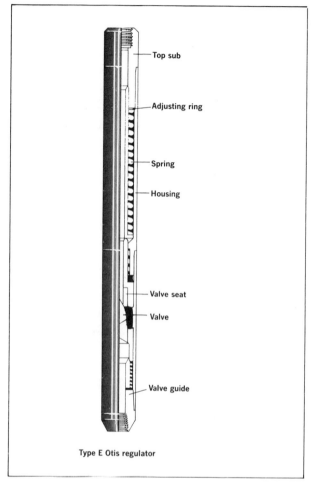

Top sub

Adjusting ring

Spring

Housing

Valve seat

Valve

Valve guide

Type E Otis regulator

FIG. *6-28—Bottom-hole regulator.*

(2) setting depth of valve, (3) surface shut-in pressure, (4) desired rate of production, (5) flowing pressure and temperature at valve depth, (6) specific gravity of produced fluids, and (7) desired pressure drop across the choke or regulator.

Bottom-hole chokes and regulators should be set in the lower portion of the wellbore. Increasing the setting depth reduces the length of tubing exposed to full flowing pressure and permits additional time and temperature for heating the fluids before they reach the surface.

Inclusion of flow coupling in the tubing string above and below landing nipples contemplated to house bottom-hole chokes and regulators should be considered. Field experience has shown that the flow restrictions from turbulence are sufficient to seriously erode the pipe. A 3-ft length has proven satisfactory in most circumstances.

Subsurface Injection Safety Valves (Figure 6-29)

There are two types of injection safety valves which will shut the well in if flow is reversed. The Otis Type T injection check valve is designed to be either made up on the end of the tubing string and run into the well with the tubing or set in a landing nipple by wireline means. The Type T valve is a ball-and-seat type valve and is designed so that injection or static pressure will hold the ball in the open position for fluid passage in the upstream direction.

Should the pressure reverse, the ball in the valve will rotate to the closed position and shut the well in. The valve will remain closed until the pressure differential across it is equalized by resuming injection. The Type T Otis input safety valve is a simple, spring-loaded valve and seat mechanism. Injection pressure forces the valve open for fluid passage.

If injection ceases or reverses for any reason, the spring tension and the flow pressure act to force the valve to the closed position. This valve may also be seated in a landing nipple or may be set on a collar stop at any point in the tubing.

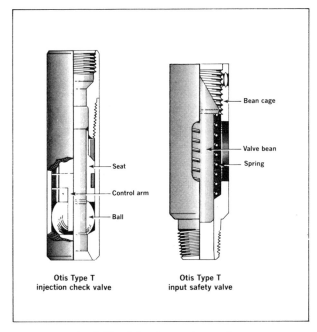

Otis Type T
injection check valve

Otis Type T
input safety valve

FIG. *6-29—Injection safety valves.*

REFERENCES

1. Lubinski, Arthur, Althouse, W. S., and Logan, J. L.: "Helical Buckling of Tubing Sealed in Packers," J. Pet. Tech., June 1962, p. 655.

2. Leutwyler, Kurt: "Completion Design for Corrosive Environment," Petroleum Engineer, February, 1970.

3. Krause, Jr., W. E. and Sizer, P. S.: "Selective Criteria for Subsurface Safety Equipment for Offshore Completion," J. Pet. Tech., July 1970, p. 793.

4. Logan, J. L.: "How to Keep Tubing Sealed in Packers," Journal of Canadian Petroleum Technology, Summer, 1963.

5. Raulins, G. M.: "Platform Safety by Downhole Well Control," J. Pet. Tech., March 1972, p. 263.

6. Lindsey, Jr., H. E.: "Deep Gas Well Completions Practices," SPE Paper 3908, September 11, 1972, SPE Amarillo, Texas.

7. Hammerlindl, D. J.: "Movement, Forces and Stresses Associated with Combination Tubing Strings Sealed in Packers," J. Pet. Tech., Feb. 1977.

8. Baker Oil Tools—Packer Calculations Handbook.

9. API Specification 14A Subsurface Safety Valve.

10. API Recommended Practice 14B, Design, Installation and Operation of Subsurface Safety Systems.

11. Purser, P. E.: "Review of Reliability and Performance of Subsurface Safety Valves," OTC Paper No. 2770, May, 1977.

12. Burley, J. D., and Holland, W. E.: "Recent Developments in Packer Seal Systems for Sour Oil and Gas Wells," SPE Paper 6762, October, 1977.

TABLE 6-8
Tubing and Packer Constants

Tubing Constants						
OD IN (Inches)	WT. IN (Lbs/Ft)	A_o IN (Sq. In.)	A_i IN (Sq. In.)	A_s IN (Sq. In.)	I IN (In.^{4})	R^2
1.660	2.40	2.164	1.496	.668	.195	1.448
1.900	2.90	2.835	2.036	.799	.310	1.393
2.000	3.40	3.142	2.190	.952	.404	1.434
2-1/16	3.40	3.341	2.405	.936	.428	1.389
2-3/8	4.70	4.430	3.126	1.304	.784	1.417
2-7/8	6.50	6.492	4.680	1.812	1.611	1.387
3-1/2	9.20	9.621	7.031	2.590	3.434	1.368

$w_s + w_i - w_o$															
Tubing OD (Inches)	Weight (Lbs/In.)	w_i and w_o (Lbs/In.)	7.0 / 52.3	8.0 / 59.8	9.0 / 67.3	10.0 / 74.8	11.0 / 82.3	12.0 / 89.8	13.0 / 97.2	14.0 / 104.7	15.0 / 112.2	16.0 / 119.7	17.0 / 127.2	18.0 / 134.6	Lbs/Gal. Lbs/Cu. Ft.
1.660	$w_s = .200$	w_i	.045	.052	.058	.065	.071	.078	.084	.091	.097	.104	.110	.116	
		w_o	.065	.075	.084	.094	.103	.112	.122	.131	.140	.150	.159	.169	
1.900	$w_s = .242$	w_i	.062	.070	.079	.088	.097	.106	.115	.123	.132	.141	.150	.159	
		w_o	.086	.098	.110	.123	.135	.147	.159	.172	.184	.196	.209	.221	
2.000	$w_s = .283$	w_i	.066	.076	.085	.095	.104	.114	.123	.133	.142	.152	.161	.171	
		w_o	.095	.109	.122	.136	.150	.163	.177	.190	.204	.218	.231	.245	
2-1/16	$w_s = .283$	w_i	.073	.083	.094	.104	.114	.125	.135	.146	.156	.167	.177	.187	
		w_o	.101	.116	.130	.145	.159	.174	.188	.202	.217	.231	.246	.260	
2-3/8	$w_s = .392$	w_i	.095	.108	.122	.135	.149	.162	.176	.189	.203	.217	.230	.243	
		w_o	.134	.153	.172	.192	.211	.230	.249	.268	.288	.307	.326	.345	
2-7/8	$w_s = .542$	w_i	.142	.162	.182	.203	.223	.243	.263	.284	.304	.324	.344	.364	
		w_o	.196	.225	.253	.281	.309	.337	.365	.393	.421	.450	.478	.506	
3-1/2	$w_s = .767$	w_i	.213	.243	.274	.304	.335	.365	.395	.426	.456	.487	.517	.548	
		w_o	.291	.333	365	.416	.458	.500	.541	.583	.625	.666	.708	.749	

Area of Packer Bores			
Bore (Inches)	Area (Sq. In.)	Bore (Inches)	Area (Sq. In.)
6.00	28.26	2.50	4.91
5.24	21.55	2.42	4.60
4.75	17.71	2.28	4.08
4.40	15.20	2.06	3.33
4.00	12.56	1.96	3.00
3.87	11.76	1.87	2.75
3.62	10.29	1.68	2.22
3.25	8.30	1.53	1.84
3.00	7.07	1.43	1.61
2.68	5.67	1.25	1.23

Chapter 7

Perforating Oil and Gas Wells

Perforating methods
Perforator performance evaluation
Factors affecting perforating results
Fluids, pressure differential, formation properties
Operational considerations
Optimum perforating practices
API RP-43 tests

INTRODUCTION

Perforating is probably the most important of all completion functions in cased holes. Adequate communication between the well bore and all desired zones, as well as isolation between zones, is essential to evaluate and to optimize production and recovery from each zone.

Research[2,3,6] by Exxon developed the significance of (a) plugging of perforations with mud or shaped charge debris, (b) perforating with a pressure differential into well bore, and (c) the effect of formation compressive strength on perforation hole size and penetration. This work led to the development of non-plugging shaped charges, through tubing perforators, improved bullet guns, and an API standard, Section 2, API RP 43, to evaluate perforators under simulated downhole flow conditions. The development of effective jet perforators has aided penetration where a high compressive strength formation, high compressive strength cement, and/or thick-walled, high-strength casing is present.

Although technology is available to insure good perforating in most wells, unsatisfactory perforating tends to be the rule in many areas. The three most prevalent causes for poor perforating probably are (1) a lack of understanding of the requirements for optimum perforating, (2) inadequate control of gun clearance, particularly with through-tubing guns, and (3) the rather widespread practice of awarding perforating jobs on the basis of price, rather than job quality.

TYPES OF PERFORATORS
Bullet Perforators

Bullet guns, 3¼ in. OD or larger, are applicable in formations with compressive strength less than about 6,000 psi. Bullet perforators in the 3¼ in. or larger size range, may provide deeper penetration than many jet guns in formations with less than about 2,000 psi compressive strength. However, tests of specific jet and bullet guns should be made in various compressive strength reservoir rocks, using tests similar to those described in Section 2, API RP 43 to validate these conclusions on a current basis.

Muzzle velocity of bullet guns is about 3,300 ft/sec. The bullet loses velocity and energy when the gun clearance exceeds 0.5 in., the clearance at which most comparative tests have been made. At zero gun clearance penetration increases about 15% over 0.5 inch clearance, along with a deburring effect. Loss in penetration with one-inch clearance is about 25% of the penetration at 0.5 in. clearance, and at 2 in. clearance the loss is 30%.

Deburring of bullet holes is not dependent on decentralization if the bullet carries a deburring device on its ogive or point. This device is more effective in deburring than using zero gun clearance.

Bullet guns can be designed to fire either selectively or simultaneously.

Jet Perforators

The jet perforating process is illustrated in Figure 7-1. An electrically-fired detonator starts a chain reaction which successively detonates the primacord, the high velocity booster in the charge, and finally, the main explosive. High pressure generated by the explosive causes the metal in the charge liner to flow, separating the inner and outer layers of the liner. Continued pressure buildup on the liner

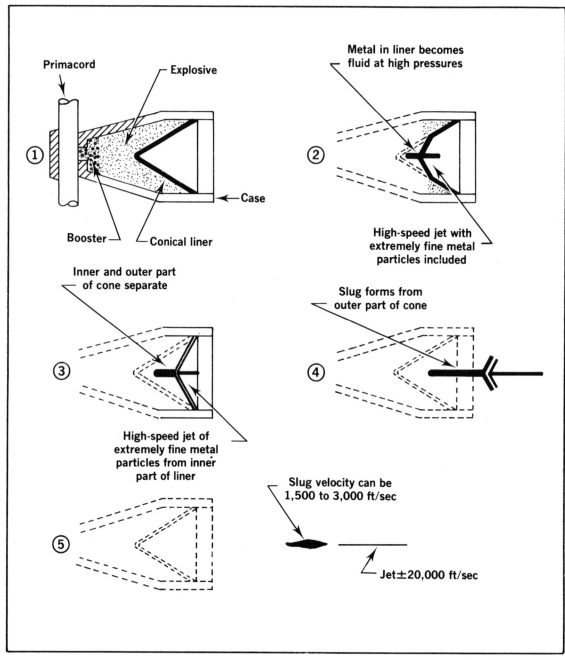

FIG. 7-1—Jet perforating process using a solid metal liner.

causes a needle-like high speed jet of fine particles to spew from the cone at a speed of about 20,000 ft/sec at its tip with a pressure at this point estimated to be 5 million psi.

The outer shell of the liner collapses to form a slower moving metal stream, with a velocity between 1,500 and 3,000 ft/sec. This outer liner residue may be in the form of a single slug sometimes called a "carrot" or a stream of metal particles. Premium-priced charges usually are relatively carrot-free, with residue being sand-sized or smaller particles.

Because of the sensitivity of the jet perforating process to a near perfect sequence of events from the firing of the detonator through the formation of the jet, any fault in the system can cause malfunction. This can result in an irregular or inadequate hole size, poor penetration, or no hole at all. Some of the causes of malfunction are insufficient electric current or voltage to detonator, poor quality or defective detonator, mashed or twisted primacord, poorly packed booster, poorly packed or poor quality main explosive, and liner incorrectly positioned or not in effective contact with the explosive.

Water or dampness in gun, primacord, or charge may cause malfunction or low order detonation. High temperature aging of explosive in primacord or charge may reduce charge effectiveness or cause low order detonation. Procedures for loading, running and firing the gun must be carefully followed to insure good job performance and a safe job.

Conventional and Through-Tubing Jet Perforators—Conventional retrievable hollow carrier steel guns normally provide adequate penetration without damaging casing. Guns are run in the hole with atmospheric pressure inside the gun carrier. Most of the explosive energy not used in producing the jet is absorbed by the gun carrier rather than the casing.

Through-tubing jet perforators, including capsule-type guns, Swing-Jet guns, wire and strip carrier guns, and thin wall or expendable hollow carrier guns are available. Their primary advantage is being able to run and retrieve through the tubing and to shoot with a pressure differential into the wellbore.

Many of the through-tubing guns give inadequate penetration and/or hole size. The swing jet provides relatively large hole size and sufficient penetration

for most wells. Its major disadvantage is the mechanical manipulation required and the large amount of debris remaining after shooting. Guns with exposed charges such as the capsule gun will swell the casing and may split the casing.

Also it is difficult to obtain proper gun clearance for effective penetration. The thin-wall expendable hollow carrier gun eliminates casing-splitting and much of the debris left inside the casing. It also overcomes the gun clearance problem if the gun is properly positioned, but sacrifices some penetration. To prevent severe swelling of the gun body, most thin-wall hollow carrier guns should normally be fired under a fluid head of at least 500 psi.

Many special-purpose jet perforators are available:

—Selectively-fired guns are usually available for both conventional and through-tubing completions.

—Four-way or five-way directional single-plane jet perforators are usually not recommended because of either inadequate penetration or casing-splitting.

—Perforators are available to penetrate only the tubing. Tubing should be centralized to prevent casing damage; however, it is acceptable to perforate just above the packer or possibly immediately above or below a tubing collar.

—Jet-type pipe cutters are also available to cut various sizes of tubing, casing, and also large-diameter steel piling.

—Open hole perforators have primary application in penetrating scale and other damage near the wellbore. Some premium-type hollow carrier guns provide satisfactory penetration with less hole size than open hole jets.

—The Jet-Vac perforator is designed to clean out perforations with a high pressure differential into the gun carrier immediately after perforating.

—Big-hole perforators, providing an entrance hole 0.75 in. or larger, have been developed for special uses such as gravel packing.

Other Perforating Methods

Hydraulic Perforators—Cutting action is obtained by jetting sand-laden fluid through an orifice against the casing. Penetration is greatly reduced as well bore pressure is increased from zero to 300 psi. Penetration (see Figure 7-7) can be increased appreciably by the addition of nitrogen to fluid stream.

Mechanical Cutters—Knives and milling tools have been used to open slots or windows to provide communication between the well bore and the formation. Milling a window in the casing, underreaming, and gravel packing is the standard procedure for sand control in some areas.

The Permeater—This was developed to eliminate cracking of cement due to perforating. Permeater units are welded over holes in the casing. These sections are then located adjacent to intended completion zones. After cement is circulated behind the casing, permeater sleeves are extended with pump pressure to the wall of the borehole. Communication with the formation is established after cement is set by dissolving soluble plugs in the Permeater sleeves with acid or caustic.

The permeater has not proved successful. It is more expensive than jet or bullet perforating and requires more rig time. Welding of permeater to casing is conducive to increased corrosion. There are difficulties in depth correlation, or if casing sticks before reaching desired depth, permeater holes are not placed opposite the correct interval. Also, cement cracking is not a problem when perforating with hollow carrier jet guns.

EVALUATION OF PERFORATOR PERFORMANCE

Prior to 1952, essentially all perforator evaluation was done in actual downhole tests in wells, or in surface tests at atmospheric temperature and pressure in casing cemented inside steel oil drums. Comparative downhole testing was generally impractical because of the difficulties in controlling well and reservoir conditions.

Surface tests at atmospheric pressure proved to be misleading for several reasons. Liner "slugs" from the shaped charge which otherwise would plug a perforation downhole tended to be deflected away from the perforation hole at atmospheric pressure. Surface tests were made in sand-cement targets instead of sandstone or carbonates. Also surface tests do not stimulate downhole flow through perforations.

Development of Flow Index System

In 1952, Exxon developed the first reliable testing procedure to simulate perforating under downhole

conditions.[3] This system initially was called the "Productivity Method of Perforator Testing" or "Well Flow Index" system. The test apparatus is illustrated in Figure 7-2.

The test plan designed to simulate actual downhole conditions, include (1) use of large diameter formation cores conditioned to specific hydrocarbon and interstitial-water saturations; (2) determination of effective formation permeability prior to perforating, and after perforating and flowing the well; (3) isolation of the formation from the wellbore by casing and a suitable cementing material; (4) perforating the casing, cementing material, and formation with various perforating fluids in the well; (5) maintenance of reservoir temperature, and reservoir and wellbore pressure during and after perforating; (6) backflowing the well to simulate producing a well to clean perforations; and (7) evaluation of test results.

Current Methods of Evaluating Perforators— APR RP 43, Sections 1 and 2 presented in Appendix I, are the primary methods of evaluating both shaped charge and bullet guns. Section 1 covers surface tests where casing is cemented inside a light steel container using two parts sand and one part cement.

Section 2 tests are a modification of the original flow index system and should be used for gun selection.

Effect of Perforating in Various Fluids

Table 7-1[3] shows the results of Well Flow Index tests of jet and bullets in various fluids. Two types

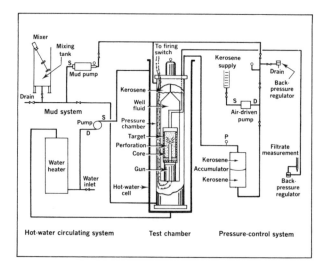

FIG. *7-2—Schematic of simulated downhole gun-perforator testing.[3] Permission to publish by API Production Department.*

TABLE 7-1
Comparative Results of Perforating in Various Fluids[3]

Type perforator, well fluid, and differential pressure	Perforation data		Average differential to initiate flow, psi	Avg. flow index	Condition of perforation after backflowing
	Penetration in.	Hole size in.			
Jet perforator					
10-lb saltwater					
200 psi into wellbore	6½–8	¼–½	0	1.00	Hole clean to total depth.
500 psi into formation for 3 to 10 hours	6½–8	¼–½	0	0.61	Hole clean or partially filled with charge debris and sand.
10-lb caustic-quebracho mud					
500 psi into formation	6½–8	¼–½	30	0.55	Partially or completely filled with mud and charge debris.
16-lb lime-base mud					
500 psi into formation	6½–8	¼–½	100	0.41	Completely filled with mud, sand, and charge debris.
Bullet perforator					
10-lb saltwater					
500 psi into formation	3–3½	½	0	0.61	Cleaned out or partially filled with sand.
10-lb caustic-quebracho mud					
500 psi into formation	3–3½	½	30	0.53	Filled with mud and sand.

of tests were made with jet perforators in salt water, one with a 200 psi differential pressure into the wellbore, and a second with a 500 psi differential pressure into the formation. Tests of bullet perforators were run with a 500 psi pressure differential into the formation. Test temperature was 180°F. All tests were conducted with the Well Flow Index system illustrated in Figure 7-2. Essentially the same test equipment is used for tests under Section 2, API RP 43.

Figure 7-3 shows a mud-plugged jet perforation; perforating was performed in 10 lb/gal mud and cleaned by backflowing. On one test perforating in 16 lb/gal lime-based mud with pressure into the formation, a drawdown pressure into wellbore of 430 psi was required to initiate flow through a single perforation.

Clean perforations result from perforating in clean oil or clean saltwater with carrot-free shaped charges when differential pressure is into the well-

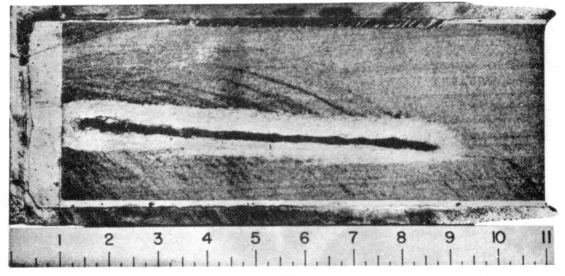

FIG. *7-3—Jet perforation made in mud with a differential pressure into formation.*[3] *Permission to publish by API Production Department.*

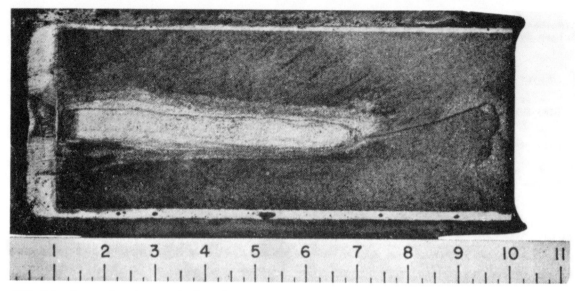

FIG. *7-4—Jet perforation made in saltwater with differential pressure into wellbore.*[3] *Permission to publish by API Production Department.*

bore, as shown in Figure 7-4.

These tests also showed that flow patterns and perforation geometry prevent the cleaning out of an appreciable percentage of mud-plugged perforations by the normal procedures of putting a well on production.

Permanent plugging of many perforations may result from killing a well with mud or dirty fluid during well completion, servicing or workover.

Effect of Formation Strength on Perforator Performance

Simulated downhole tests[3,6] run using test apparatus shown in Figure 7-2, showed that perforator penetration varies with formation compressive strength as measured by ASTM C-190 tests of cores. The ASTM test is a crushing test of an unsupported column at atmospheric pressure. The crushing strength or compressive strength of a rock, as shown by this test, may not be representative of downhole conditions, particularly in the case of relatively unconsolidated sandstones. Crushing strength measured on the surface is less than the rock crushing strength downhole in the zone to be perforated.

As shown in Figure 7-5, jets penetrate deeper than bullets in hard formation. However, some premium bullet guns may penetrate deeper than some jets in low compressive strength formations, particularly if guns are fired at zero clearance.

Figure 7-6 shows that penetration of jets, bullets, and hydraulic perforators was reduced with increased compressive strength of formation penetrated. However, bullet penetration declined at a more rapid rate as rock strength increased. One set of these tests was run in Berea sandstone which has an average compressive strength of about 6,500 psi. Tests run under Section 2, API RP 43 also use Berea sandstone targets. The curve slope derived from Figure 7-6, thus can be used to correct Section 2, APR RP 43 perforator penetration test data to specific formations if the formation com-

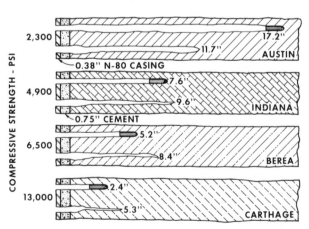

FIG. *7-5—Effect of formation compressive strength on penetrating efficiency of bullet and jet perforators.*[6] *Permission to publish by API Production Department.*

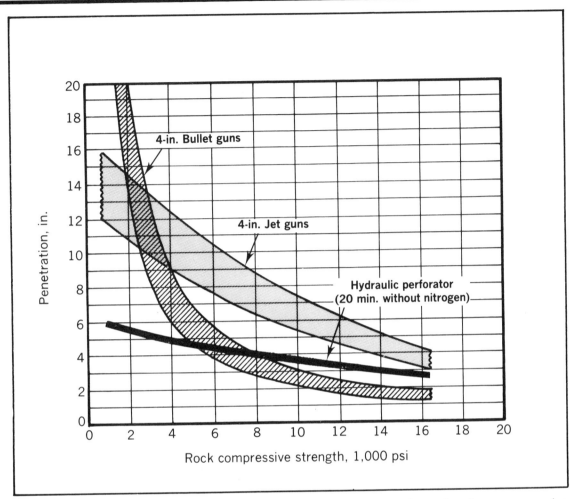

FIG. *7-6—Bullet, jet, and hydraulic perforator performance in formations of various compressive strength.[6] Permission to publish by API Production Department.*

pressive strength can be reliably estimated.

If representative rock compressive strengths are available from a specific well, perforator penetration projections can be made for comparison with Berea sandstone test data, or for comparison of different guns in the same formation using Figure 7-6. It may be possible to estimate strength based on cores from nearby wells. Drilling speed, under properly controlled conditions, can be used to roughly correlate compressive strength of formations penetrated in uncored wells with those from previously cored wells in the same formation and geologic trend.

Figure 7-7 shows semilog plot of rock compressive strength vs. penetration with a hydraulic perforator with and without nitrogen in the hydraulic jet streams.[6]

Table 7-2 shows permeability, porosity and compressive strength of representative formations in Texas and Louisiana, U.S.A. Data from Table 7-2 can be used where applicable to determine probable penetration using perforator data from Section 2, API RP 43 and curves shown in Figures 7-6 and 7-7.

Downhole Evaluation of Perforations

A recent method of evaluating actual perforations and perforation plugging downhole involves running a soft rubber impression packer opposite perforations. The packer is then hydraulically expanded. If a perforation is open, packer rubber will extrude into the perforation. If the perforation is completely sealed, no extrusion will take place.

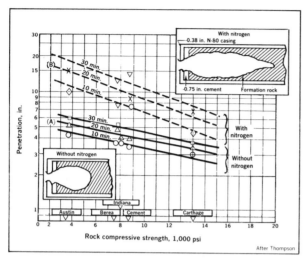

FIG. 7-7—Effect of rock compressive strength on penetration with hydraulic perforator.[6] Permission to publish by API Production Department.

FACTORS AFFECTING GUN PERFORATING RESULTS

Perforation Plugging—Plugging of perforators with pieces of charge liner residue slugs can be very severe. Through the use of powdered metal cones in charges, large residue slugs have been eliminated in a number of the premium-priced charges. The slug or liner residue is still formed but is carried to the bottom of the hole as sand-sized or smaller particles. Surface tests under atmospheric pressure are not reliable to evaluate this type of perforation plugging, because "slugs" are frequently deflected from the perforation at atmospheric pressure.

Perforations tend to be filled with crushed formation rock, mud solids, and charge debris when perforating in mud. These plugs are not readily removed by backflowing. Crushed and compacted formation around the perforation further reduces the probability of perforation cleanout. High weight muds, mixed from high density solids, cause high density plugs in perforations.

Differential pressure, from the formation into the wellbore, required to initiate flow through each plugged perforation varies. When a few perforations requiring low differential pressure open up, flow through these perforations make it difficult to create the greater drawdown needed to open

TABLE 7-2
Physical Characteristics of Some Oil-producing Formations[6]

Formation	Field*	Approx. depth, ft	Avg. perm., md	Average porosity, %	Avg. comp.** strength, psi
Woodbine	Van	2,700	—	—	370
Frio	Tom O'Connor	5,500	1,725	30	245
Frio	Seeligson	5,900	6	18	1,850
Frio	Trull	12,000	50	20	3,620
		12,000	20	18	6,615
Cockfield	Conroe	5,000	—	—	60
Yegua	South Dayton	8,600	—	—	2,400
		8,600	1	13	6,600
Wilcox	Barbston	6,300	500	32	6,000
Paluxy	Merit	10,500	57	9	8,040
		10,500	2	7	13,860
Rob-7	Bayou Sale	13,300	129	23	1,550
		13,300	63	19	2,490
		13,300	12	16	4,940
Grayburg	—	4,000	—	—	12,600
San Andres	—	4,200	—	—	15,800
Wolfcamp	—	8,500	—	—	11,400
Pennsylvanian	—	8,950	—	—	12,400
Devonian	—	10,450	—	—	8,215
Ellenburger	—	12,220	—	—	16,000

*Fields located in the U.S.
**Crushing strength using ASTM C-190 tests.

more perforations. A large number of plugged perforations remain and may result in failure to drain specific zones in stratified formations such as sand-shale sequences. When one or more zones in a layered multizone well are plugged or partially plugged, drill stem tests, production tests, and PI measurement may provide an erroneous evaluation of well damage, well productivity, and recovery.

Cleanout of Plugged Perforations—For unconsolidated sands, backsurge tools and perforation washers have been successfully used to clean out perforations in many areas. If perforations in sandstone wells cannot be cleaned out with perforation washers or backsurge tools, the next approach should usually be to break down each perforation with clean water or oil using ball sealers. This procedure results in mud being driven into formation fractures. Normally these fractures will heal a short time after frac pressure is removed.

Acidizing sandstone wells will not usually clean out all mud-plugged perforations, unless each perforation is isolated and fractured, and the mud driven into the formation fracture.

Mud plugs are somewhat easier to remove from perforations in carbonate formations, because acid entering a few perforations can usually dissolve enough rock to open other perforations. However, limestone or dolomite wells are frequently perforated in acid with a slight differential into formation. Perforating in clean water or clean oil with a pressure differential into wellbore, however, is quite satisfactory.

If a section of perforated casing is poorly cemented, providing vertical communication behind the pipe between perforations, the resulting condition is similar to a screened open hole. If any flow occurs from the formation, all casing perforations will usually be cleaned out. However, the perforated hole in the formation may or may not clean out.

Plugging of perforations during production with paraffin, asphaltenes, or scale is a major problem in many areas of the world. Solvent soaking will usually remove paraffin or asphaltenes. If perforations are sealed with acid-soluble or acid-insoluble scale, it is often advisable to reperforate and then to treat with acid or other chemicals.

Effect of Pressure Differential—When perforating in mud with a pressure differential into formation, perforations are filled with mud solids, charge debris, and formation particles. Mud plugs are difficult to remove, often resulting in some permanent perforation plugging and reduced well productivity.

Even when perforating in clean fluids such as water or oil, having high filtration rates, particles of charge and debris, clay, or other fines may cause some plugging of perforations and deep damage in the formation. As a rule of thumb, permeabilities of about 250 md or more will allow clay size particles to be carried into formation pores or fractures.

For carbonate formations it is often possible to obtain high well productivities and low perforation breakdown pressure when perforating in HCl or Acetic acid with a small differential pressure into the formation. Because of the slow reaction of Acetic acid on limestone, it is usually desirable to leave the Acetic acid in the perforations for about 12 hours after perforating. Mud solids should not be allowed to enter the acidized perforations.

Perforating with a differential pressure into the wellbore in clean fluids aids in perforation cleanout. This is the preferred method of perforating sandstone wells and some carbonate formations.

Effect of Clean Fluids—If a particular perforator provides adequate hole size and penetration under given well conditions, well productivities will be maximized by perforating in clean oil or saltwater with pressure differential into wellbore during perforating and well clean-up period.

Effect of Compressive Strength[6,15]—Penetration and hole size of jet perforators are reduced as compressive strength of casing, cement, and formation rock is increased. Penetration of bullets is severely decreased with increases in strength of casing, cement, and formation rock.

Perforation Density—Shot density usually depends on required production rate, formation permeability, and length of perforated interval. For high volume oil or gas wells, perforation density must permit the desired flow with reasonable drawdown. Four 0.5-in. or larger holes/ft are usually adequate, with one or two shots per foot being satisfactory for most low volume wells. In wells to be fractured, perforating is planned to allow communication with all desired zones. For sand consolidation, four shots/ft of large diameter perforations is usually preferred. For gravel packing, four to eight shots/ft of very large perforations, 0.75 in. or greater in diameter, is preferred.

Perforating with four or more shots/ft in low-strength, small-diameter casing, with the exposed charge perforator such as the capsule gun, may cause casing splitting. Also, cement behind pipe may be so badly cracked that squeeze cementing is required to seal off undesired water or gas. High-strength casing collars may be damaged by multiple perforations in the collars.

Cost—Perforating prices vary from area to area; however, reduced perforation density usually results in lower job costs. Selectively-fired guns can save appreciable rig time where productive zones are separated by nonproductive intervals.

The use of through-tubing guns can frequently save rig time if tubing is run open-ended and set above all zones to be perforated. On new wells, tubing may be run within a few hours after the cement is in place. Then through-tubing perforating may be carried out without a rig on the well, resulting in little or no rig time being charged to well completion.

Pressure and Temperature Limitations—Pressure and temperature ratings are available on all perforators. Bottom hole pressure may impose limitations on some exposed charge guns. However, there are few wells being perforated where pressure is a problem with most conventional hollow carrier casing-type guns.

Figure 7-8 shows self-detonation time-tempera-ture curves for (RDX) cyclonite powder used in conventional charges, and for high temperature explosives used in very hot wells.

As a general rule, high temperature charges should not be employed even in wells in the 300–400°F range. This recommendation is based on the following: (1) most high temperature charges provide less penetration, (2) high temperature powder is less sensitive, resulting in increased misfires, (3) high temperature charges are more expensive, and (4) there is much less choice in charge selection.

When operating near the upper limit of low temperature charges, these approaches may be used:

1. Wells can be circulated with low temperature fluids to lower the bottom hole temperature. This is especially applicable for through-tubing guns which can be run soon after fluid circulation has been stopped.

2. When there is some question as to whether the temperature limit of the gun will be reached prior to firing the gun, high temperature detonators may be employed in guns equipped with low temperature charges. This approach will prevent accidental perforating due to high temperature, because shaped charges will "fuse off" or burn with no resulting perforation unless fired with the perforating gun detonator.

For very high temperature wells there may be no alternative but to run the entire high temperature perforating package. This includes the detonator, the primacord, the booster charge and the main powder charge. As previously noted, the detonator is the key to the system. Unless the detonator is fired, the shaped charge will not function as intended to perforate. Table 7-3 provides an example of temperature and pressure ratings on selected perforators.

Well Control—Low pressure oil wells can be perforated with oil or water in the casing with little surface control beyond a wiper type of packing gland. It is always good practice, however, to use a wireline BOP. Normal pressure oil wells can be perforated with oil or water in the hole with through-tubing guns using conventional wellhead control and the labyrinth type of "flow-tube" or "blow-by" packing gland.

A grease seal lubricator should be used on all gas wells, and on all wells if greater than 1,000 psi surface pressure is anticipated. Abnormally high

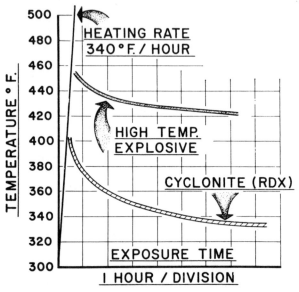

FIG. 7-8—Self-detonation curves of 26-gm encapsulated charges.[17] Permission to publish by The Society of Petroleum Engineers.

TABLE 7-3
Pressure and Temperature Ratings on Selected Perforators

Gun types	OD in.	Pressure rating psi	Temp. rating °F
Low temperature powder			
Bullet	3¼, 3⅜	15,000	250
Jet (Std. hollow carrier)	3⅜, 4	20,000	340
Jet (Big hole-hollow carrier)	4	20,000	325
Jet (Thin wall hollow carrier)	1⅜, 1¹¹/₁₆, 2⅛, 2⅞	15,000	340
Swing Jet (Expendable thru tubing)	1¹¹/₁₆	12,000	300
Jet (Partially expendable thru tubing)	1¹¹/₁₆, 2⅛	20,000	300
High temperature powder			
Jet (Std. hollow carrier)	3⅜, 4	25,000	470
Jet (Thin wall hollow carrier)	1¹¹/₁₆, 2⅛	25,000	470
Jet (Thin wall hollow carrier)	1⁹/₁₆	25,000	600

pressure wells can be safely perforated with saltwater in the hole with through-tubing guns using high pressure wellhead control equipment. Equipment with working pressure ratings in excess of 10,000 psi is available.

Casing and Cement Damage[7,11]—Hollow carrier jet guns absorb unused energy from charge detonation. This prevents casing splitting and virtually eliminates cement cracking. Little casing damage occurs with conventional bullet guns. Shooting with zero gun clearance tends to eliminate burrs inside the casing.

Jet guns with exposed charges, such as strip or capsule-type guns, can cause deformation, splitting, and rupture of the casing, and appreciable cracking of cement. Explosive weight, degree of casing support with cement, perforation density, casing diameter and casing "mass-strength" are factors in casing splitting with exposed jet charges. Casing "mass-strength" has been defined as the product of wt/ft and yield strength. Tests shown in Figure 7-9 indicate that unsupported and semi-supported 5½" J-55 casing can be safely perforated with exposed-charge guns using 20 grams or less of RDX powder.

Need for Control of Gun Clearance—Excessive gun clearance with any jet gun and especially with some as illustrated in Figure 7-10, can result in inadequate penetration, inadequate hole size, and irregularly shaped or "keyed" holes. Bullet guns should usually be fired at zero or 0.5 in. gun clearance to avoid appreciable loss of penetration. There is usually little problem with most large

diameter conventional hollow carrier jet guns, except when perforating in 9⅝-in. OD or larger casing.

Clearance control can be achieved through spring-type deflectors, magnets, and other methods. Two magnets, one located at the top and at the bottom of a through-tubing gun are usually needed to increase the probability of proper gun clearance. Depending on charge and gun design, 0 or ½ in. clearance for jet guns usually provides maximum penetration and hole size. In some hollow carrier casing-type guns, significant changes in hole size result as gun clearance is increased from 0 to 2 in. In such cases, centralization may produce a satisfactory and more consistent hole size. With gun clearance above 2 in. it is usually desirable to decentralize and to orient the direction of fire from the gun.

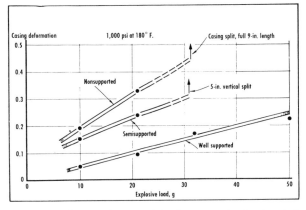

FIG. *7-9—Casing damaged by large explosive loads.*[7]
Permission to publish by API Production Department.

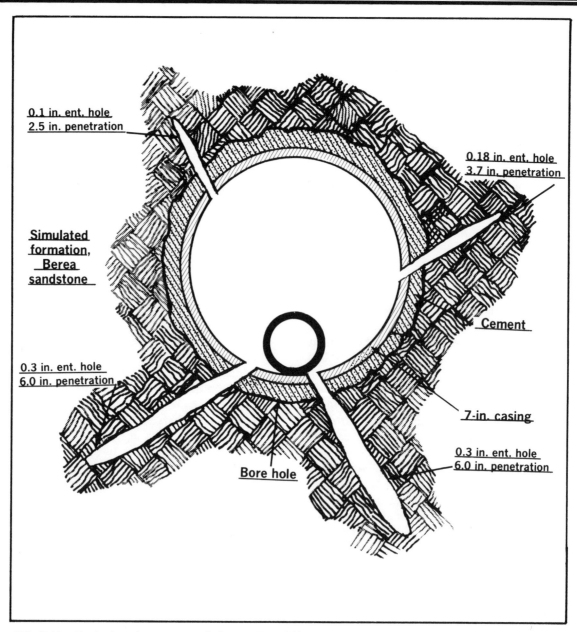

FIG. *7-10—Typical performance variations with a 90° phased 1¹¹/₁₆-in. perforator inside 7-in. casing.*[17] *Permission to publish by The Society of Petroleum Engineers.*

Centralizing is not a valid approach for most through-tubing jet guns which are usually designed to be perforated with zero gun clearance. Swing-Jets can usually alleviate the clearance problem for through-tubing guns. However, debris and mechanical problems can be quite severe.

Depth Measurements—The accepted method of insuring accurate perforation depth control is to run a collar locator with the perforator, and to make measurement from casing collars which have been previously located with respect to formations using radioactive logs. Radioactive tags can be inserted inside selected shaped charges to assist in locating exact depth of perforations. Collar logs may show location of old or new perforations made with exposed charges such as those used in capsule guns. In this case the collar log records casing swelling or bulging from detonation of the exposed charges.

Oriented Perforating—Oriented perforating is required for multiple casing strings, or through-tubing perforating in multiple completions where adjacent tubing strings are present, as illustrated in Figure 7-11.

Mechanical, radioactive, and electromagnetic[13] gun orientation devices are available. When using oriented perforators in multiple through-tubing completions, as shown in Figure 7-11, a thin wall hollow carrier gun should always be used. Capsule-type guns can cause collapse of an adjacent tubing string.

To avoid perforating adjacent strings of casing cemented in the same borehole, the most prevalent practice is to run the radioactive source and detector on the same electric cable as the perforating gun and then to rotate the gun to avoid perforating adjacent strings of casing. If there is doubt in interpretation, a radioactive pill is run in adjacent casing to aid in location of these strings.

Penetration vs. Hole Size—In the design of any shaped charge, greater penetration can be achieved by sacrificing hole size. Because maximum penetration appears to be important on the basis of theoretical flow calculation,[1] the oil industry has frequently requested, and often received, greater penetration at the sacrifice of hole size. When perforating thick-walled high strength casing, dense high strength formations, maximum penetration is probably required even if hole size is reduced to 0.4 in.

However, for more normal situations, because of the difficulties in removing mud, shaped charge debris, sand, and carbonate particles from a long

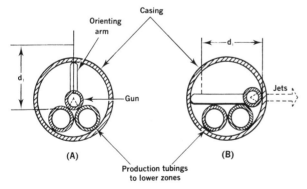

FIG. 7-11—*Oriented perforating of zones between packers. In A, gun cannot be fired, as orienting arm does not extend beyond d-1 arming distance. In B, firing position: gun armed.*

small-diameter perforation[2,3], the hole into the casing, cement, and formation should normally have a minimum entrance diameter of 0.5 in. with a uniform size smooth bore to maximum depth.

For specific perforating situations points should be considered.

1. For gravel packing, there should be large round holes with a minimum diameter of 0.75 in. and a shot density up to 8/ft. When perforations are to be cleaned with a perforation washer, 4 in. of penetration into the formation should be adequate.

2. In sand consolidation with plastic, perforation should be a minimum of 0.5 in. in diameter, along with maximum obtainable depth. Perforate with an underbalance condition, and/or clean with a perforation washer or surge tool.

3. When ball sealers are to be used as a diverting device in acid fracturing or hydraulic fracturing, entrance holes should be round and smooth. A hole size of 0.75 in. is desirable where practical.

4. When perforating carbonates in acid, hole size should be about 0.75 in. in diameter. Four to 6 in. of penetration into formation is probably adequate because acid will provide the required penetration into carbonates.

5. In wells where scale plugging is anticipated, hole size should be as large as practical, probably 0.75 in. to reduce plugging tendencies and to aid in perforation cleanout.

Plugging of perforations with scale in producing, shut-in, or injection wells appear to be related to hole diameter, and the period of time perforations are exposed to a particular scale precipitation. Pressure drop through small diameter perforations increases scaling of $CaCO_3$ and $CaSO_4$. Temperature drop through small diameter perforations in gas wells will increase scaling of $BaSO_4$.

Limitations in the Use of Exposed-Charge Jet Perforators

Most exposed-charge guns are more difficult to decentralize or centralize for proper gun clearance than are hollow carrier guns.

There is usually no way to determine whether all exposed charges produced perforations. With top fired guns, charges may drop off the gun after some charges have been fired, with no way to determine number of charges fired. This type of gun should always be bottom fired.

With an exposed charge gun, approximately 93% of the total energy resulting from charge detonation must be absorbed by the casing and fluid system; thus, the chance of splitting the casing is great. With the hollow carrier gun, this energy is absorbed by the steel gun carrier.

Charges in exposed charge guns are in direct contact with well fluids and well pressure. Many cases of leaky or cracked charge cases have been reported with exposed charges in water, or gas filled holes. Aluminum or ceramic charge cases have leaked in gas-filled holes, reducing perforating effectiveness as much as 50%. As a general rule, exposed charge guns should not be run in gas filled wells. Aluminum charge cases must be coated with a sealer if they are to be run in acid.

PERFORATING IN A CLEAN FLUID WITH A DIFFERENTIAL PRESSURE INTO THE WELLBORE

For most formations, perforating in a clean solids-free fluid with a differential pressure into wellbore results in the highest well productivity. The ideal method would be to perforate with a through-tubing gun with zero gun clearance in clean fluids with a differential pressure into the wellbore. In many cases, however, because of diameter limitations of tubing, landing nipples, and "No-Go" nipples, available small-diameter through-tubing guns frequently do not provide adequate hole size and penetration. Also, the effectiveness of some decentralizing devices required for zero gun clearance may be questionable.

If available through tubing guns are inadequate for a particular perforating job. There are several other possible means of optimizing results from perforating.

1. One approach, currently being used, requires the use of a large diameter lubricator on the wellhead and employs the following procedures:

Perforate with a full-sized casing gun in clean fluid with a differential pressure into wellbore and then retrieve the gun under well pressure. For normal pressure wells a differential pressure of 200 to 300 psi into the wellbore during perforating is quite adequate. Required differential is usually obtained by partially or completely filling the casing with salt water, fresh water, or oil. A permanent type packer is then on an electric line run through a lubricator under well pressure and set at the desired depth. After setting the packer (equipped

with a positive seal) pressure is bled off the casing above the packer. One of several available systems may then be employed to complete the well, depending on packer design.

With one type of packer having a retrievable inside blanking plug on the lower end of the packer, tubing is run to bottom and set with tubing seals positioned inside the packer body. Then the blanking plug in the lower part of the packer unit is retrieved through the tubing with a wireline to provide pressure communications between the wellbore and the tubing.

With another type of packer, the lower part of the packer body is equipped with a frangible seal. After the tubing has been run and tubing seals are positioned inside the packer body, continued weight on the tubing is employed to push out the frangible disc, thus opening the tubing to well pressure. If the tubing is equipped with an inside blanking plug, this plug is pulled with a wireline when the well is ready to be produced.

This system is limited by the ability of the stuffing box on the electric cable lubricator to seal around the cable. Therefore, completions of this type have been limited to cases where relatively low pressures are anticipated on wellhead after perforating. A number of low pressure wells have been swabbed down to provide a differential into wellbore during perforating; on these wells pressure on the wellhead is usually zero or quite low after the reservoir fluids are allowed to rise to a balanced pressure condition in the well casing.

Perforating with large diameter guns, while maintaining a differential pressure into the wellbore offers considerable potential in improving well completions if the lubricator stuffing boxes can be designed to handle higher pressures while pulling the casing-type perforator and running the packer.

2. A second approach would be to perforate casing with a conventional casing gun in clean fluid with a pressure differential into wellbore, to pull the gun under pressure, and to produce the well without tubing. Producing wells without tubing has been the usual procedure for a number of years on many wells in the Middle East. This technique has also been a part of the tubingless completion for a number of years.

3. A third approach is to perforate with a conventional casing gun, with a pressure differential into wellbore. Pull the gun under full well pressure

and snub tubing in the well.

4. In some high volume wells equipped with $3\frac{1}{2}$ to 7-in. diameter tubing, a conventional hollow carrier may be run instead of a smaller diameter through-tubing type of gun.

SUMMARY OF OPTIMUM PERFORATING PRACTICES

1. Select a perforator on the basis of test data from Section 2, API RP 43, Third Edition, October, 1974. Correct results of API test data for compressive strength of formation to be perforated. Surface tests made under Section 1, API RP 43 are of very limited value in perforator selection.

2. Gun clearance must be considered for each job to optimize penetration and hole size. Through-tubing perforators are normally designed for zero phasing to be fired with zero gun clearance. If through-tubing guns are fired at any gun clearance other than zero or possibly $\frac{1}{2}$-in. anticipated penetration and hole size should be corrected for gun clearance as well as rock compressive strength.

3. The preferred perforating method is usually to perforate in clean, nondamaging, solids-free fluid with differential pressure into the wellbore. A pressure differential into the wellbore of 200-500 psi is normally sufficient.

4. In limestone or dolomite, it may be desirable to perforate in HCl or Acetic acid with the differential pressure into the formation if clean oil or water will provide required hydrostatic head to control the well.

5. Perforating in oil, water, or acid below a mud column is not recommended.

6. When perforating in mud or relatively dirty fluids, it should be recognized that:

—It is virtually impossible to remove mud or silt plugs from all perforations by backflowing or swabbing.

—Mud or silt plugs are *not* readily removed with acid or other chemicals from perforations unless each perforation is "broken down" or fractured with ball sealers.

—Backsurge tools and perforation wash tools have proved to be effective in removing mud plugs from perforations in some wells completed in unconsolidated sands.

7. Drilling mud and dirty completion fluid should not be allowed to enter perforations throughout the life of the well. Dirty water or dirty oil may be very damaging due to perforation plugging or plugging of the formation with solids.

8. Mud-plugged perforations contribute to these problems:

—Well productivity can be appreciably reduced.

—Depending on type of reservoir drive and completion practices, oil or gas recovery can be appreciably reduced.

—Efficiency of waterflooding or other improved recovery methods can be greatly reduced.

—Wildcat wells may be abandoned as a result of erroneously indicated poor well productivities during drill stem tests or production tests.

—Effectiveness of squeeze cementing, gravel packing, and sand consolidation can be appreciably lowered.

—Well sanding problems often result from creating high rates of flow through a few perforations when most perforations are plugged.

—Blast joints in dual-completed or triple-completed gas wells usually cut out because most perforations are plugged. This is very apparent when perforations have been made in mud or fluids containing appreciable solids.

—Screens will cut out very quickly in oil or gas wells flowing at relatively high rates if most casing perforations are plugged.

—The probability of gas or water coning or fingering is increased if a high percentage of perforations are plugged.

REFERENCES

1. McDowell, J. M. and Muskat, M.: "The Effect on Well Productivity of Formation Penetration Beyond Perforated Casing," Transactions, AIME, 1950.

2. Allen T. O., and Atterbury, J. H.: "Effectiveness of Gun Perforating," Transactions, AIME, Vol. 201, 1954.

3. Allen, T. O., and Worzel, H. C.: "Productivity Method of Evaluating Gun Perforating," Drill. and Prod. Practice, API, 1956.

4. Delacour, J., Lebourg, M. P. and Bell, W. T.: "A New Approach Toward Elimination of Slug in Shaped Charge Perforating," J. Pet. Tech., March 1958, p. 15.

5. Lebourg, M. P., and Bell, W. T.: "Perforating of Multiple Tubingless Completions," J. Pet. Tech., May 1960, p. 88.

6. Thompson, G. D.: "Effects of Formation Compressive Strength on Perforator Performance," Drill. and Prod. Practice, API, 1962, 225, 191–197.

7. Bell, W. T., and Shore, J. B.: "Preliminary Studies of Casing Damage From Gun Perforators," API Paper 906-8-He, Fort Worth, March, 1963.

8. Harris, M. H.: "The Effect of Perforating on Well Productivity," Transactions, AIME, Vol. 237, 1966.

9. Porter, W.:"Recent Improvements in Perforating Tools," API Paper 906-8-Hc, March, 1965.

10. Bell, W. T.: "Recent Developments in Perforating Techniques," Seventh World Petroleum Congress, Mexico City, 1967.

11. Godfrey, W. K., and Methven, N. E.: "Casing Damage Caused By Jet Perforating," SPE 3043, October, 1970.

12. "Standard Procedure for Evaluation of Well Perforators," API RP 43, Third Edition, October, 1974.

13. Stroud, S. G., and DeGough, K. G.: "An Electromagnetic Method of Orienting a Gun Perforator in Multiple Tubingless Completions," SPE 3446, October, 1971.

14. Bell, W. T., Brieger, E. F., and Harrigan, Jr.: "Laboratory Flow Characteristics of Gun Perforators," J. Pet. Tech., Sept. 1972, p. 1,095.

15. Weeks, S. G.: "Formation Damage of Limited Perforating Penetration? Test-Well Shooting May Give a Clue, J. Pet. Tech., Sept. 1974, p. 979.

16. Keese, J. A., and Oden, A. L.: "A Comparison of Jet Perforating Services, Kern River Field," SPE 5690, January, 1976.

17. Bell, W. T., and Auberlinder, G. A.: "Perforating High-Temperature Wells," J. Pet. Tech., March, 1961, p. 211.

18. Wade, R. T., Pohoriles, E. M., and Bell, W. T.: "Field Tests Indicate New Perforating Devices Improve Efficiency in Casing Completion Operations," J. Pet. Tech., Oct. 1962, p. 1069.

19. Saucier, R. J., and Lands, J. F.: "A Laboratory Study of Perforations in Stressed Formation Rocks," SPE-6758, October, 1977.

Appendix

API Tests for Evaluation of Perforators

Tests under Section 1, API RP 43, Third Edition, October 1974 [12], is a revision of surface tests system developed in the early 1940's. These tests were designed to evaluate multishot bullet and jet perforators at surface conditions in a concrete test target.

To prepare the target, casing is cemented inside a steel form using one part, or 94 lb (1 sack), of Class A cement, and two parts, or 188 lb, dry sand, mixed with 0.45 part, or 43.3 lb (5.1 gal), of potable water. At the time of perforating, cement-sand mixture must have set minimum of 28 days and have a minimum tensile strength of 400 psi.

Bullet guns are shot in air and jets in water. The outside steel form may either be left on the target or removed when firing shaped charges; it must remain in place when testing bullet guns. Hole size, burr height, and penetration are measured and recorded on API Form 43D.

Value of Section 1, API RP 43—These surface tests probably have value for field testing to evaluate changes in hole size, due to changes in gun clearance, and consistency of performance. It may be useful outside the U.S. to determine whether the shaped charges have been damaged in transit. Results are *not* indicative of flow potential through perforations or susceptibility of perforations to be cleaned out.

These surface tests do not evaluate plugging with liner residue slugs, or "carrots," because the slugs tend to tumble and be diverted away from perforations during surface tests at atmospheric pressure, particularly with exposed charge perforators. Also debris is not flushed out of perforations as in the flow test. Data from tests under Sections 1 and 2, API RP 43, Third Edition, October 1974, are reported on API Form 43D. Appendix I provides examples of test data reported on 43D for specific jet guns.

Tests under Section 2, API RP 43, Third Edition, October, 1974 [12] are designed to evaluate, under simulated downhole wellbore and reservoir conditions, the hole size, penetration, and ability of specific perforators to provide perforations with high flow efficiency. The test apparatus and procedure developed by Exxon,[3,6] and described earlier in the chapter, are used in tests under Section 2, API RP 43. However, a change was made in method of calculating flow efficiency in the 1971 revised API Standards. The current evaluation procedure is described here.

Evaluation of Flow From Perforations—Original effective permeability (k_o), to kerosene of a "restored state" Berea sandstone core is measured prior to preparing the target.

Perforated effective permeability (k_p) to kerosene of the Berea sandstone target is measured after perforating and cleaning out by backflowing.

(k_p/k_o), the ratio of the perforated effective permeability to original effective core permeability, is then calculated. This experimental permeability ratio was originally called WFI or Well Flow Index under Section 2, API RP 43, October 1962.

k_i is the ideal perforated permeability of an ideal clean undamaged hole with the same depth as the perforated hole used to obtain k_p and is based on a 0.4-in. diameter perforation. k_i for any given depth perforation in a given core length has been developed from a computer program and is available in Appendix A, API RP 43, Third Edition, October, 1974.

(k_i/k_o) is the ratio of the ideal perforated effective

permeability to original core permeability.

Core Flow Efficiency, CFE $= (k_p/k_o)/(k_i/k_o)$. CFE represents the relative effectiveness of the perforation for conducting fluid compared to an ideal undamaged 0.4-in. diameter perforation.

Total core penetration in inches is designated as TCP. It is always $1\frac{1}{8}$-in. less than TTP, Total Target Penetration.

Effective Core Penetration, ECP, is core flow efficiency, CFE, multiplied by total *core* penetration, TCP. ECP $=$ (CFE)(TCP). This provides a means of comparing fluid flow capacity of perforations with different penetration (TCP) and different core flow efficiency (CFE). The perforation with the highest ECP, effective core penetration, in the same length *core* target, should have the highest flow rate under clean downhole conditions at a given differential pressure.

All perforator tests reported by service companies in Section 2, API Form 43D, are made in Berea sandstone with an average compressive strength of about 6,500 psi, unless a different compressive strength is noted. Therefore, total target penetration (TTP) for each gun should be corrected for compressive strength of the particular downhole formation if the API test data is to be used for perforator selection for a specific well.

For example, Section 2, API RP 43 test data reported on Form 43D for a specific $1\frac{11}{16}$ in. through-tubing perforator showed the TTP to be 5.03 in. To illustrate the effect of formation compressive strength on perforator selection, assume the compressive strength of one formation to be 2,000 psi and a second formation to be 14,000 psi. Referring to Figure 7-6, (perforator performance on formations of various compressive strengths), the corrected TTP for the 2,000 psi formation is about 8.4 in.; and for the 14,000 psi formation about 1.4 in.

To obtain the correct TCP, subtract the $1\frac{1}{8}$-in. thickness of the casing and cement from the TTP, leaving about 7.3 in. of penetration in 2,000 psi rock and about 0.3 in. of penetration in 14,000 compressive strength rock. Penetration in the low compressive strength formation would be adequate, whereas penetration in the 14,000 psi formation is completely unsatisfactory.

Recommended Use for API RP 43 Test Data

Concrete Test Target, Section 1, API RP 43—The entry hole diameter obtained in Section 1, Concrete Target Tests, is a most essential part of this test because entry hole diameter is measured in casing, a curved surface, backed up by cement.

Section 1, Concrete Target Data, should not be used for evaluating penetration because the sand-cement target is not representative of the downhole formation.

Berea Sandstone Core Target, Section 2, API RP 43—A more reliable indication of penetration can be obtained by using the data given in the Berea target. Berea sandstone core targets are more uniform and do not have the wide variation in strength as experienced with concrete targets. The Berea target data will provide more useful information relative to penetration, flow characteristics of perforations, and well productivity.

Entry hole diameters measured in a Berea target is measured in a flat surface, a mild steel face-plate. Entry hole diameters in Berea targets will be slightly larger than those obtained in the concrete target.

Use of Data on Form API 43D—Data from both Section 1 and Section 2 is reported on Form 43D.

Figure 7-12 shows test data on the Jet Research Center (JRC) charge 4 in. XLH (Extra Large Hole). The scale drawing of penetration on the API Form shows a penetration of 4.91 in. and hole size of 0.93 in. in the Berea flow target. This hole is large and relatively uniform. It should have good application for perforating zones to be gravel packed or carbonate wells perforated in acid where scaling is a major problem.

Figure 7-13 shows test data on Schlumberger's $2\frac{7}{8}$-in. Domed Scallop gun. The average penetration in the Berea flow target is 10.63 in., which is excellent penetration for a through-tubing gun. This gun should have application in any well where large diameter tubing is run to obtain high productivity oil or gas wells.

Figure 7-14 shows test data on the Dresser-Atlas 4 NCF V Jumbo Jet Charge used in the 4-in. Konshot gun. Average penetration in the Berea flow target is 14.01 in. with an average hole size of 0.44 in. This charge is representative of many competitive charges used in conventional casing guns. This type of charge should be a preferred charge when penetration is a problem. Because of the probability of downhole plugging with scale, paraffin, asphaltenes and other debris, a hole diameter of at least $\frac{1}{2}$ in. would be desirable for most perforating.

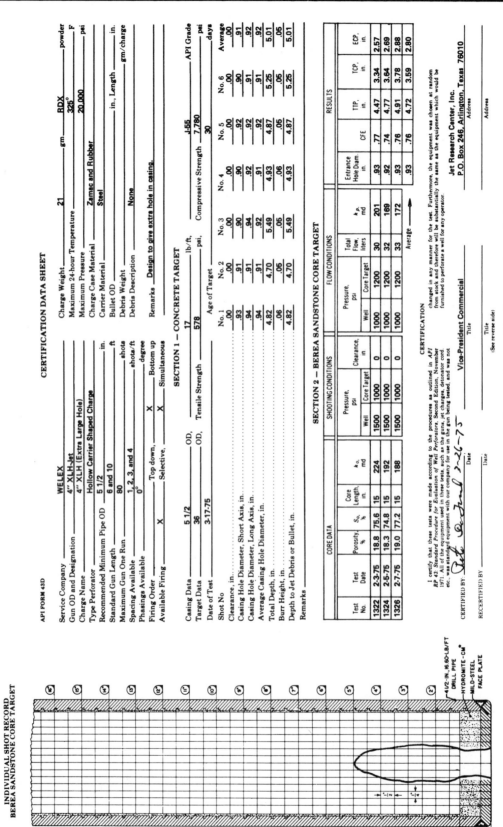

FIG. 7-12—Test data on the Jet Research Center charge 4 in. by extra large hole.

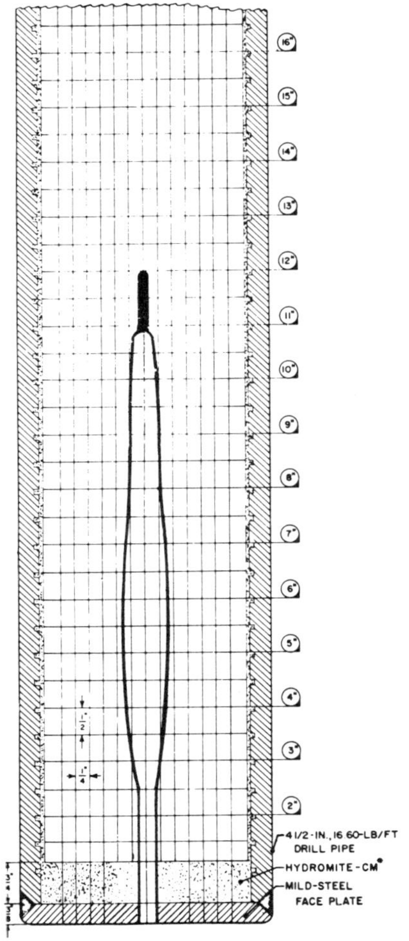

API FORM 43D

CERTIFICATION DATA SHEET

Service Company	Schlumberger Well Services
Gun OD and Designation	2-7/8" Domed Scallop
Charge Name	2-7/8" Hyper Dome
Type Perforator	Shaped Charge
Recommended Minimum Pipe OD	___ in.
Standard Gun Length	Random ___ ft
Maximum Gun One Run	___ shots
Spacing Available	4 shots/ft.
Phasings Available	0 and 180 degree
Firing Order	Top down X, Bottom up
Available Firing	X Selective, X Simultaneous

Charge Weight	14.0 gm
Maximum 24-hour Temperature	340° F
Maximum Pressure	___ psi
Charge Case Material	Steel
Carrier Material	Steel
Bullet OD	- in., Length - in.
Debris Weight	None gm/charge
Debris Description	-
Powder	RDX
Remarks	___

Casing Data	4.50 OD
Target Data	60 OD
Date of Test	3-20-74

SECTION 1 – CONCRETE TARGET

Tensile Strength 446 psi   11.6 lb/ft   Compressive Strength 5,597 psi   Age of Target 28 days   API Grade J-55

	No. 1	No. 2	No. 3	No. 4	No. 5	No. 6	Average
Shot No							
Clearance, in.	.51	.51	.51	.51	.51	.51	.51
Casing Hole Diameter, Short Axis, in.	.36	.34	.35	.35	.35	.36	.35
Casing Hole Diameter, Long Axis, in.	.37	.38	.35	.35	.38	.37	.37
Average Casing Hole Diameter, in.	.37	.36	.37	.35	.37	.37	.37
Total Depth, in.	16.25	16.50	14.13	17.38	16.38	16.13	16.12
Burr Height, in.	.09	.08	.09	.07	.07	.09	.08
Depth to Jet Debris or Bullet, in.	16.25	16.50	14.13	17.38	16.38	16.13	16.12

Remarks ___

SECTION 2 – BEREA SANDSTONE CORE TARGET

CORE DATA						SHOOTING CONDITIONS			FLOW CONDITIONS				RESULTS				
Test No	Test Date	Porosity %	S_{u} %	Core Length, in	k_{u}, md	Pressure, psi Well	Core Target	Clearance, in.	Pressure, psi Well	Core Target	Total Flow, liters	k_{p} md	Entrance Hole Diam in	CFE	TTP in	TCP in	ECP in
3142	2-28-74	19.3	79.4	18	217.7	1500	1000	0.50	1000	1200	40	320	.40	.76	10.90	9.77	7.42
3168	3-1-74	18.6	68.7	18	219.5	1500	1000	0.50	1000	1200	40	309	.42	.78	10.25	9.12	7.11
3169	3-1-74	19.2	68.6	18	258.5	1500	1000	0.50	1000	1200	40	396	.39	.81	10.75	9.62	7.79
											Average →		.40	.78	10.63	9.50	7.44

CERTIFICATION

I certify that these tests were made according to the procedures as outlined in API RP 43 Standard Procedure for Evaluation of Well Perforators, Second Edition, November 1971. All of the equipment used in these tests, such as the guns, jet charges, detonator cord etc. was standard equipment with our company for use in the gun being tested, and was not changed in any manner for the test. Furthermore, the equipment was chosen at random from stock and therefore will be substantially the same as the equipment which would be furnished to perforate a well for any operator.

CERTIFIED BY: A. Mahi   10/8/74 Date

Vice President, Engineering   Title

P.O. Box 2175, Houston, Texas 77001   Address

RE-CERTIFIED BY: ___ Date   ___ Title (See reverse side)   ___ Address

INDIVIDUAL SHOT RECORD
BEREA SANDSTONE CORE TARGET

4-1/2-IN. J6 60-LB/FT DRILL PIPE · HYDROMITE-CM · MILD-STEEL · FACE PLATE

FIG. 7-13—Test data on Schlumberger's 2⅞-in. Dome Scallop gun.

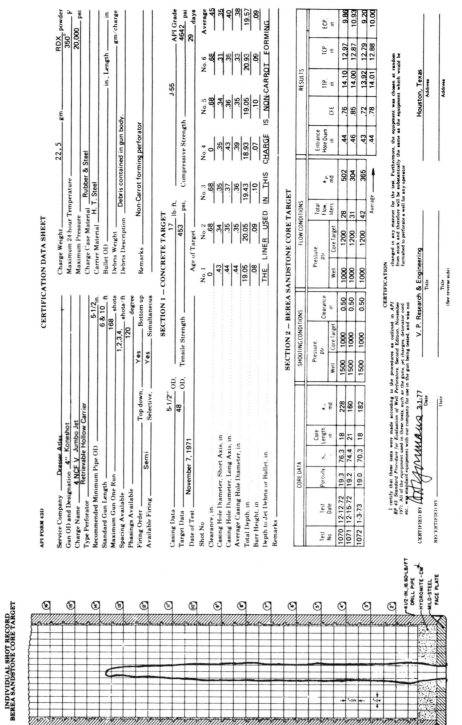

FIG. 7-14—Test data on the Dresser-Atlas 4 NCF V Jumbo Jet charge used in the 4-in. Konshot gun.

Chapter 8 Completion and Workover Fluids

Functions, requirements, selection criteria
Formation damage considerations
Oil fluids, practical applications
Clear-water fluids, practical applications
Clay, emulsion, and wettability problems
Viscosity, fluid loss, and density control
Maintenance of clean fluids
Conventional drilling fluids
Aerated fluids
Perforating fluids, packer fluids

FUNCTIONS—REQUIREMENTS—SELECTION CRITERIA

By definition a completion or workover fluid is a fluid that is placed against the producing formation while conducting such operations as well killing, cleaning out, drilling in, plugging back, controlling sand, or perforating. Basic fluid functions are to facilitate movement of treating fluids to a particular point downhole, to remove solids from the well, and to control formation pressures.

Required fluid properties vary depending on the operation—but the possibility of formation damage should always be an important concern. In certain operations, such as sand consolidation or gravel packing, sand-face or perforation plugging is a prime concern. In recent years many new fluid systems have appeared, most due to the recognition of the high risk of reducing the productivity, or completely plugging certain sections of the producing zone, through contact with a foreign fluid.

These points should be considered in selecting a workover or completion fluid:

Fluid Density—Fluid density should be no higher than needed to control formation pressure. With reasonable precautions a hydrostatic pressure of 100–200 psi over formation pressure should be adequate. Balanced pressure workovers are ideal from the standpoint of formation damage and, with proper equipment to contain the surface pressure, are practical for some operations.

Solids Content—Ideally, the fluid should contain no solids to avoid formation and perforation plugging. Figure 8-1 shows plugging of Cypress sandstone (450-md brine permeability) with salt water fluids containing various sizes and concentrations of solids. Particles up to 5 micron size caused significantly more plugging than particles less than 2 micron size. In both cases plugging occurred within the core pore channels.

Particles larger than about one-half the average pore diameter should bridge at the entrance to the pore. These larger particles are probably not detrimental if they are removed by backflow or degraded by acid or crude oil.

Particles which plate out to plug the face of the formation or a perforation, obviously obstruct operations, such as sand consolidation, gravel packing, or squeeze cementing.

Filtrate Characteristics—Characteristics of the filtrate should be tailored to minimize formation damage considering swelling or dispersion of clays, wettability changes, and emulsion stabilization. Many times this means that the fluid should contain the proper surfactant as well as the proper electrolite.

Fluid Loss—Fluid loss characteristics may have to be tailored to prevent loss of excessive quantities of fluid to the formation, or to permit application of "hydraulic stress" to an unconsolidated sand formation. Bridging at the formation face by prop-

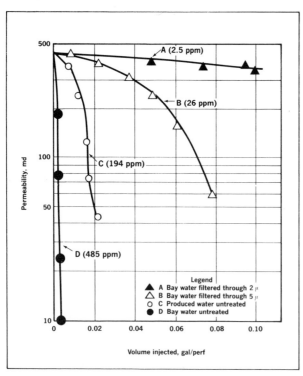

FIG. 8-1—Permeability reduction in Cypress sand-stone cores.[7] Permission to publish by The Society of Petroleum Engineers.

erly sized acid-soluble particles (calcium carbonate) is a desirable approach to fluid loss control. Where limitations permit, oil soluble resin particles may substitute for the calcium carbonate. In either case colloidal particles are also required for an effective seal.

Viscosity-Related Characteristics—Viscosity-related characteristics, such as yield point, plastic viscosity, and gel strength, may have to be tailored to provide fluid lifting capacity required to bring sand or cuttings to the surface at reasonable circulating rates. Lab tests show that many viscosity builders cause permanent reduction in permeability. This can be minimized by careful polymer selection along with adequate fluid loss control to limit invasion.

Corrosion Products—The fluid should be chemically stable so that reaction of free oxygen with tubular steels is minimized, and that iron in solution is sequestered and not permitted to precipitate in the formation.

Mechanical Considerations—Rig equipment available for mixing, storage, solids removal, and circulating is often a factor in fluid selection.

Economics—The most economical fluid commensurate with the well's susceptibility to damage should be selected.

FORMATION DAMAGE RELATED TO SOLIDS

There are two basic approaches to minimize formation damage due to solids entrained in the completion fluid.

Complete Solids Removal—To be effective fluid in contact with the formation must not contain any solids larger than 2 micron size. These points are involved:

1. Eliminate all solids to the greatest extent possible through use of two micron surface filters, backed up by other measures to minimize solids pickup downstream from the filter: i.e., control of oxygen to minimize iron oxide, careful use of thread dope, removal of rust, scale, etc. from down hole tubulars using HCl, IPA, or sand scouring techniques.

2. Accept loss of fluid to the formation and, as a practical matter, movement of very small fines into the formation. The quantity of fines is limited by minimizing differential pressure into formation. Removal of fines from the pore system after the job is maximized by returning the well to production through gradual increases in flow rate.

3. Accept possible difficulties in removing large particles from the hole due to low viscosity and carrying capacity. Many times velocity can be substituted for viscosity in lifting particles. A rising velocity of 150 ft/min should be sufficient to remove formation cuttings or sand even with 1.0 cp viscosity clear salt water.

This velocity is within the limits of many workover rig pumping systems: i.e., a circulating rate of 5 bbl/min gives 150 ft/min annular velocity with $2^{7}/_{8}$ in. tubing inside 7-in. casing or $2^{3}/_{8}$ in. tubing inside $5^{1}/_{2}$-in. casing. A no-solids fluid viscosity greater than 1.0 cp may be possible with filtered crude oil or perhaps brine viscosifiers having no residue.

Complete Fluid Loss Control—To be effective particles must not be allowed to move past the face of the formation into the pore system. These points are involved:

1. Stop all solids at the formation face by carrying in the workover fluid solid particles properly sized

to bridge quickly, and colloids to maximize the effectiveness of the seal.

2. Removal of formation-face plug after the job by backflow—and/or by degradation of the calcium carbonate solid particles and colloids with acid. With certain limitations resin solids dissolvable in crude oil could be used rather than calcium carbonate.

3. Accept the possibility that after the job pressure differential needed to unplug all sections of the zone will not be available and that it will not be possible to contact all the plugged section of the formation with acid due to bypassing tendencies.

The optimum approach depends on specific well conditions and operations. For certain critical operations; such as, sand consolidation and gravel packing through perforations, the "complete-solids-removal" approach often provides highest productivity and minimum cost.

OIL FLUIDS—PRACTICAL APPLICATION

Crude Oil—Availability makes crude oil a logical choice where its density is sufficient. Density considerations may make it particularly desirable in low-pressure formations. A low-viscosity crude has limited carrying capacity and no gel strength, and thus should drop out non-hydrocarbon solids in surface pits. Oil is an excellent packer fluid from the standpoint of minimizing corrosion, and gel strength can be provided to limit solids settling. Crude obtained from the stock tank has usually weathered enough to reduce (but not eliminate) the fire hazard.

Loss of oil to the formation is usually not harmful from the standpoint of clay disturbance or from saturation effects, as might be the case with salt water in a low pressure formation. It has no fluid loss control thus any entrained fine solids could be carried into the pore system.

Crude oil should always be checked for the presence of asphaltenes or paraffins that could plug the formation. This can be done in the field using API fluid loss test equipment to observe the quantity of solids collected on the filter paper.

Also, crude oil should be checked for possibility of emulsions with formation water. Techniques of the API RP 42 test are suitable for field use. If stable emulsions are formed, a suitable surfactant should be added.

Diesel Oil—This may be ideal where an especially clean fluid is required for operations; such as, sand

consolidation. It may even be advantageous to work under pressure at the surface where the density of diesel oil is not sufficient to overcome formation pressure.

Depending on hauling and handling practices, diesel oil should also be checked for solids. Emulsion and wettability problems should be nonexistent if the diesel is obtained at the refinery before certain motor fuel additives are included.

CLEAR WATER FLUIDS—PRACTICAL APPLICATION
Source of Water

Formation Salt Water—When available, formation salt water is a common workover fluid since the cost is low. If it is clean, formation salt water is ideal from the standpoint of minimizing formation damage due to swelling or dispersion of clays in sandstone formations.

Although "gun barrel" salt water is frequently considered to be natural water from the formation, it often contains treating chemicals, fine particles of oil, clay, silt, paraffin, asphalt, or scale and it therefore, may cause appreciable formation damage. (See Figure 8-1).

Even filtered formation salt water may contain oil treating surfactants (cationic emulsion breakers) which may cause wettability or emulsion problems. Field checks can be run using API RP 42 procedures.

Seawater or Bay Water—Due to availability, is often used in coastal areas. Again, frequently it contains clays and other fines that cause plugging.

As shown in Figure 8-1, untreated bay water caused serious plugging of Cypress sandstone cores. Depending on the salinity of bay water, it may be necessary to add NaCl or KCl to prevent clay disturbance.

Prepared Salt Water—Fresh water is often desirable as a basic fluid due to the difficulty of obtaining clean sea or formation water. Desired type and amount of salt is then added. Where clean brine is available at low cost, it may be preferable to purchase brine rather than mix it on location.

Salt Type, Concentration for Prepared Salt Water

Practicalities—From the standpoint of preventing formation damage in sandstones due to disturbance of montmorillonite or mixed-layer clays, the prepared salt water should, theoretically, match the formation water in cation type and concentration.

It is difficult to match formation brine, however, and laboratory results show that 3% to 5% sodium chloride, 1% calcium chloride, or 1% potassium chloride will limit swelling of clays in most formations. In practice these concentrations are often doubled.

Limitations of CaCl₂—In certain formations sodium montmorillonite can be flocculated (shrunk) by contact with calcium ions even in low concentrations. Thus, the clay may become mobile and could cause permeability reduction.

Where this is the case, 1% or 2% potassium chloride should be used rather than calcium chloride since the potassium ion will prevent swelling; and, in addition, low concentrations will not flocculate the sodium montmorillonite.

Additional objections to use of calcium chloride result from the observation that field mixed solutions of CaCl₂ usually exhibit pH of 10–10.5 which may disperse formation clays. CaCl₂ is also incompatible with some viscosifiers.

Extreme Water Sensitivity—In some very water sensitive formations, 2% ammonium chloride brine (while quite expensive) seems to stabilize formation clays. Some small number of sand formations should not be contacted by water of any ionic characteristics.

Emulsion—Wettability Problems

When the brine fluid base is clean fresh water, wettability and emulsion problems theoretically should not be a concern. However, even here contamination from any one of many sources often occurs.

Field Checks—Best practice dictates that the actual workover fluid be checked on location to insure that it does not form a stable emulsion with the reservoir oil, or that it does not oil wet the reservoir rock.

This is particularly true where formation salt water is used, or where corrosion inhibitors or biocides are used. Field checks can be run using the simple techniques of the API RP 42 Visual Wettability and Emulsion Breakout tests.

Prevention is the Key—Usually an unsatisfactory emulsifying or wettability situation can be corrected by the addition of a small amount (0.1%) of the *proper* surfactant.

As a general rule, workover fluids for sandstone formations where productivity is important should contain the *proper* surfactant to prevent any possibility of emulsion in the formation, and to leave the formation around wellbore strongly water wet.

Viscosity Control—Fluid Loss Control

A number of additives are available to provide "viscosity," thereby increasing the lifting, carrying, and suspending capacity of the fluid. In the Bingham plastic representation of viscosity, "plastic viscosity" relates to flow resistance between particles as well as the viscosity of the continuous fluid phase; and "yield point" relates to suspending capability when the fluid is at rest.

Completion fluid viscosity builders are all long chain polymers or colloids. They also provide fluid loss control by an indepth plugging mechanism which extends some distance back within the radial pore system. Plugging is subsequently reduced by backflow and degradation, but this process is usually not complete, and formation damage remains.

Ideally fluid loss control should be obtained strictly by a bridging mechanism at the face of the formation. This can be done effectively by use of properly sized particles. Particles larger than one-half the pore size should bridge at the pore entrance.

However, a range of particle sizes is required to reduce bridge permeability. Colloids or "plastic particles" are needed to complete the plug and further reduce permeability. The bridge should form and stabilize quickly to minimize movement of fines into the pore system.

"Viscosity" Builders

Both natural and synthetic polymers have been used in completion fluid formulations. Among them are: guar gum, hydroxyethyl cellulose (HEC), biopolymer (Xanthan gum), polyacrylamide, calcium lignosulfonate, and starch.

Natural Polymers—Guar gum is a hydrocolloid that swells on contact with water to provide viscosity and fluid loss control. Typical fluid properties are shown in Table 8-1. A filter cake is deposited, which may interfere even with squeeze cementing.

Figure 8-2 shows plugging resulting from injection of about twenty pore volumes of unbroken guar gum into a sandstone core. Permeability regain after backflow was only 25%.

In a radial flow system reduction in productivity

TABLE 8-1
Properties, Costs of Completion Fluids

	Polymer	Starch	Guar
Plastic viscosity, cp	27	20	25
Yield point, lb/100 ft^2	31	10	65
10-sec gel, lb/100 ft^2	2	1	6
10-min gel, lb/100 ft^2	5	2	12
API fluid loss, cc/30 min	14.6	10.0	42.0
Cost, $/bbl (relative value)	4.50	3.00	3.90

Composition of the fluid is:

Polymer: 2 lb polysaccharide, 1 lb polyanonic cellulose per bbl salt water.
Starch: 10 lb starch, 1 lb polyanonic cellulos per bbl salt water.
Guar gum: 2.5 lb of guar gum per bbl salt water.

would depend on the depth of the reduced permeability zone back from the wellbore. If, for example, the depth of damage was 12 in., productivity would be reduced to about 65% of the undamaged productivity.

Guar gum stability is affected by changes in pH. It forms an insoluble floc in contact with isopropyl alcohol, thus guar should not be used where sand consolidation with Eposand will be carried out. Industry trend is away from guar gum for workover fluids.

Starch primarily is used to provide fluid loss control. Other polymers may be needed for carrying capacity. Table 8-1 shows typical properties. Overall cost of starch fluids is significantly lower than guar gum or synthetic polymers, but higher concentrations of starch are required. Starch has no inherent bacterial control. Permeability loss due to plugging is significant, thus starch is losing popularity.

Bio-polymer (Xanthan gum) provides good carrying capacity and fluid loss control. Gel strength properties provide stable suspensions of calcium carbonate bridging (or weighting) particles, but may make removal of undesirable fine solids more difficult. Bio-polymer is not completely removed by HCl acid.

Synthetic Polymers—HEC, (hydroxyethyl cellulose) sometimes combined with calcium lignosulfonate, has many desirable properties and is becoming more popular. It provides:

—Good carrying capacity for hole cleaning.

—Good fluid loss control (in combination with bridging solids)

—Low gel strength to drop out undesirable solids in surface pits.

—Is degradable in HCl—or with enzyme breakers.

Unbroken HEC (without bridging particles) causes significant permeability reduction even after backflow. Damage is not as great as with unbroken guar. Acid broken HEC as shown in Figure 8-3

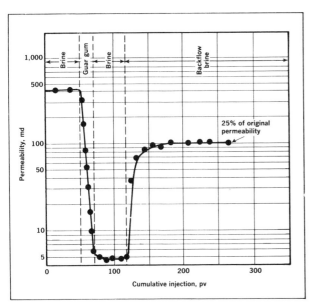

FIG. *8-2—Unbroken Guar gum.*[8] *Permission to publish by The Society of Petroleum Engineers.*

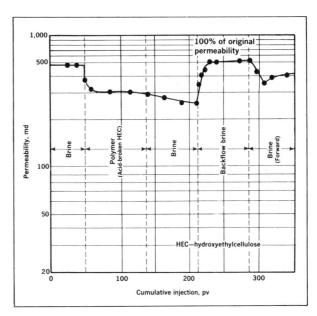

FIG. *8-3—Acid-broken HEC.*[8] *Permission to publish by The Society of Petroleum Engineers.*

shows little fluid loss control (without bridging solids) but complete permeability regain.

CMC (carboxymethyl cellulose) of the commercial grade used in drilling fluids should never be used in contact with a producing zone, due to irreparable formation damage.

Comparison of Solids-Free Fluid Properties— Typical fluid properties and costs of three solids-free completion fluids are shown in Table 8-1.

Fluid Loss Control

Due to unregained permeability loss, currently available viscosity builders should not be used without proper bridging particles to prevent movement of the viscosity colloids into the formation pore system. Bridging particles must meet two criteria:

—Form a stable, low-permeability bridge quickly.

—Be removable by degradation or backflow.

Calcium Carbonate—This material is available in several size ranges as shown in Table 8-2. For most formation pore sizes "200 mesh" particle range should be used. CaCO$_3$ is completely soluble in HCl.

Used in conjunction with HEC to provide colloids for carrying capacity and to further reduce the permeability of the "bridge," excellent fluid loss control is provided with almost perfect permeability regain if contacted with HCl acid.

Figure 8-4 shows results of lab tests using an unfiltered bay water (136 ppm fine solids) with 5% NaCl, 1.0 lb/bbl HEC and 10 lb/bbl CaCO$_3$. Bridging was almost immediate, fluid loss control excellent, and after contact with HCl acid regain was 93%.

It should be noted that in a field situation acid contact with the bridged CaCO$_3$ cannot always be assured and some permeability loss may remain.

In gravel-packing at least partial backflow of "200 mesh" bridging solids through the pack is possible

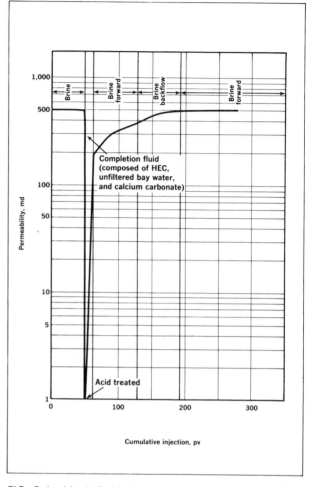

FIG. *8-4—Ideal fluid loss control and permeability regain in lab test.*[8] *Permission to publish by The Society of Petroleum Engineers.*

with 10–20 mesh gravel. Backflow probably would not occur through a tight pack of 20–40 mesh gravel.

Oil-Soluble Resins—These are available in graded size ranges needed for effective bridging action. While quite effective in brine water fluids they are quickly removed by low oil concentrations (2%). Also temperature stability is a concern since some resins tend to melt even at relatively low temperature.

Within these limitations they appear to be an excellent approach to the problems of effective fluid loss control with effective permeability regain.

Field Applications

To facilitate field use several suppliers combine polymer viscosifiers and graded calcium carbonate

TABLE 8-2
Calcium Carbonate Particle Sizes

| Designation | Diameter—Microns | |
	Mean	Maximum
Micro (400 mesh)	3.2	18
Fine (200 mesh)	60	160
Medium (70 mesh)	213	420

into one-package completion fluid systems. Poly-brine, Workover One, and Solukleen are typical of these systems. Solukleen contains oil soluble resin rather than $CaCO_3$. Usually 1–4 lbs per bbl of these products are required to prepare a satisfactory fluid. Commercially available products are shown in the appendix.

Polybrine (Magcobar) is a prepared mixture of polymers with a low concentration of 200-mesh calcium carbonate particles. All components are acid soluble. In 8.5–10.0 lb/gal brine (KCl, NaCl or $CaCl_2$) 1.0 to 4.0 lb/bbl Polybrine provides Marsh Funnel viscosity of 33 to 50 sec/qt and API fluid loss of 13–18 cc. Aluminum stearate (0.25 to 0.50 lb/bbl) must be added to brine prior to mixing Polybrine to prevent foaming.

High shear is needed to effectively mix the polymer. Low pressure centrifugal mixing systems are marginal. Dry salt added to Polybrine mixture causes severe foaming. Polybrine must be degraded with HCl before being contacted with isopropyl alcohol (IPA) to prevent formation of insoluble flocs. Above 250°F Polybrine becomes unstable.

A satisfactory calcium carbonate fluid can be prepared by adding 5 to 15 lb/bbl of 200-mesh calcium carbonate to solids-free salt solution containing 0.25 to 0.50 lb/bbl biopolymer. This should provide effective seepage control, and sufficient viscosity to circulate out sand or silt. Higher concentrations of polymer may be required to lift large cuttings or shale if the rig circulating capacity is limited.

In well killing, a technique that is used effectively is to circulate a "pill" (10–15 bbl) of fluid containing a high concentration of polymer and calcium carbonate particles. This establishes an initial bridge. Polymer and particle concentration can then be reduced. When and if additional cleaning or chip lifting capacity is needed, another pill can be circulated.

Loss of Circulation problems can usually be solved by the pill technique using additional polymer for carrying capacity with a coarser grade of calcium carbonate to bridge.

Salt Solutions Where Increased Density is Required

Table 8-3 shows the approximate density range of solids-free potassium, sodium and calcium chloride solutions:

Potassium chloride can be mixed to provide

TABLE 8-3
Density Range of Salt Solutions

Density (lb/gal)	Salt solutions
8.3– 9.7	Potassium chloride
8.3– 9.8	Sodium chloride
9.8–11.0	Sodium chloride—calcium chloride
11.0–11.7	Calcium chloride
11.7–15.1	Calcium chloride-calcium bromide

densities up to about 9.7 lb/gal at 85°F, as shown in Table 8-4.

Sodium chloride can be mixed to provide densities up to 9.8 lb per gal. Figure 8-5 shows quantities of water and salt required for 100 barrels of solution. Sodium chloride-calcium chloride mixtures can provide densities from 10.0 to 11.0 lb per gal. Calcium chloride could be used alone, but addition of sodium chloride reduces cost. Material requirements are shown in Figure 8-6.

Calcium chloride can be used for weights up to 11.7 lb/gal with material requirements as shown in Figure 8-7.

Formulations of calcium chloride and calcium bromide can provide densities up to 15.1 lb/gal without the use of solids. Although expensive, and still experimental, the fluid is much less corrosive than zinc chloride formulations.

Freezing can be a problem with high density calcium chloride solutions, or calcium chloride-calcium bromide combinations as shown in Figure 8-8.

A calcium chloride-zinc chloride solution can be weighted to 17.0 lb/gal. The mixture is expensive, and present organic corrosion inhibitors will not provide effective inhibition over extended periods

TABLE 8-4
Density of KCl Fluids

% KCl	Lb KCl per bbl	Sp Gr of solution	Density lb/gal
1	3.52	1.0046	8.37
2	7.03	1.0110	8.42
3	10.56	1.0175	8.48
4	14.08	1.0239	8.53
6	21.12	1.0369	8.64
8	28.16	1.0500	8.75
10	35.20	1.0633	8.86
15	52.79	1.0986	9.15
20	70.39	1.1328	9.44
25	87.98	1.1670	9.72

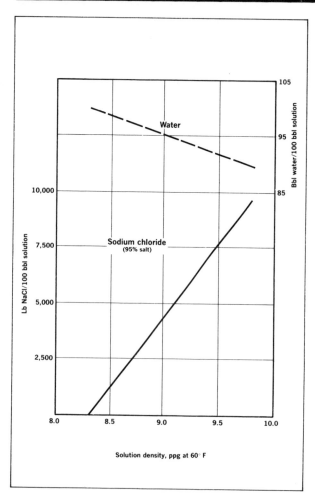

FIG. *8-5—Material requirements for preparing sodium chloride solutions.*

TABLE 8-5
Fluid Density Adjustment For Downhole Temperature Effect

Surface-measured density		Loss in density per 100°F rise in average circulating temperature above surface-measured temperature	
lb/gal	sp gr	lb/gal	sp gr
8.5	1.020	0.35	0.042
9	1.080	0.29	0.035
10	1.201	0.26	0.031
11	1.321	0.23	0.028
12	1.441	0.20	0.024
13	1.561	0.16	0.019
14	1.681	0.13	0.016
15	1.801	0.12	0.014

solids. The following equations estimate the weight of solids needed to obtain a certain fluid density, and the resulting increase in fluid volume.

Weighting material required:

$$W = \frac{K(\rho_f - \rho_i)}{C - \rho_f}$$

of time. (See Fig. 8-9.). In no case should it be used as packer fluid.

Temperature Effects on Brine Density—As the temperature of any water solution is raised, the volume is increased, and density is reduced. Where minimum formation pressure overload is maintained this effect can be important. Table 8-5 can be used to make the needed adjustment. The decrease in density is more important in the lower weight fluids.

Use of Solid Particles to Provide Density—Calcium carbonate (200 mesh) can be used to obtain higher density fluids. Although viscosity and gel strength will be needed to suspend the solid material, this approach may be desirable where densities exceed 11.0 lb/gal.

Iron carbonate, barium carbonate, and ferric oxide (all acid soluble) also can be used as weighting

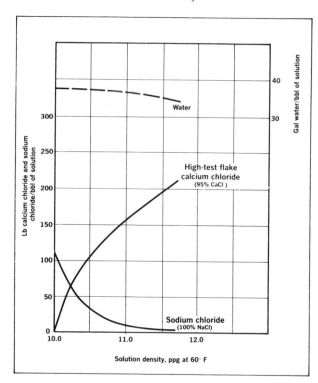

FIG. *8-6—Material requirements for preparing calcium chloride-sodium chloride solutions.*

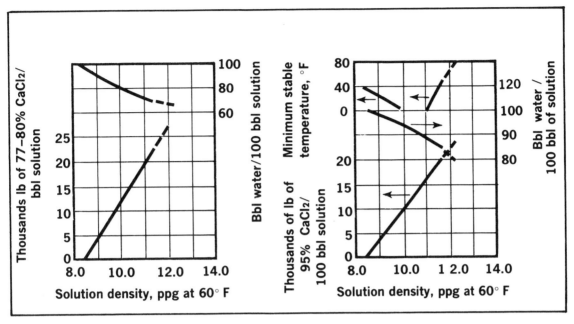

FIG. *8-7—Material requirements for calcium chloride solutions.*

Increase in volume: $\Delta V_i = W/K$

W = weighting material needed, lb/bbl of initial fluid.
ρ_f = fluid density desired, lb/gal.
ρ_i = density of available brine, lb/gal.
ΔV_i = volume increase, bbl/bbl initial fluid.

K, C = constants for weighting material (from Table 8-6).

Based on above relations, 11.5 lb/gal $CaCl_2$ brine would require 150 lb/bbl of calcium carbonate to prepare a 13.0 lb/gal fluid. Volume increase would be 16 bbl/100 bbl of initial brine. Adequate suspension qualities would probably require 0.5 to 1.0 lb/bbl Bio-Polymer. Settling should be checked before putting fluid in the hole. If settling occurs, more polymer is required.

For effective fluid loss control some of the calcium carbonate should be larger than the 200-mesh size range. Suspension improves, however, as particle size decreases. As an alternate approach

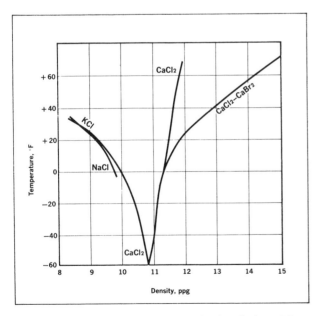

FIG. *8-8—Freezing temperature of salt solutions.*[9] *Permission to publish by Gulf Publishing Co.*

TABLE 8-6

Weighting material	Specific Gravity sp gr	Density increase obtainable, lb/gal	Equation constants	
			K	C
Calcium carbonate	2.7	3.5	945	22.5
Iron carbonate	3.85	6.5	1348	32.1
Barium carbonate	4.43	8.0	1551	37.0
Ferric oxide	5.24	10.0	1834	43.7

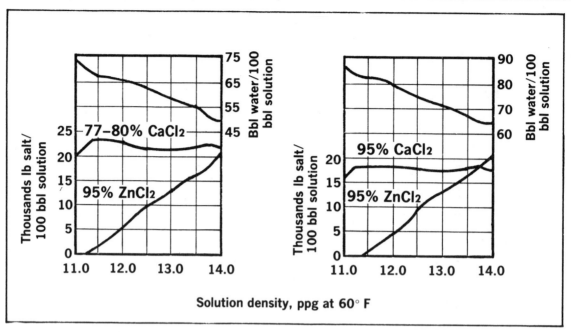

FIG. *8-9—Material requirements for calcium chloride-zinc chloride solutions.*

initial fluid loss control could be established with a "pill" of graded calcium carbonate, after which 200-mesh size range could be used for density requirements.

Care and Maintenance of Clean Salt Water Fluids

These points are important in the care and maintenance of clean salt-water fluids:

1. Dirty mixing and storage tanks or dirty vacuum tanks are a common source of contamination to a clean workover fluid system. Tanks must be thoroughly cleaned before use.

2. Workover rig tanks should be equipped with sumps and bottom baffles to contain settlings. Suction should be about 18 in. off bottom. Tanks should have easily accessible clean-out plates. Rounded corners aid cleaning.

3. Settlings in workover tanks should be checked hourly and cleaned as needed. Samples from the pump discharge are helpful in checking for undesirable solids.

4. Dirty tubing strings are often a source of rust, scale, pipe dope, etc. They can best be cleaned in the well by setting a wireline plug at the bottom, running a string of 1-in. and circulating HCl, I.P.A. or simply water with about one lb/gal frac sand for scouring. If a sliding sleeve is available above

the packer, HCl can be circulated to the bottom of the tubing, then reversed out; however, a bottom tubing plug is desirable.

5. A 4-in. cone desilter properly operated should take out a high percentage of solids down to 10–20 microns, with decreasing percentage removal of solids down to 2–3 microns.

6. For situations where clean no-solids fluids are critical (i.e., sand control), filters should be used. One type of filter frequently used is the Peco filter having 2–5 micron sock-type cartridge. Skid-mounted unit with two cartridge containers will filter water with 2-micron cartridge filter at 72 gal/min through one filter unit, or 144 gal/min through both. With 10 micron cartridge, capacity is five times as great.

7. Recent Shell work has shown that surface filtration alone is not enough to insure a no-solids fluid at the bottom of the hole. Fluid, previously filtered through a 2-micron unit was circulated through a tubing string having additional filters at the top and bottom. The top tubing filter remained clean, but the bottom filter consistently plugged with chemically-formed iron oxide, and particulate matter from the tubing, i.e., pipe dope, scale, and rust.

8. Laboratory work by Union has emphasized the point that iron oxide particles are a serious plugging material. The reaction between oxygen and

iron can be prevented by addition of sodium sulfite (and cobalt sulfate as a catalyst) to scavenge the oxygen; and sodium citrate to sequester the iron.

CONVENTIONAL WATER-BASE MUD

Economics and availability sometimes suggest use of water-base mud rather than weighted salt water where weights above 10.5 to 11.0 lb/gal are required. However, the filtrate contains clay thinners and dispersants and a high concentration of fine solids certain to cause irreparable formation damage both inside and on the face of the formation.

Thus, water mud should never be used except in zones to be abandoned.

In new wells the presence of mud can be avoided economically by pumping the primary cement plug down with salt water or oil. Mud can be circulated out before perforating using production tubing string; however, once mud solids are inside the production casing complete removal may be difficult.

OIL BASE OR INVERT-EMULSION MUDS

These muds are usually less damaging from the standpoint of clay problems than conventional water-base muds since filtrate is oil and very low filtration rates can be obtained. Most oil-base systems contain strong emulsifiers which may oil wet the formation, and blown asphalt which can plug the formation as well as present an oil wet surface. Thus, emulsion blockage could be severe.

Some recent invert-emulsion systems utilize emulsifiers chosen to minimize this problem. Perforating in weighted oil muds could form mud plugs which would not be removed by backflowing. Although lab tests show flow index is not reduced, the mud plugs are detrimental to subsequent squeeze cement jobs.

The cost of oil-base and invert-emulsion muds is relatively high, and usually can be justified only in cases where formation clays would be seriously damaged by conventional water-base mud.

Oil-base or invert-emulsion muds are better packer fluids than water-base muds from the standpoint of corrosion and settling of solids. Temperature stability is very good. Solids do tend to settle over long periods; unless the mud is properly treated with oil-dispersible clay to develop gel strength.

FOAM

In low fluid level wells where circulation of solids-free oil or water-base fluids would not be possible, foams can be used for certain workover operations such as washing out sand, drilling in or deepening. Depending on the ratio of air to foam water circulated, flow gradients as low as 0.1 to 0.2 psi/ft are possible.

Foam is a simple mechanical mixture of air or gas dispersed in clean fresh water or field brine containing a small amount of surfactant. Surfactant type and concentration should be selected to develop a stable foam with the specific well fluids encountered.

Equipment requirements include an air compressor or source of measured gas, mixing tanks for foamer solution, a liquid pump, metering facilities for air and liquid volumes, and a foam generator to provide good dispersion of the air in the foam solution.

Equipment needed to handle foam returns include a tubing rotating head or stripper assembly at the wellhead to divert the foam returns into a blooie line and to a disposal pit. At the pit, a water spray system may be required to break the foam. Aluminum Stearate acts as a good defoamer.

Typical air compressor requirements are 500 to 1000 cfm at a pressure of about 500 psi. Water and foaming agent are mixed and injected into the air stream at a rate of 10 to 20 gal/min. Foaming agent concentrations of 0.5 to 1.0% are typical. Bentonite or polymers are added to the water to produce a "stiff foam" having greater carrying capacity.

The prime advantage of foam is the combination of low density and high lifting capacity at moderate flow velocities. Bottom-hole pressures as low as 50 psi have been measured at 2,900 ft while circulating foam. Use of foam in sand washing is justified on the basis of a much faster operation and more complete sand removal.

Foam generated with natural gas or nitrogen has been used in connection with small diameter reelable tubing or snubbing equipment to clean out higher pressure wells without killing them. Foam returns in these cases are directed through the normal flowline system to production separation facilities.

Being a compressible two-phase fluid, the rheology of a foam system is complicated. One major company has developed a computer program to determine injection pressures, bottomhole circulating pressure, annular velocity, and foam-lifting ability at various gas and foamer solution rates. The program considers liquid and gas entry from the formation, temperature gradients, hole deviation, etc.

PERFORATING FLUIDS

Perforating fluids are not necessarily a distinct type of fluid, but are distinguished here to emphasize the importance of perforating in a no-solids fluid.

Salt Water or Oil—When clean, these do not cause mud plugging of perforations, but if the pressure differential is into the formation, fine particles of charge debris will be carried into the perforation.

Acetic Acid—This is an excellent perforating fluid under most conditions. In the absence of H_2S, acetic acid can be inhibited against any type of steel corrosion for long periods at high temperatures. Normally a ten percent solution is used. Acetic acid plus H_2S is very difficult to inhibit against embrittlement. Acetic acid will put iron sulfide and mineral carbonate in solution. These may result in added corrosion problems.

Nitrogen—This has advantages as a perforating fluid in low pressure formations, or where rig time or swabbing costs are very high, or where special test programs make it imperative that formation contamination be avoided.

Gas Wells—These can be completed economically in "clean fluid" by perforating one or two holes, bringing the well in and cleaning to remove as much wellbore fluid as possible, then perforating the remaining zones as desired.

PACKER FLUIDS

Criteria—Water-base drilling muds as used today are generally not good packer muds. An acceptable packer fluid must meet two major criteria:

1. Limit settling of mud solids and/or development of high gelation characteristics.

2. Provide protection from corrosion or embrittlement.

Corrosion Protection.

Laboratory tests have shown that many drilling mud additives degrade upon prolonged exposure to high temperatures to form carbon dioxide and hydrogen sulfide. Both are corrosive in water-base mud. Bacterial activity in water-base mud can cause breakdown of organic additives to form organic acids, carbon dioxide and soluble sulfide. Lignosulfonate solutions can react electrochemically at a metal surface to form sulfide even at moderate temperatures.

Thermal degradation of drilling fluid additives and the action of sulfate-reducing bacteria have been recognized for some time. Normally, bacterial action does not occur above temperatures of 150° to 175°F or in fluids treated with sufficient biocide, or where pH is maintained above 12.

Normally serious thermal degradation of drilling mud chemicals will not occur at temperatures below 300°F. Previously recommendations for packer fluids were based on avoiding these conditions. In special cases low temperature degradation may be important and would limit the use of some chemicals in packer fluid environments.

Case histories have shown that, even when thermal degradation and bacterial activity were eliminated, serious corrosion and embrittlement of high strength tubing still occurred. Laboratory evidence shows that sulfide can be formed under the influence of electrochemical action where even small amounts of sulfur-containing organics are present.

If a corrosion cell is formed on tubing or casing with enough potential to release hydrogen (0.4 volts in lab experiments), then a sulfur-containing organic mud additive could be reduced to sulfide even at room temperature and with no bacterial activity. These conditions can exist downhole since the potential between mill scale and pipe wall has been reported to be 0.5 volts and under some conditions as high as 0.9 volts.

A number of detailed case histories are available from deep wells in South Louisiana which indicate electrochemical reactions may have been the primary cause of failures of N-80 and harder tubing and casing. Even where chemically-treated drilling muds are replaced by a CMC/Bentonite weighted fluid corrosion occurred since the chemically-treated fluid was not successfully removed by circulation.

Where oil muds were used in drilling and as the packer fluid, local cell electrochemical action apparently did not occur. Casing potential surveys have shown that current is not entering or leaving the pipe, indicating corrosion cell action is not occurring.

Although there is still much to be learned about corrosion and embrittlement due to packer fluids, the following recommendations suggested by Baroid should be considered:

PACKER FLUID RECOMMENDATIONS

Condition A

No high strength pipe involved in completion (N-80 is borderline case).

Packer fluid density of less than 11.5 ppg required.

Recommendation:

1. Use diesel oil or sweet crude treated with an inhibitor where density requirements permit.
2. Use clear water or brine with an inhibitor and a biocide. Inhibitor and biocide must be compatible.

Condition B

No high strength pipe involved in completion.

Fluid density greater than 11.5 ppg required.

Bottom-hole temperature does *not* exceed 300°F.

Recommendation:

1. Economics of workover must be considered. Where workovers are inexpensive, a water-base mud treated with a biocide might be economical. Tests should be made to ascertain that mud does not contain soluble sulfide. pH should be maintained at 11.5 for a few days prior to completion if possible. Solids should be kept to a minimum to avoid gelation with high pH.
2. In remote locations where workovers are expensive or where workover frequency has been found to be high with water-base muds, use a properly formulated oil mud.

Condition C

No high strength pipe involved in completion.

Density of more than 11.5 ppg required.

Bottom-hole temperature exceeds 300°F.

Recommendation:

1. Use properly formulated oil mud.

Condition D

High strength pipe to be used under any condition of fluid density or bottom-hole temperature.

Recommendation:

1. Where fluid density requirements permit, use oil treated with both an oil-soluble and a brine-dispersible corrosion inhibitor.
2. Use oil mud formulated to meet density and temperature requirements.

In a conductive water-base packer fluid no sulfide can be tolerated without risk of stress cracking of high strength pipe. If the well has been drilled with water-base muds as presently formulated, some of this mud would remain and electrochemical reactions might result in sulfide being formed unless corrosion inhibition were 100% effective.

Amine compounds in the oil-mud packer fluid should oil wet the surface of the pipe and protect against this type of stress corrosion cracking. Maximum protection would be obtained by using an oil mud for drilling and then gelling the oil mud to be left as the packer fluid.

WELL KILLING

Circulation rather than bullheading is the preferable way to kill conventional completions.

An adjustable choke should be used to hold casing back pressure on the formation when killing a well by circulation. For high pressure well a Swaco well control choke may be desirable.

For single completions on a packer, recommended procedure is as follows:

1. Fill the annulus.
2. Open circulating port in tubing or punch hole in tubing above packer.
3. Pump slowly down casing-tubing annulus ($1/4$–$1/2$ BPM) as wireline tools are retrieved to build up a back pressure on formation.
4. After wireline tools are retrieved, pump at a constant rate of 2–3 BPM to build up 200–300 psi on tubing.
5. Maintain a constant pump rate and manipulate the adjustable choke, controlling tubing returns to keep casing pressure constant.

For a tubingless completion—or where circulation is not possible—bullheading a non-damaging fluid is best if formation will take fluid without "breakdown" or fracture. Here are four important points.

1. Breaking down the formation may cause difficult squeeze cementing and producing problems.
2. For "bullhead" well killing the surface pressure plus fluid gradient times depth should be less than formation breakdown pressures.
3. It may be necessary to have a surface pressure regulator to prevent over-pressuring.
4. If it is necessary to break down the formation, the size of the resulting fracture can be minimized by low injection rates and high fluid loss.

REFERENCES

1. Glenn, E. E. and Slusser, M. L.: "Factors Affecting Well Productivity: I, Drilling Fluid Filtration; and II, Drilling Fluid Particle Invasion into Porous Media," Transactions AIME (1957) 210, 126 and 132.

2. Monaghan, P. H., Salathiel, R. A., Morgan, B. E., and Kaiser, A. D., Jr.: "Laboratory Studies of Formation Damage in Sands Containing Clays," Transactions AIME (1959) 216, 209.

3. Black, H. N. and Hower, W. F.: "Advantageous Use of Potassium Chloride Water for Fracturing Water-Sensitive Formations," API Paper 850-39-F (1965).

4. Simpson, J. P. and Barbee, R. D.: "Corrosivity of Water-Base Completion Fluids," 23rd Annual NACE Conference, March 3, 1967, Los Angeles, California.

5. Hutchinson, S. O.: "Foam Workovers Cut Costs 50%," World Oil, November 1969.

6. Christensen, R. J., Connor, R. K., and Millhone, R. S.: "Applications of Stable Foam in Canada," Oilweek, September 20, 1971.

7. Bruist, E. H.: "Better Performance of Gulf Coast Wells," SPE No. 4777, New Orleans, February 1974.

8. Tuttle, R. N. and Barkman, J. H.: "The Need for Non-Damaging Drilling and Completion Fluids," J. Pet. Tech., Nov. 1974, p. 1,221.

9. Suman, George, O., Jr.: Sand Control Handbook, Gulf Publishing Company, Houston, Texas, 1975.

Appendix

Commercial Completion Fluid Products:

Product Name   *Description or Use*

Brinadd Co.

Barium Carbonate—weight material (W.M.)

BEX—stabilized HEC and calcium carbonate ($CaCO_3$) for viscosity

BREAK—higher alcohol base defoamer

BRIGEHEAL—lignosulfonate (lgs.) complex, $CaCO_3$ and polymers for fluid loss control (F.L.C.)

BRINEFOAM—anionic surfactant foamer

CARBWATE—$CaCO_3$ W.M.

CHEMCIDE #3—amine salt corrosion inhibitor, bactericide

DEFOAM—metallic stearate in diesel defoamer

EMULGO—$CaCO_3$ fatty acid and surfactant oil base emulsion

EMULGO PILL—$CaCO_3$ fatty acid and surfactant lost circulation (L.C.) pill

HEAL—S—lgs. complex and $CaCO_3$ for F.L.C.

HEAL—S—PILL—$CaCO_3$, lgs. complex and stabilized HEC L.C. pill

HEAL—S—PILL HT—as above with other polymers L.C. pill

HTHP PILL—$CaCO_3$, stabilized polymers evaluated temp. L.C. pill

IRONWATE—iron oxide (W.M.)

KARI—stabilized HEC viscosifier

OSS PILL—oil soluble resin (O.S.R.), lgs. complex, stabilized HEC, other polymers L.C. pill

PERFHEAL—lgs. complex, $CaCO_3$, stabilized HEC, other polymers F.L.C.

Q—PILL—$CaCO_3$, stabilized HEC and other polymers L.C. pill

Q-R-PILL—O.S.R., stabilized HEC and other polymers L.C. pill

SANHEAL PILL—$CaCO_3$, lgs. complex, stabilized HEC, other polymers L.C. pill

SLUGGIT—$CaCO_3$, Also available in MAX, COARSE and GRANDE grades

SLUGHEAL—stabilized HEC, lgs. complex, $CaCO_3$, other polymers viscosity and F.L.C.

SOLUBRIGE—oil soluble resin (O.S.R.) available as FINE, MEDIUM and COARSE

SOLUKLEEN—O.S.R. lgs. complex, stabilized HEC and other polymers

SOLUPAK—O.S.R., stabilized HEC, other polymers and enzyme breakers

SOLAQUIK—fatty acid, lime hydrate and surfactant oil base emulsion

SOLVAQUIK PILL—$CaCO_3$, lime hydrate, fatty acid and surfactant L.C. pill

THIX—stabilized polymers for weighting brines with acid soluble W.M.

THIX PAK—polymer for treating weighted packer fluid

ZEQ—stabilized HEC, other polymers for viscosity and F.L.C.

SAFE-VIS X—synthetic polymers for viscosity and carrying capacity when F.L.C. is not required

WATE—ground and sized $CaCO_3$ W.M.

Dowell Division of The Dow Chemical Co.

A-181—bisulfite oxygen scavenger

D-47—defoamer

F-68-Eze Flo—non-ionic surfactant foaming agent

G-2—non-ionic surfactant foaming agent

J-158—fast hydrating HEC

J-168—soluble bridging material

J-211—lgs., $CaCO_3$

J-212—HEC, lgs., $CaCO_3$

J-213—HEC, lgs., $CaCO_3$ large particles

J-214—$CaCO_3$

J-215—HEC and $CaCO_3$

J-237, J-275—liquid oil soluble F.L. additive

J-333—BRINE SAVER liquid oil soluble F.L. additive

M-76—Bactericide 400 water soluble, cationic bactericide

M-129—Oxygen scavenger

M-133—Bactericide 300 water soluble, cationic bactericide

M-155—Bactericide 600 water soluble cationic bactericide

S-54—calcium bromide brine

S-55—dry sacked calcium bromide

DOWCIDE G—sodium pentachlorphenate compound bactericide

Imco Services, A Division of Halliburton Co.

DRIL—S—Polymers, biocides and F.L.C. materials for drilling

FOAMBAN—surface active, readily dispersible, liquid defoamer

PRESERVALOID—paraformaldehyde to prevent starch fermentation

SAFE—PAC—selected polymers and other compounds

SAFE-PERFSEAL—lgs., synthetic polymers and sized carbonates for L.C.

SAFE-SEAL—sized carbonates

SAFE-SEAL-X—sized carbonates

SAFE-TROL—filtrate control additives

SAFE-VIS—synthetic polymers for viscosity and F.L.C.

Magcobar Operations, Oilfield Products Division, Dresser Industries, Inc.

MAGCONAL—aluminum stearate defoamer

CEASCAL—lgs. and $CaCO_3$ F.L.C.

CEASTOP—Polymers, $CaCO_3$, lgs and complex heavy metals L.C. pill

MIXICAL—sized $CaCO_3$ and surface active agents bridging material

POLYBRINE—polymers and carbonates viscosifier and F.L.C.

IRON CARBONATE—W.M.

Milchem

AEP-132—bactericide

W.O. DEFOAM—alcohol-base defoamer

W.O.20—polymeric viscosifier and F.L. additive

W.O.30—$CaCO_3$ in Fine and Coarse grades

W.O.50—$CaCO_3$

Symbols and Abbreviations, Vol. 1

Reservoir Engineering and Well Testing

A = area

AOF = absolute open flow potential

B = formation volume factor, res bbl/stb

c = compressibility psi^{-1}

DR = damage ratio

E_i = exponential integral

FE = flow efficiency

GOR = gas-oil ratio, cu ft/bbl

h = formation thickness, ft

J or PI = productivity index (stbd)/psi

J' = modified productivity index for deliverability test

IPR = inflow performance relation

k = permeability, md

K = permeability, darcys

k_f = fracture permeability, md

k_o = permeability to oil, md

k_{rg} = relative permeability to gas, fraction

k_{ro} = relative permeability to oil, fraction

k_{rw} = relative permeability to water, fraction

log = logarithm base 10

ln = logarithm base e (2.7182)

L = length

m = slope of linear portion of transient pressure semilog plot, psi/cycle

n = power in productivity index formula or AOF plot

p = pressure, psi

p_c = critical pressure, psia

p_{ff} = final flowing pressure on DST, psi

p_{fhm} = final hydrostatic mud column pressure on DST, psi

p_{ihm} = initial hydrostatic mud column pressure on DST, psi

p_i = initial pressure, psi

p_{wf} = flowing bottomhole pressure, psi

p_{ws} = shut-in bottomhole pressure, psi

$p_{1\,hr}$ = pressure on straight line portion of semilog plot 1 hr after beginning of transient test, psi

p_e = pressure at external radius, psi

$\bar{p}$ = average reservoir pressure, psi

Δp = pressure change, psi

q = flow rate stbd for liquid, Mcfd for gas

q_g = gas flow rate, Mcfd

q_o = oil flow rate, stbd

r = radius, ft

r_e = external radius or drainage radius, ft
r_w = well bore radius, ft
s = van Everdingen-Hurst skin factor
S_g = gas saturation, fraction
S_o = oil saturation, fraction
t = time, hours
t' = time, minutes
t_p = time well on production before shut-in, hours
t'_p = time well on production before shut-in, minutes
t_s = stabilization time, hours
Δt = shut-in time, hours
$\Delta t'$ = shut-in time, minutes
V = volume, bbls
WOR = water-oil ratio, % of water in total flow stream
μ = viscosity, cp
μ_o = viscosity oil, cp
μ_g = viscosity gas, cp
μ_w = viscosity water, cp
ρ = density lb_m/cu ft
ϕ = porosity, fraction
Z = gas deviation factor

Cementing

B_c = Bearden units of consistency—dimensionless
S_s = shear stress, lb/ft^2
S_r = shear rate, sec^{-1}
n' = flow behavior index, power law fluid
K' = consistency index, power law fluid
N = range extension factor for Fann viscometer (usually 1.0)
PV = plastic viscosity, Bingham plastic fluid
YP = yield point, Bingham plastic fluid
Q_b = pumping rate bbl/min
Q_{cf} = pumping rate cu ft/min
D = inside diameter of pipe, in.
D_o = outer pipe id or hole-size, in.
D_i = inner pipe od, in.
N_{Re} = Reynolds Number
f = Fanning friction factor
ρ = density, lb/gal
ΔP_f = frictional pressure drop, psi
V_c = critical velocity, ft/sec
P_h = hydrostatic pressure, psi
H = height of fluid column, ft

Downhole Production Equipment

A_i = cross-sectional area based on inside diameter of tubing, in.^{2}
A_o = cross-sectional area based on outside diameter of tubing, in.^{2}
A_p = cross-sectional area through the packer seal, in.^{2}
A_s = cross-sectional area of steel in tubing, in.^{2}
C = coefficient of expansion of steel per °F
d = inside diameter, in.
D = outside diameter, in.
F = force, lb

I	= moment of inertia, in.^{4}
E	= modulus of elasticity (30×10^6 psi, for steel)
μ	= Poisson's Ratio (0.3 for steel)
L	= length, in.
P_i	= pressure inside tubing at the packer seal, psi
P_o	= pressure outside tubing above the packer seal, psi
p_i	= pressure inside tubing at surface, psi
p_o	= pressure outside tubing at surface, psi
r	= radial clearance between concentric tubulars, in.
R	= ratio of od to id of tubular
ρ_i	= density of fluid inside tubing, lb/cu in.
ρ_o	= density of fluid outside tubing, lb/cu in.
δ	= pressure drop in tubing due to flow, psi/in.
S_i	= stress in inside fiber of tubular, psi
S_o	= stress in outside fiber of tubular, psi
Δ	= change from initial packer setting conditions
T	= temperature, °F
w_s	= weight of tubing, lb/in.
w_i	= weight of fluid contained inside tubing, lb/in.
w_o	= weight of annulus fluid displaced by bulk volume of tubing, lb/in.

Completion Fluids

W	= weighting material needed, lb/bbl of initial fluid
ρ_f	= fluid density desired, lb/gal
ρ_i	= density of available brine, lb/gal
ΔV_i	= volume increase, bbl per bbl initial fluid
K, C	= constants for weighting material